BIRKHÄUSER

To the memory of Andreas Speiser and Clifford Ambrose Truesdell,
and to my dear wife Ruth

Discovering the Principles of Mechanics 1600–1800

Essays by David Speiser

Kim Williams
Sandro Caparrini
Editors

Birkhäuser
Basel · Boston · Berlin

Editors:

Kim Williams
Kim Williams Books
Via Cavour, 8
10123 Turin (Torino)
Italy
e-mail: kwb@kimwilliamsbooks.com

Sandro Caparrini
Via Simeto, 3
10043 Orbassano (Torino)
Italy
e-mail: caparrini@libero.it

2000 Mathematical Subject Classification: 001A45, 001A50

Library of Congress Control Number: 2008933952

Bibliographic information published by Die Deutsche Bibliothek. Die Deutsche Bibliothek lists this publication in the Deutsche Nationalbibliografie; detailed bibliographic data is available in the Internet at http://dnb.ddb.de

ISBN 978-3-7643-8564-4 Birkhäuser Verlag AG, Basel - Boston - Berlin

Basel · Boston · Berlin
P.O. Box 133, CH-4010 Basel, Switzerland
Part of Springer Science+Business Media
Printed on acid-free paper produced from chlorine-free pulp. TCF ∞
Cover illustration: Jacob Bernoulli's solution to the problem of the elastica, Jacob Bernoulli, "Op. LVIII. Curvatura Laminae Elasticae. Ejus Identitas cum Curvatura Lintei a pondere inclusi fluidi quibusdam Theorematis huc pertinentibus &c.", *Acta eruditorum*, Junii 1694.
Cover photo: Peterskirche, Basel, by Karin Neidhart

ISBN 978-3-7643-8564-4

9 8 7 6 5 4 3 2 1 www.birkhauser.ch

Contents

Foreword

Kim Williams invited me to publish a selection of conferences which I had presented at various occasions during my stay at the UCL and elsewhere. She selected some papers which refer to topics on the history of science, especially on the history of mechanics.

My occupation with the history of science began through the mathematician Andreas Speiser, my uncle, at the time the General Editor of the *Opera Omnia* of Leonhard Euler, who invited me to edit volume III, 5 of the *Opera*, containing Euler's work on physical optics. I accepted this offer with great pride and indeed, the task gave me much pleasure and satisfaction.

Of all the scientists with whom I entered into contact, aside from Andreas Speiser, I must mention especially Alexandre Koyré. I had the great fortune to meet him at the Institute for Advanced Study in Princeton where we were both guests during the same period. Koyré also introduced me into questions dealing with philosophy of science, an important field, though one that plays a minor role in this present volume.

Later I was invited more and more often to give lectures concerning a given subject, or one I working on at the moment.

The other scientist whom I must mention here is Clifford Ambrose Truesdell. While Andreas Speiser was a mathematician and a historian of mathematics, Truesdell's science was classical mechanics. But his work as a scientist as well as a historian was by no means restricted to point mechanics and to hydrodynamics, as is the case with most physicists whose main field is quantum mechanics. Truesdell worked in all sectors of classical mechanics: point mechanics, rigid body dynamics, hydrodynamics, the theory of elasticity and statistical mechanics: he had a view over the development of classical mechanics from its beginning until the twentieth century.

Truesdell's work as a historian has opened our eyes to the work of the Bernoullis, especially to the oldest of them, Jacob I, whose importance he really rediscovered. He was probably also the first scientist in the twentieth century to rediscover Euler's work in mechanics in its fullness. And he was also deeply acquainted with most of the work done in mechanics in the later centuries.

To Truesdell we owe especially a great deal of our modern understanding of the work in mechanics of Lagrange and many other scientists of the eighteenth century. His work as a historian, it may be said, complemented his work as a scientist in the field of mechanics, especially mechanics of the continua and its axiomatics.

The third scientist and historian whom I must mention, is J. O. Fleckenstein, who provided me with much historical information concerning the Bernoullis.

I should like to stress that the collection contained in this book does not pretend to be a part of a "history of science": each lecture is focused on one question or on one or several persons, irrespective of all the other lectures.

The question "Why do we need history of science at a university?" is often asked and discussed. Why indeed?

First, the importance of the role which science played and still plays today – for example, in our economy – is clear to everybody. But the question: how must the teaching of science, e.g., at a university, be conceived and organized, is answered in many very different ways. Even a superficial acquaintance with the history of science can teach much that helps to answer this question. This holds as well for the perennial question: how important is mathematics for physics, and how much time must be reserved for its teaching? History of science provides answers here that would surprise deeply many physicists, experimentalists and theoreticians alike.

One of the most interesting, indeed one of the most beautiful aspects that the history of science presents, is that we see how each progress of science began as a specific step taken in a single branch of the science. But these steps then lead often to a fusion of some of the numerous branches of physics, so that we can proudly say that today there remain only a few large separate fields! This growing together is one of the most interesting and indeed the most beautiful aspect of science and by its very nature, its understanding and its appreciation demands from us that we think about science itself as well as about its history! For many scientists the observation of this process which we observe in the history of science supports the hope that one day, possibly together with branches that we do not know yet, as well as branches of the sciences which are separated today from physics, will form one well-connected science with one common basis, just as physics became united with chemistry through quantum mechanics!

I should like to express here my deep gratitude to Kim Williams for the initiative to publish these lectures and for all the hard work she put into it. My gratefulness also goes to Sandro Caparrini, who devoted many hours to the edition of these lectures. My wife always supported me morally throughout this adventure, not least by typing and e-mailing the corrections to Italy! I am extremely grateful to all three for their help!

David Speiser
Arlesheim, May 2008

Introduction

When Dr. Speiser asked me in 2006 to edit an anthology of his essays on the history of science and the relationships between science and art, my gratitude was almost equalled by my trepidation. I had just finished editing a *Festschrift* in honor of Dr. Speiser's eightieth birthday (*Two Cultures. Essays in Honour of David Speiser*, Birkhäuser, 2006) that contained papers dealing with pure physics as well as other, interdisciplinary papers, and this had been no easy task. Knowing that I would not able to do justice to a book of scientific papers on my own, I invited historian Sandro Caparrini to work with me.

Our work consisted first in making a thoughtful selection from among the thirty-nine papers first given to me by Dr. Speiser in order to focus the work on an over-riding theme. We chose that dealt with by the most part of these papers, that is, the history of mechanics in the seventeenth and eighteenth centuries, beginning with Galileo and going through Huygens, Newton, the Bernoullis, Euler, Maupertuis and Lambert in the fields of rigid body dynamics, hydrodynamics, elasticity, optics, magnetism, electricity, and celestial mechanics. We arranged the papers in a roughly chronological historical order rather than according to the original dates of publication, which range from the early 1960s through the first decade of the twenty-first century. It is remarkable how well this group of papers, written over a forty-year period, comes together to recount the foundations of mechanics, a history which has yet to be examined definitively. We have brought this history to life by including portraits of the protagonists of this exciting period, whose work not only concerns science but betrays their personalities, strengths and weaknesses. We have also included reproductions of the frontispieces of many of the key publications of the age, books so famous as Newton's *Principia* and Euler's *Lettres à une princess d'Alegmagne* as to become protagonists in their own right.

Of the eighteen papers in this collection, seven were originally in languages other than English. I translated the papers from Italian and German; Dr. Caparrini and I together translated the papers from French. All general editing on the translations was carried out jointly.

The next part of our work consisted in editing the papers as required to minimize repetitions and overlaps in the material. For each single paper the full bibliographical reference for the original publication is provided; all of these papers can be found in their original form should the reader wish to do so. We have cross-referenced sections where further discussion of an individual topic or person is carried on in another paper in the collection.

Dr. Caparrini, and to a lesser extent myself, have confirmed all of the original references and updated them to the newest, most widely available editions. References throughout the book follow the [author, date: page] format, with the full publication information given in the bibliography. For quotations, we have provided English translations within the text and placed the original citation (in Latin, French, or German as the case may be) and its reference in footnotes. We have constructed two indexes, of names and of subjects. We chose to use original names rather than anglicized versions (Johann rather than John Bernoulli, for example.

The "Editor's notes" were compiled by Sandro Caparrini, as was the comprehensive bibliography.

In short, our goal was to make Dr. Speiser's papers on the early years in the science mechanics available in a single source, in a single language, accessible for specialists and non-specialists alike.

The remaining papers in the original group that were given to me which were not included in this book are in preparation for a forthcoming volume.

I wish to thank Thomas Hempfling and Karin Neidhart at Birkhäuser for their support of this publication. I am very grateful to Sandro Caparrini, without whose valuable collaboration I could not have edited this book, and who not only worked unflaggingly for over a year to track down references and citations both ancient and modern, and who proofread the text with me many a time, but who also patiently explained the difficult concepts to me in a way that I could understand. Galileo, Newton, the Bernoullis, Euler and others have now become familiar figures to me, and I have even grown fond of them, but not nearly as fond as I am of David Speiser and his loving wife, Ruth, who have encouraged, inspired and supported me since I first met them in 1993

Kim Williams
Turin, May 2008

I was greatly honoured by the invitation to be co-editor of a collection of papers by Dr. David Speiser. When it turned out that the book as originally conceived would have been too bulky, it was a natural choice to select the papers on the heroic age of the history of mechanics. In his Foreword, Dr. Speiser describes how the writings of C.A. Truesdell influenced his work. (Indeed, every research in the history of eighteenth-century mechanics necessarily bears the mark of the "lion's paw.") Since I too have been heavily influenced by Truesdell's ideas, my editing work became a welcome return to a much-loved subject.

My "Editor's notes" which appear at the back of the book, separately numbered so that they cannot be confused with Dr. Speiser's original notes, were written to provide auxiliary information for the reader who may lack a background in the sciences, and more precise information for readers who are specialists. They do not pretend to be anything more than a collection of useful remarks, mainly taken from the works of well-known historians.

It is a pleasant duty to add my thanks to Ruth and David Speiser to those already expressed by Kim Williams. The responsibility of the editing was divided between the two co-editors according to the sphere of competence of each: thus it is clear that I must also thank Kim Williams for her linguistic advice.

Sandro Caparrini
Orbassano, May 2008

No.LVII. quod solutio hujus Problematis abstrusæ Methodi Tangentium inversæ plurimum perficiendæ & promovendæ magnum lumen præbere possit.

N°. LVIII.

CURVATURA LAMINÆ ELASTICÆ.

Ejus Identitas cum Curvatura Lintei a pondere inclusi fluidi expansi.

Radii Circulorum Osculantium in terminis simplicissimis exhibiti;

Una cum novis quibusdam Theorematis huc pertinentibus, &c.

Acta Erud. Lips. 1694. Jun. p. 262. POst triennale silentium promissi tandem fidem libero; sed ita, ut moram, quam Lector alias inique ferre posset, nonnullo fœnore compensem, dum Elaterum curvaturam non in una sola, [ut initio fueram pollicitus] sed generaliter in quacunque extensionum hypothesi constructam exhibeo; quod primus, ni fallor, exequor, postquam a multis inutiliter tentatum Problema fuisset. Extitit enim jam inde a GALILÆI quoque temporibus celebre, qui curvam hanc, perinde ut Funiculariam, Parabolam esse suspicatus erat. Eandem denique sententiam tuitus est Pater *Gasto* PARDIES Gallus, in exiguo tractatu de

staticis,

First page, Jacob Bernoulli, Curvatura laminæ Elasticæ

The Beginning of the Theory of Elasticity from Galileo to Jacob Bernoulli[1]

Introduction

The discoveries of Galileo that had the greatest, or certainly the most important, impact on science were his empirical discoveries, made in Padua and Venice, of the composition of the milky way, the moons of Jupiter, the craters of the moon and, a bit later, the phases of Venus. As a result of these discoveries, men looked at the world with very different eyes. The awareness of how small the earth is compared to the ever growing and by now almost immense cosmos is perhaps the strongest feeling that imbues us today.

More than 360 years have passed since Galileo's last and greatest work, the *Discorsi e dimonstrazioni intorno à nuove Scienze* [1638] was printed. It is correspondingly difficult to affirm something solid about the influence of this book, an influence which often enough was only indirect, let alone to speak about his general impact, that is, about the most important of his ideas and how they propelled science. Thus here I pick out only one special, very narrow domain, and within this domain, one particular question in order to present Galileo's results and then pursue the problem exactly how Galileo's impulse stimulated his successors. The part of the *Discorsi* most often presented and discussed is his theory of gravitation and the falling body, which he treated during the third and fourth days of the *Discorsi*. Instead, here I choose a subject from the second day, where he treats the breaking of a beam. This choice has the advantage that whereas Newton, Galileo's most important successor in the theory of gravitation in the seventeenth century, was almost certainly not influenced by the *Discorsi* when he wrote the *Principia* [1687], and knew of his discoveries only indirectly, mainly through Huygens, we know a small series of three scientists who knew and quoted Galileo's results in elasticity: Mariotte, Leibniz and Jacob Bernoulli. This allows us to see how they reacted to his ideas, how they were stimulated by them and how the evolution of his ideas may be followed in their work. We

---

[1] Lecture given at the 4th International Symposium on Neutrino Telescopes on 13 March 1992 in the Palazzo Loredan in Venice, originally entitled "Galileo and the Beginning of the Theory of Elasticity". First published in: *Galileo Scientist, his years at Padua and Venice*, Ed. Milla Baldo Ceolin. Istituto Nazionale di Fisica Nucleare, Istituto Veneto di Scienze, Lettere ed Arti, Dipartimento di Fisica "Galileo Galilei" dell'Università degli Studi di Padova, 1992, pp. 87-112.

can follow this development up to 1800, a century and two-thirds after the publication of the *Discorsi*.

The other advantage is that during this period only very few worked in this domain. For example, Newton did not publish anything about it. Thus for the period I have indicated the three names form an almost complete list. But even later there would not be many who did so: Leonhard Euler, Charles Augustin Coulomb and Augustin-Louis Cauchy are the most important ones; with Cauchy we arrive already at 1826, when he gave the definitive foundations of the theory of elasticity, or if you wish, of the whole of continuum mechanics. The work of all of them can be traced back to Jacob Bernoulli and through him back to Galileo.[2] Looking over those first 167 years after the *Discorsi* we can see part of what Galileo's impact on science was. The *Discorsi* is known mostly for Galileo's treatment of the ballistic problem,[3] but it is much less known that Galileo does not open his book with this problem. First he examines a branch of science that we may call "the solidity of matter and the rupture of bodies". This cannot be easily found in today's books of physics; in textbooks it is only briefly mentioned, if at all. So the reader may ask where its legacy is? As it happened, a scientific development outside physics, which I shall mention, guided and stimulated physics into an unexpected new direction. Indeed, fifty years after the publication of the *Discorsi* a new branch of physics came into existence: the theory of elasticity. This new science looks at first glance quite different from what we find in the *Discorsi*. Yet Galileo's question, together with his main idea, survived this break with tradition and undoubtedly stimulated the discovery made by Jacob Bernoulli in 1691. Precisely for this reason, when we try to pin down this stimulus, we can measure the depth of Galileo's vision and the penetration of his grasp of mechanics.

The breaking of a beam

A physicist, upon learning for the first time that Galileo intended to produce *a theory of breaking*, can only shake his head out of puzzlement, if not out of disbelief. If there are two things which any physicist learns consciously or by instinct they are first, that every experiment must be reproducible, that is, following a number of indications any colleague must be able to replicate it, and second, if you want to find a new law, which is always done by guessing, you must consider a situation as simple as possible. Now, if you have ever built a fire in the woods, you know that only the driest wood breaks cleanly, the green one splinters, and no two branches splinter the same way. Someone who had heard of "preparation of a state" might try to prepare an experiment by buying, in the same shop of course, a dozen or so equal broomsticks. He may break them one after another, all over the same knee (provided the knee can stand it), but he will hardly be left with a series of experimental results out of which, by sheer power of exact observation, he can extract a new law of nature, as restricted and modest as you allow it to be. So after this stick breaking experiment, the physicist will exclaim in exasperation, "what the h... induced Galileo to choose that subject of research, and what the d... did he possibly get out of it?" We shall see that Galileo, imaginative as he is, begins by formulating what we call today a

[2] For the history of the theory of elasticity from 1638 to 1788 see [Truesdell 1960a].

[3] See "The Second New Science in Galileo's *Discorsi* " in this volume, pp. 17-38.

"model". In this case, it is the model of what we may call the "clean break of a rigid body": the beam is conceived as rigid and breaks along a straight line. That means that Galileo's result, which I shall show, applies only to bodies that are at least almost rigid and only if the break has this simple form, that is, if it does not splinter.

Is that ever the case? As we shall hear in a moment, for metal beams as for wooden beams, it never really is. Do marble columns or stone pillars or gargoyles always break this way? I do not know. However, I will mention one object that does indeed, as we all know, break in this way. This is the professor's very weapon in his daily intellectual fight with his students, that is, his chalk when he teaches. Indeed, some professors begin every lecture with the solemn ritual of "breaking the chalk".

Fig. 1

But Galileo does not just observe. He wants to formulate a law, and as such laws are not only in a vague poetical way "written in the language of mathematics", this law too must be an equation, for he was one of the first to understand that clearly. In mechanics, as elsewhere, an equation connects two quantities. In order that the relation also be valid for readers who may want to test it under different conditions, it must fulfill certain requirements, as Galileo saw more clearly than anyone before him. For instance, what we call today the "physical dimension" of the two quantities must be the same on both sides of the equation. Indeed, the first day of the *Discorsi* is largely dedicated to what we call "scaling properties", that is, the question of how the solidity of a construction – a bridge, for instance – changes if all its dimensions are multiplied by a common factor. The answer is by far not as simple as Simplicio, another participant in the dialogues, thinks. One by-product of the inquiry is precisely dimensional analysis.[i]

Further, an equation in physics must connect quantities that have the same character, or as we say today, the same covariance property under the Galileo group, otherwise it does not have an invariant meaning and does not lead to a reproducible experiment. Of course, only a first, dim understanding of the Galileo group is present, but rotational invariance lead Galileo in his endeavor to formulate equations that could be put to an experimental test. That will bring us to the central point.

But the greatest difficulty still remains. The equation which a scientist writes down must be the answer to a *good question*, that is, it must be a nontrivial relation between fairly independent quantities. And such a good question is difficult to ask even today, in spite of all the theories that are available to us. So we must not be surprised if Galileo's question seems a very modest one. Its importance can only be judged by the impact it had on subsequent research. The question he asks that I am going to pursue here is the following: how does the transversal breaking of the beam described above compare with the longitudinal break, both defined now by the same idealized model?[ii]

Is it easier to break the beam transversally or longitudinally? Here "easier" means "with a smaller weight". Intuition tells us transversally, but exactly how much smaller can the weight be?

Before I give you Galileo's answer I should like to emphasize three facts. The first is the apparent modesty of the possible results. "That's all he tells us?" you will ask, and say "Big deal! It even applies only to highly unlikely cases". But it is precisely the modesty of the question that guarantees its success.

Second, Galileo is answering a question that he himself thought up. No one before had asked such a question, or at any rate, no one had formulated it so carefully, nor indicated the limitations of the validity of the answer so strictly.[iii] Moreover, the question is formulated by notions which, if not coined by Galileo, are combined by him in an original way, and these notions now connect mathematics and experiments.

Third, and this is perhaps the most important and original characteristic, Galileo was driven and supported by the sheer belief that, whatever the material and whatever the form of the beam, the answer was given by *one universal law of nature*; one only, of course, within the limit of his model!

I will restrict myself to the most important point. Consider the beam to be a lever with its fulcrum at the point *B*. If we assume (and this is the weak point in the reasoning) that the resistance may be represented as a force attached to the mid-point *M*, then for the transversal break Archimedes' law of the lever tells us

$$l \cdot W_t = R \cdot \frac{b}{2} \tag{1}$$

On the other hand the W_l is equal to the resistance R itself (he calls it the "absolute resistance")

$$R = W_l \tag{2}$$

This is a special case of what was to become, when stated in full generality, Newton's third axiom!

It follows that

$$\frac{W_t}{W_l} = \frac{b/2}{l} \quad \text{or} \quad W_t = \frac{1}{2}\frac{b}{l} \cdot W_l \tag{3}$$

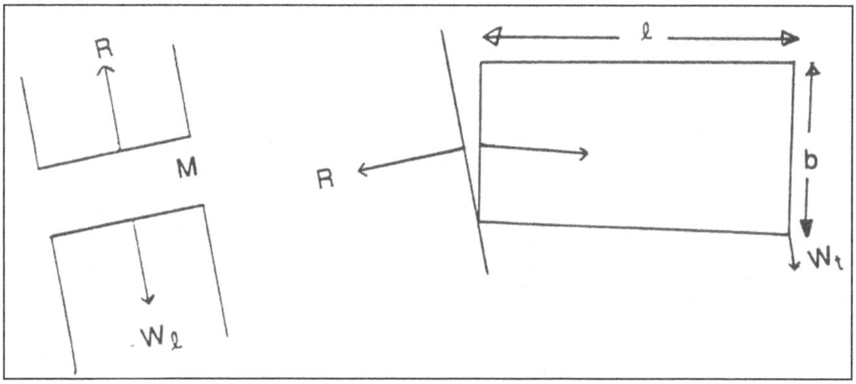

Fig. 2

This is Galileo's law of a break: it simply relates two ways of breaking a beam. Looking now at what Galileo did we may say that it was not the observation that inspired the discovery. Rather, Galileo thought out an application of Archimedes' law in a new domain.

The experiments of Edme Mariotte

The first to put Galileo's theory to the test and make systematic measurements was the Catholic priest Edme Mariotte. This is the same Mariotte whose name is associated with the law of Townley and Boyle. In the history of physics Mariotte's name does not belong to the greatest ones, but his writing nevertheless exerted considerable influence. Mariotte's book, *Traite du mouvement des eaux et des autres corps fluides*, appeared posthumously in 1686. It contains a *Discours de la force des tuyeaux de conduire et de l'épaisseur qu'ils doivent avoir suivant leur matières et la hauteur des réservoirs*. The location of this digression on the strength of materials – for that's what the discourse is – in a book on hydraulics shows that the purpose of this book was a practical one. With Mariotte we are still in the tradition of practical engineering, and we do not find much mathematics. Indeed, in this respect Galileo's book, written years before, is more modern. Moreover, Mariotte was not a very systematic writer and as such he stands no comparison with his contemporary, Christiaan Huygens. His aim is to solve problems presented by engineering praxis. His use of mathematics is minimal, and in his "proofs" the hypothesis are not always stated explicitly. For example, he uses what we call Hooke's law, but he does not state the law explicitly. But while in many ways more old-fashioned than Galileo, he was a careful observer and experimenter at a time when measurements were very hard to perform, and some of his results are valuable indeed. His measurements relate directly to Galileo's formula. His most important results for us are: 1) there is indeed a universal connection between longitudinal and transversal break, so that Galileo's formula – his theory if you wish – is sound in principle; 2) the numerical factor 1/2 seems to be wrong and should rather be 1/3 or may be 1/4; and 3) the reason for this failure seems to be the fact that beams do not break suddenly, but always bend before breaking, even if only slightly.

We may connect these three findings with both what Galileo had done before and what would be done after Mariotte. First, that he compares his findings with Galileo's formula is proof of his predecessor's importance. Indeed no other name is mentioned, and already this should warn us that Galileo's discovery is not to be underestimated. Second, Galileo's erroneous numerical factor due to the use of an oversimplified model brings to mind a remark made by another Italian. Enrico Fermi once said: "I can calculate anything in physics within a factor of two on a few sheets. To get the numerical factor in front of the formula right may well take a physicist a year to calculate, but I am not interested in that."[4] Finally, Mariotte's experimental discovery of the importance of bending was indeed pioneering. But as always happens with a pioneering discovery, it led further development, including theoretical development, in a direction that no one could have foreseen, because the reason for this sudden new turn lay outside physics; it came from another science. Indeed, the account of our story will now jump into a radically new period of physics.

Leibniz's mathematical theory of elasticity

At about this time, physics was enriched by the most powerful stimulus it has ever received from an outside agent probably until the present. I am referring, of course, to the invention of infinitesimal calculus independently by Newton and by Leibniz. This new calculus not only brought to the physicist new tools for computing, it actually changed the character of his science fundamentally. It made it possible to attack problems that had been out of reach, to formulate them much more adequately than had ever been possible, and even to formulate problems that no one had even thought of yet. But, of course, not everything was accessible even to this powerful calculus. Thus people jumped to the problems to which it was best suited: smooth motions and smooth deformations. Physics now became the investigation of such processes, while those problems which were not of this kind were either neglected or reformulated. Hard spheres were now thought of as ever so slightly deformable, the corners of an orbit were rounded off, etc. In short, differentiable motions and what we today call differentiable manifolds became the focus of physicists' attention. Small wonder then that few bothered any longer with breaking beams (a perfect example of what is now called a catastrophe!). But bending was another matter: it was accessible to the new calculus. The first to realize this was Leibniz himself.

In his paper *Demonstrationes de resistentia solidorum* [1684a] Leibniz says: (I abbreviate somewhat):

> Mechanics seems to have two parts, one [that deals] with the power of acting and moving, the other with the power of resisting, i.e., with the solidity of a body. Archimedes who alone among the ancients introduced geometry into mechanics did not touch on this second part. Thus no one [after him] did anything in mechanics until Galileo, who, trained through his sharp judgment and his great knowledge of geometry was the first to push the boundaries of science outward, and thus began to trace back the resistance of solid bodies to the laws of geometry [Leibniz 1684a: 319].

[4] Quoted in [Bretscher & Cockcroft 1955: 74] and in [David Speiser 1962: XXVI].

Leibniz makes it clear that this introduction of geometry must always be the goal of science.

It follows that if the findings of Mariotte, whom Leibniz quotes, do not confirm Galileo's theory, the reason cannot be Galileo's endeavor itself (that is, to find laws that are geometrically formulated), but rather the fact that he had considered too simple a model. In total agreement with his whole philosophy, Leibniz reasons that there cannot be a sudden break of a beam, just as there is never a sudden break of a string or an instantaneous repulsion of an infinitely hard body by another one. Before the repulsion there is always a slight deformation, and breaking as well is always preceded by either bending or stretching. If we consider a beam as made up of fibers, we see that each fiber is stretched in proportion to its distance from the fulcrum *f*. If we now assume that the stress in the fiber is proportional to the elongation, we obtain another factor x, and the stretching moment m acting on each fiber is proportional to x^2.

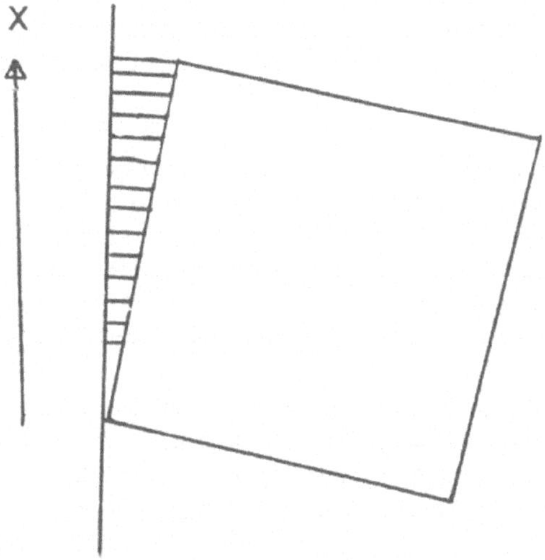

Fig. 3

If we integrate over the cross section (a computation Galileo and Mariotte could not yet do) we get a factor of 1/3, since

$$M_{tot} = \int m \, dx = \int x^2 \, dx = \frac{x^3}{3}$$

What about the break itself? Leibniz does not say, and I can only submit a guess. Leibniz's central mechanical notion is energy. Thus I suspect that in an intuitive way he may have felt that the energy needed for the break proper is negligible compared with the energy needed for the bending. Be that as it may, more important are the following observations. First, contrary to Mariotte's, Leibniz's paper is logically and mathematically rigorous, in so far as the physical hypotheses and the mathematical operations that lead to the final result are fully stated. Second, Leibniz, as we shall see, was fully aware that he had really treated the elongation and not the bending of the beam. Of this we shall hear more.

Jacob Bernoulli's foundation of the modern theory of elasticity

Leibniz's paper, printed in 1684 in the *Acta Eruditorum*, was read by Jacob Bernoulli, the oldest and greatest of the Bernoulli dynasty. By that time at the latest his attention had been drawn to the *Discorsi*, but probably he knew them earlier. Although very little is known about Jacob's early education, we know that this paper aroused his curiosity. But only in 1687 did this diffident and introverted man dare to write to Leibniz asking him for explanations of points he could not understand and expressing his disagreement with others. For in the meantime he had undertaken a check by experiment of Hooke's law (which he called Leibniz's) and found it in error. According to his measurements, it took more than twice the weight to elongate a string of gut to twice its length, more than three times to elongate it to three times its length, etc. Thus, on this point he contradicted Leibniz. Jacob's experimental research into principles of statics proved to be useful later, as we shall see.

With his letter to Leibniz, he was unlucky. Leibniz had left shortly before the letter arrived and returned three years later. Only then did he find Jacob's letter, and he hastened to answer it, without going into many details since he was now busy with other, mostly historical subjects [1855]. He complimented Jacob by saying that from his latest article in the *Acta* he had seen that he had understood everything about his own calculus. And since Leibniz recognized at once the true rank of Jacob Bernoulli as a scientist, he suggested to him "a new and not inelegant problem", namely, the exact shape assumed by a rod that is bent under a weight.[iv]

Exactly when Jacob began to work and when he achieved his goal we do not know, but in June 1691 he announced, again in the *Acta*, that he had found its solution [1691a]. He invited everybody else to send in a solution too, "before the harvest festival", when he would publish his own. Alas, no one sent a solution, nor was one sent during the three years before Jacob eventually published his results. Compare this with what happened when Johann Bernoulli invited "all Mathematicians of his Age" to find the shape of the brachystochrone, when four mathematicians immediately sent a solution: Newton, Leibniz, his brother Jacob, and the Marquis de L'Hôpital. A bent beam was a much more formidable problem. Certainly no one found any help or hint in the works of either Huygens or Newton, and both, as we know, kept silent.

In a first paper, "Curvatura laminae elasticae" [1694], Jacob indicated the solution and gave the main new mathematical theorem for finding it. But as neither Huygens nor Leibniz could understand it, he was forced to write a second one, Explicationes, Annotationes et Additiones ad ea, quae in Actis sup. anni de Curva Elastica…" [1695] which contained a more detailed explanation. Moreover, he corrected a mistake which he had made following Leibniz and of which I shall speak later. I shall now sketch the main ideas of his solution; we shall see that quite a few can be traced back to Galileo.

First, Jacob formulates the problem by asking exactly what curve is described by a rod put vertically into the ground but then bent by a weight such that its uppermost part is exactly horizontal (fig. 4).

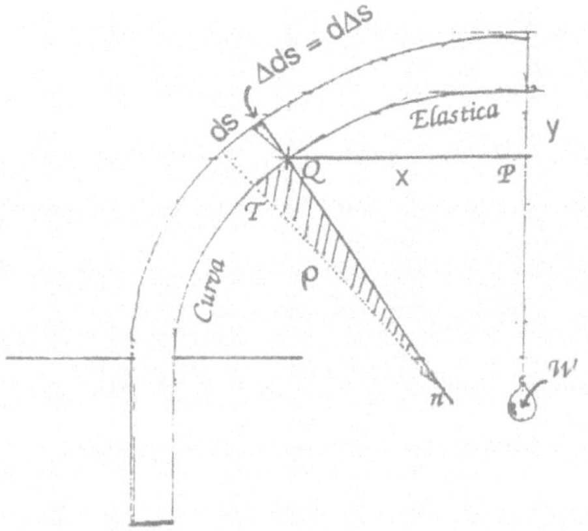

Fig. 4

He derives the answer from four assumptions:

1. The dynamical principle is Archimedes' law of the lever, which Bernoulli uses following Galileo. The weight W induces through its moment (longer arm of the lever) $x \cdot W$, a stretching force F, which is counter-balanced by the restoring force $R = F$. The force R induces the moment $c \cdot r$, where c (the shorter arm of the lever) is the thickness of the beam. The fulcrum of the lever is at Q. Thus the law of the lever yields:

$$x \cdot W = c \cdot F = c \cdot R \qquad (4)$$

and

$$F = \frac{w}{c} \cdot x .$$

From this alone it follows immediately that the beam is curved most where it is stuck into the ground, since the weight's moment is greatest there, and that it is not curved at all at the upper end where $x = 0$.

2. If we assume that a plane section of the beam remains a straight section after bending, we see that the two shaded triangles are similar, thus:

$$\frac{d\,\Delta s}{c} = \frac{ds}{\rho} \quad or \quad \frac{d\,\Delta s}{ds} = \frac{c}{\rho} \qquad (5)$$

3. Since, unlike Galileo, Jacob applies (4) to an infinitesimal part of the beam, he needs an integrating theorem. For this he uses one of the forms of the formulae for the radius of curvature which he had discovered and published in this paper for the first time:

$$\frac{1}{\rho} = \frac{d^2 y}{dx\,ds} . \qquad (6)$$

4. Into (4) enters the stretching force $F = xW/c$ where W and c are constant. But for determining the shape of the beam we need the elongation $d\,\Delta s/ds$ rather than the stretching force. Thus Jacob needs a "constitutive relation", that is, the stress-strain relation. The simplest one is Hooke's law, where the elongation is proportional to F:

$$\frac{d\,\Delta s}{ds} = F \text{ i.e. } \frac{d\,\Delta s}{ds} = \lambda F\,, \tag{7}$$

and with the help of (4):

$$\frac{d\,\Delta s}{ds} = \frac{\lambda W}{c} \cdot x = 2Hx\,;\; 2Hx \equiv \frac{\lambda W}{c}\,.$$

If (5) and (7) are inserted into (6) we obtain

$$d^2 y = \frac{2Hx}{c} dx \cdot ds\,. \tag{8}$$

After integration over dx and a slight manipulation we obtain

$$dy = \frac{x^2 dx}{\left(a^4 - x^4\right)^{1/2}}\,, \tag{9}$$

where

$$a^{-2} \equiv \frac{H}{c} = \frac{\lambda W}{2c^2}\,.$$

We see here how Galileo's initial stimulus is combined with the new analysis to produce an unexpected mathematical triumph: (9) is the first indefinite elliptic integral. Together with the corresponding "integral of the first kind" for the line element

$$ds = \frac{dx}{\left(a^4 - x^4\right)^{1/2}}\,,$$

it is thoroughly discussed by Jacob and computed numerically.

Of course, this curve was not to be determined by measurement, which would not help anyway, since, as we shall see, the rod describes a curve which no mathematician of that day could define. Rather, the questions as Jacob saw them were, what the mechanical principles that determine this form are, and only then, what its exact form under some special given circumstances is. The analogy to Newton's procedure for solving the problem of the solar system is clear. It will be recalled that Newton had not asked "Given the law of motion and the law of the gravitational force, what are the orbits?" but rather, "Given the orbit and assuming the law of motion, what is the exact law of the gravitational force?"[5]

Two remarks may be added. First, as I said, Jacob had to explain in a second paper how he had used Galileo's principle, since no one was able to follow the compact and crowded style of the first. Besides, Huygens had correctly criticized Jacob for having placed the fulcrum of his lever on the concave boundary of the beam, following Leibniz, rather than in the middle. This meant that this inside boundary is the neutral fiber, while in reality it is compressed. The neutral fiber is somewhere in the middle of the beam. This error of Leibniz and Jacob Bernoulli shows how closely both followed Galileo, for Galileo, who had considered a rigid rather than an elastic beam, had correctly placed the fulcrum of his lever on the boundary of the beam!

Second, in 1690 when Jacob announced his discovery, he had found the solution assuming only a linear stress-strain relation, that is, Hooke's law. But his own careful

[5] See "The Kepler Problem from Newton to Johann Bernoulli" in this volume, pp. 123-134.

experiments had convinced him that this relation could not always be true. Moreover, Galileo himself had used the principle of the lever to tackle the problem of the break of a stone that was rigid rather than elastic. So Jacob generalized his formula for whatever the stress-strain relation may be. If we set

$$\int \frac{d\,\Delta s}{ds}\,dx \equiv S$$

one finds

$$dy = \frac{S\,dx}{\left(a^4 - S^2\right)^{1/2}}$$

For Hooke's law, $S = x^2$. This may be the first time that a problem was solved for a whole class of constitutive equations.

According to Truesdell, the distinction between universal dynamical principles and constitutive equations that describe a special material, fundamental for classical mechanics, can be traced back to this paper.[6] Jacob had understood this distinction through his experiments as well as through his study of Galileo's book.

How deeply Galileo's problem had occupied Jacob is shown by his very last paper, "Véritable hypothèse de la résistance des solides, avec la démonstration de la courbure des corps qui font ressort", written shortly before his death [1705]. He again takes up the problem of the break, wanting now to prove Galileo's result rigorously. His proof, alas, was correctly criticised by the French physicist Antoine Parent, who pointed out a fallacy in Jacob's reasoning. Suffice it to say that even today the problem is not fully solved. Yet Jacob's effort was not fruitless. According to Truesdell [1960a: 106], it is in this paper, stimulated directly by Galileo, that we find the first clear definitions of the notions of "strain", "mean elastic stress", and "stress-strain relation", and the concept of a modulus of elasticity begins to emerge.

The ulterior history

We can follow Galileo's impulse beyond Jacob Bernoulli. The latter exerted great influence on Euler. Euler first continued Jacob's theory of the beam in his first appendix, *Additamentum I de curvis elasticis*, to his *Methodus inveniendi* [1744], deriving the results from a variational principle suggested to him by Jacob's nephew, Daniel.[7]

I have not found an explicit reference to Galileo's work in Euler,[v] but that the Galileo-Bernoulli idea of the lever was a potent stimulus for Euler is shown by the following. Recall that Newton's *Principia* dealt with point mechanics and hydrodynamics but with neither rigid nor elastic bodies. Therefore Newton's axiom, $\dot{\mathbf{P}} = \mathbf{F}$, published in 1687, suffices for his goals. Starting from it, sixty-five years later, Euler could formulate the definite foundations of hydrodynamics.

Euler then struggled for over twenty-five years in his attempt to derive a theory of rigid bodies – that is, the spinning top – from Newton's law. Eventually he succeeded, but only at a price. His theory was not fully general and the derivation was unnecessary

---

[6] The reader of Truesdell's book will see immediately how much I am indebted to his account.

[7] See [Truesdell 1960a: § 28].

complicated. Eventually – he was by then over sixty – his work in elasticity, especially his discovery of the shear force, showed him the right way: the law of the lever of Archimedes and Galileo, which had become in the hands of Jacob Bernoulli the momentum principle for rigid bodies. Euler's last derivation of the spinning top equation starts from Newton's and his equation, $\dot{\mathbf{J}} = \mathbf{M}$. The derivation now becomes short and transparent, and is fully general.

But that is not yet the end of Galileo's impulse, which, let me repeat, bypassed Newton. After Euler, Coulomb discovered the shear stress. And eventually, by combining all these elements – the law of the lever, the momentum principle, the shear force, and the shear stress – Cauchy, through his formulation of the stress-tensor, will establish the definite foundation of the theory of elasticity and thereby of the whole of continuum mechanics [1823; 1826]. Galileo's shadow was *that long* in the history of mechanics!

But I would like to finish this historic account in Italy. In 1785 a book was published in Milan with the title *Nuove Ricerche sull'equilibrio delle volte*. *Volte*, that is, vaults, and domes are typically Italian, and so was its author. He is the priest Lorenzo Mascheroni from Bergamo, known from the Euler-Mascheroni constant, and his book, like the *Discorsi*, deals with mathematics, physics and engineering. Jacob's formula plays a prominent role in Mascheroni's book, and in spirit we are not far from Galileo.

Conclusion: Galileo's impulse on mechanics

In Galileo we may see how the medieval tradition of artisanship and engineering handed down from master to master merged with the much older Greek tradition laid down in their writings. The man who now must effect this fusion is also mentioned at the beginning of the *Discorsi*: he is an *intelletto speculativo*, a speculating intellect. As the content of the book makes clear, he speculates about the laws of nature. Perhaps, using a philosophical term, he has "productive imagination"; *intelletto speculativo* may also be translated as "imaginative mind". No one will deny that Galileo had this faculty in enormous measure. Apart from the new results, what is new in his procedure?

What strikes us most in the *Discorsi* is Galileo's explicit and systematic use of mathematics. When he says, "Mathematics is the language of Nature", this is not a mere metaphor. He proceeds by rigorous deductions and proofs anchored in axioms. Never mind that the axioms formulated are not yet dynamical but merely kinematical; his desire for a rigorous foundation *more geometrico*, "by means of geometry", is there for everyone to see. He sets forth mechanics in the dual way the Greeks had done: mechanics is presented as the science of the space and time in which we live, of the bodies that move in them, and of the forces that act between them, but the theory is presented as a coherent set of theorems, all proved by starting from a few rules and with no internal contradictions. All those theorems proved mathematically must also be confirmed by experiment, even if this is difficult and tedious. If Galileo's presentation of mathematics is not always systematic, we nevertheless see that he uses it everywhere. But this use of mathematics is made possible only by a clear and often explicit use of concepts, since the observation and experiments presented must be reproducible by anyone. For this reason, the prescriptions transmitted in the *Discorsi* for performing them with special tools and apparatuses are extensively described and always expressed in unambiguous notions. It was clear to

Galileo as it was clear to probably only a few before him that the intercourse between theory and experiment needs not only elaborate mathematics but clear, explicitly defined concepts. And, as we saw, his use of the moment of a force survived in the theory of elasticity and was to stimulate a long ulterior development.

It is often said that Galileo always proceeded from observation and experiment to theory. That he did observe constantly and was stimulated by observations, and that he sometimes performed experiments, is true enough. But observations had been made in the Middle Ages, long before him. In the example we have examined here, we have seen that observation and experiment were the least successful part of what he actually achieved, and there are other examples which show the same. What is really new is that Galileo considered – or better, constructed – through mathematical formulation intellectually ideal objects, which are called today "models". Galileo's models are simplified objects which, however, retain all the essential properties of the object or system studied, just as we still construct them today, even if ours are perhaps in general more refined and more complicated. Here too Galileo was following the old recipe of the engineers and artisans. Had they not built fortresses, canals, dams and, above all, cathedrals using geometrically simple elements whose static properties they had tried out? And did they not repeat again and again simple constructive connections of which they had the well acquired certitude that they would last and not crumble?

What we observe in nature is infinitely complicated and diversified, while the theories which we construct are always simple and oversimplified. (Hence, by the way, the drama of the uniformity of modern technology.)

But Galileo introduced mathematics not just for computing, but for penetrating into the laws and formulating concepts and equations. I cannot better describe the difference between what his predecessors, the artisans and engineers, had done and what he did with the help of his mainly still Greek mathematics than to quote the motto which the twentieth-century artist, musician and pedagogue Alfred Cortot adopted for the institution which he had founded in Paris: *N'étudiez jamais un passage difficile, mais toujours la difficulté en elle même!* Translated into science: "Never be content with an ad hoc solution of some problem, but try to find *the laws* themselves that govern this situation as they also do many others!"

Indeed, while before, people had solved the problems of the construction of solid fortifications, had successfully designed canals here and there, improved on guns, and surmounted the immense difficulties of building cathedrals and towers, now they began to study the *laws* of statics and solidity, the *laws* of hydraulics and hydrodynamics, the *laws* of ballistics, of elasticity, etc. Once these laws were understood in a certain domain, people could solve problems which they had invented themselves, and apply them for practical purposes. In short, technology took off on its own grand flight. But underlying technology are *the laws of nature*, as Galileo saw, and this is perhaps what puzzled and occupied him most. Indeed, if I were asked to identify what I think was his greatest legacy for us, I would answer that no one contributed so much to how we think today of *a law of nature*. To be truthful, since Archimedes, or even earlier, very few before Galileo had done so; and after him came others (though in fact not so many), who understood the essence of laws of nature better and went much beyond what he could grasp. To Galileo, the flow of water and the phenomena of sound seemed separate fields, but today they are connected through the laws – that is, the equations – of hydrodynamics. But if his followers could see more clearly, we must not forget that it took almost two hundred years after the appearance of the *Discorsi*, from 1638 to 1826, to lay the foundations of classical mechanics alone.

In science, developments are slow and it is difficult to reduce periods. Still if we look over the past, the *Discorsi* is probably the document that best and most clearly testifies to the merging of Greek mathematics and medieval artisanship and engineering, out of which sprang modern natural science.

Acknowledgments

Part of this conference was written at the I.H.E.S. at Bures-sur-Yvette (France) where, thanks to Profs. M. Berger and L. Michel, I could discuss these topics with L.A. Radicati; it was finished at the Scuola Normale Superiore di Pisa. To both institutions I should like to express my gratitude.

Galileo Galilei

The Second New Science in Galileo's *Discorsi*[1]

To Pisa and all the friends I made in that beautiful city

Introduction

Towards the end of the first period of the development of mechanics, during which the fundamental principles of this discipline were made and tied together, Joseph-Louis Lagrange, in his famous *Méchanique analitique*, wrote:

> Galileo was the first to take this important step, and so opened a new and enormous road for the advancement of mechanics. This discovery is set forth and developed and the work entitled *Dialoghi delle scienze nuove* etc. [*sic*], which appeared for the first time in Leyden in 1637. These did not earn as much celebrity for Galileo during his lifetime as those that he wrote on the system of the world, but they are today the most solid and real part of this man's glory.[2]

Lagrange underlines an important point: although the *Sidereus nuncius* [1610] earned Galileo immediate celebrity, it is in the *Discorsi e dimostrazioni matematiche intorno à due nuove scienze* [1638] that today we can identify his real legacy, which is unfortunately unknown to most people (fig. 1).

The aim of this article is to illustrate the essence of Galileo's achievement in order to render at least the part of this extraordinary book written in Italian accessible to the reader, to pique his curiosity, and to explain the effects that it had on the development of science.

The *Discorsi*

As was the *Dialogo [...] sopra i due massimi sistemi del mondo* [1632], the *Discorsi* is written in the form of "dialogues" that take place over the course of four days (*giornate*) at the Arsenal in Venice.

---

[1] Originally published as "La seconda Nuova Scienza nei "Discorsi" di Galileo", pp. 251-290 in *Principio di Secol Nuovo,2 Saggi su Galileo*, Ed. L.A. Radicati di Brozolo, Pisa: Cassa di Risparmio di Pisa, 1999.

[2] *Galilée a fait le premier ce pas important, & a ouvert par-là une carriere nouvelle & immense à l'avancement de la Méchanique. Ces découvertes sont exposées et développées dans l'ouvrage intitulé :* Dialoghi delle scienze nuove, *&c. lequel parut pour la premiere fois à Leyde en 1637. Elles ne procurerent pas à Galilée, de son vivant, autant de célébrité que celles qu'il avoit faites sur le systême du monde, mais elle sont aujourd'hui la partie la plus solide & la plus réelle de la gloire de ce grand homme* [Lagrange 1788: partie II, sect. I].

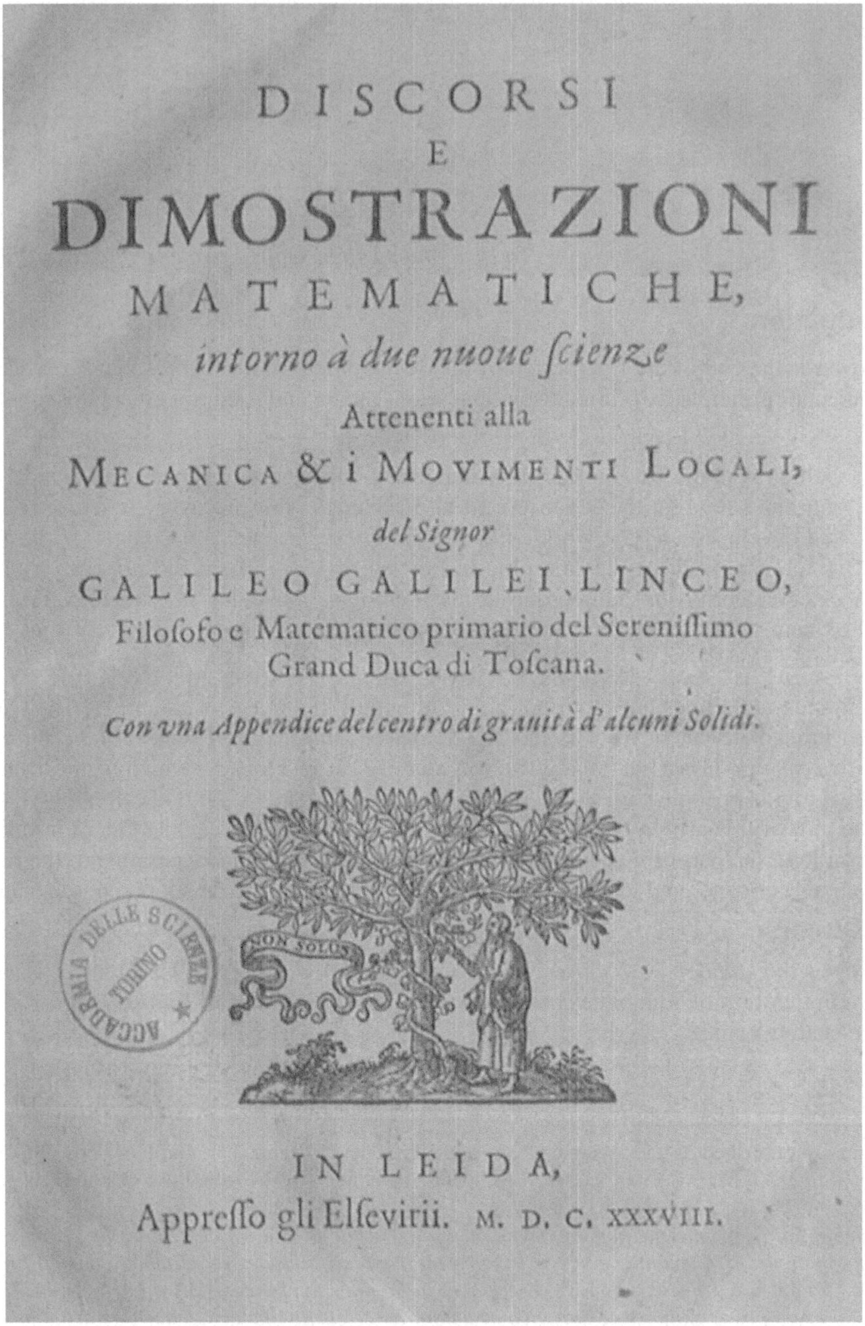

DISCORSI
E
DIMOSTRAZIONI
MATEMATICHE,
intorno à due nuoue ſcienze
Attenenti alla
MECANICA & i MOVIMENTI LOCALI,
del Signor
GALILEO GALILEI, LINCEO,
Filoſofo e Matematico primario del Sereniſſimo
Grand Duca di Toſcana.
Con vna Appendice del centro di grauità d'alcuni Solidi.

NON SOLUS

IN LEIDA,
Appreſſo gli Elſevirii. M. D. C. XXXVIII.

Fig. 1. Frontispiece of the first edition of the *Discorsi*

The protagonists are the same as the *Dialogo*: Salviati, Galileo's Florentine friend and disciple, and his spokesman; the Venetian Sagredo, another friend;[3] and Simplicio, a figure that Galileo invented. Simplicio is less learned than Sagredo, especially regarding the fields of modern science and philosophy, but he is not the simpleton often described in the literature; his objections are often interesting and are always taken seriously by the other two.

Of the four dialogues, the first two are dedicated to the solidity of bodies and materials, or as Galileo says, to their "resistance" (Cf. [Villaggio 1999]); the other two are dedicated to "local motion" and the "motion of projectiles" and deal with what we would today call dynamics, that is, the science that Galileo was the first to set forth systematically. The first and third dialogues fill some hundred pages each, while the other two are only half as long, an important fact, as we shall see. The book is written partly in Italian and partly in Latin, but the two parts do not play the same role, and their proportions vary greatly from dialogue to dialogue. The first two are written entirely in Italian. In the two dialogues that interest us here, that is, the third and the fourth, both languages are used, but in varying proportions. About three-quarters of the third dialogue are written in Latin, while less than half of the fourth is. The Latin text contains a principle, a succession of axioms, definitions, problems, theorems and corollaries that are rigorously proven. This parts provide the book the systematic rigor for which it is justly famous, and to which I intend to return. The text is presented as a treatise written by "our Academic" and Galileo, in laying out the *Discorsi*, probably resorted to his old manuscripts, in particular, his notes from his time in Padua.

But it is above all in the dialogues in Italian, in keeping with a tradition that goes back to Plato, that scientific problems are discussed, objections are formulated, and results evaluated. Often these discussions are genuinely "philosophical", a term used at the beginning by Galileo himself, and these make the book especially attractive and fascinating. In some parts of the dialogue of the fourth day, Galileo mentions the sources of his knowledge, a subject that I will discuss first.

The sources of Galileo's new science

The four most important threads contributing to the foundation of the new science are medieval craftsmanship, medieval science, the influence of Greek science, and the Renaissance.

1. Medieval craftsmanship

The first thread is medieval craftsmanship and engineering. At the beginning of the *Discorsi*, Galileo says explicitly, in the words of his spokesman Salviati:[ii]

> It seems to me, good Venetian sirs, that frequent visits to your famous shipyard (*arsenale*) offer ample opportunity for philosophizing to an enquiring mind, especially that part of enquiry involving mechanics; since

---

[3] It is probable that it was from Sagredo that Galileo learned of the Copernican system through the mathematician Christian Wurtysen of Basel (not of Rostock).[i]

> here all types of instruments and machines are constantly being put to work by a great number of artisans, some of whom, partly through what they have learned from those who came before them, and partly through their own observations from their own experience, are necessarily greatly expert and highly intelligent.[4]

That this is not pure rhetoric is proven by the fact that in the "Table of the most noteworthy things" at the end of the book is the item *Arsenale di Venezia, gran campo di filosofare à gl'ingegni* "Venetian Shipyard, great arena of philosophizing for scholars", surely placed there by Galileo himself, or at least in obeyance to his explicit instructions.

Lynn White [1969] has underlined the importance of the report of ambassadors of the Byzantine Empire, in which they express their wonder at the new inventions that they observed in Italy: the Venetian ships, the canals, the fortresses, and especially, as the fourth day attests, the extremely developed art of artillery. Even today we marvel at the clean monuments of Pisa's Piazza dei Miracoli, at Giotto's Campanile and Brunelleschi's dome in Florence, and at many other wonders of engineering. And it was precisely in Italy that, according to Guy Beaujouan, that there arose a new profession: engineering [1957: 572]. An engineer, the Picard Pierre le Pélerin de Maricourt, wrote a treatise in Lucera on magnetism that was probably based in part on what he had learned from the Arabs, and which is still admirable for its truly modern organization of the subject.[iii] Thus Galileo was basing his work on a very old tradition.

D'Alembert wrote in his *Discours preliminaire à l'Encyclopedie*:

> Regarding the distinction between the liberal and mechanical arts and the superiority attributed to the second with respect to the first. This superiority is without a doubt unjust in several ways … it may be that the most admirable demonstrations of sagacity, spirit, patience and its resources should be looked for among the artisans …[5]

2. Medieval science

A second thread, which unfortunately has not so far been studied thoroughly enough, is medieval culture, and in particular, the way in which philosophy and science were taught in the universities. Medieval science, as we know today, had a greater influence on Galileo

---

[4] *Largo campo di filosofare a gl'intelletti specolativi parmi che porga la frequente pratica del famoso arsenale di voi, Signori Veneziani, ed in particolare in quella parte che mecanica si domanda; atteso che quivi ogni sorte di strumento e di machina vien continuamente posta in opera da numero grande d'artefici, tra i quali, e per l'osservazioni fatte da i loro antecessori, e per quelle che di propria avvertenza vanno continuamente per se stessi facendo, è forza che ve ne siano de i peritissimi e di finissimo discorso* [Galileo 1638 in *Opere*, VIII: 49].

[5] *De-là la distinction des Arts en libéraux & en méchaniques, & la supériorité qu'on accorde aux premiers sur les secondes. Cette supériorité est sans doute injuste à plusieurs égards (...) ... c'est peut-être chez les Artisans qu'il faut aller chercher les preuves les plus admirables de la sagacité de l'esprit, de sa patience et de ses ressources ...* [1751 : XIII]. I am indebted to Mr A. Englebert (Bruxelles) for showing me d'Alembert's judgment, which is much less known than it deserves.

than historians of the nineteenth century thought.[6] My hypothesis – or perhaps it is a speculation – is that this is illustrated by a characteristic aspect of Galileo's way of thinking: his preference for the circle and circular motion, a preference that the French scholar Alexandre Koyré traced to the *hantise de la circularité* (the spell of circularity).

In addition to his treatise on magnetism, Pierre le Pélerin wrote two other treatises, one of which has been lost, on concave – and probably convex– mirrors, and on the astrolabe. In all three treatises the circle plays a fundamental role and the theorems regarding circles are essential. For this reason, among others, I suspect that medieval geometry – which, as a discipline, was only set forth approximately and not axiomatically – was not geared towards Euclidean geometry and its applications but rather towards what we would today call "conformal geometry". In conformal geometry circles are the fundamental curves, while straight lines are considered degenerations of circles. Angles are well-defined, as are the relationships of line segments and circular arcs, but the lengths of line segments and circular arcs are not well-defined. However, this geometry suits two natural sciences – optics and astronomy – in many ways, because in these sciences it was possible to take exact measurements, precisely, those of angles, which furnished an empirical basis for every "system of the world". It is sufficient to mention the best known of these, the great geometrical construction of Dante's cosmos, which is based on circles and the geometry of which can be justifiably defined as "conformal". For example, in the *Divine Comedy* we always know exactly what direction Virgil and Dante take, but not what distances they travel. The distances are never given; at most the ratios between them are given. It would be a serious error indeed to neglect the influence of one Tuscan on another, especially since Galileo presumably knew extensive parts of Dante's poem by heart.

3. The influence of Greek science

Greek science was known throughout the medieval period, particularly in Pisa, through the Arabs. This is shown by the works of Fibonacci (Leonardo da Pisa). Personally, I am convinced that the Baptistery and the Leaning Tower testify to the geometric knowledge of the architects in Pisa [Speiser 1994: 511]. When in 1453 the Byzantine court found refuge in Ferrara, it brought with it many works which if not completely unknown up to that time had been only vaguely heard of in the West, and the effect that these had on Western science was great and lasting.

The strongest and most significant influence on Galileo was Archimedes, founder of the science of hydrostatics and discoverer of the laws of the lever. Both of these intellectual achievements were reprised by Galileo, who especially studied hydrostatics from the point of view of the construction of ships with an eye to both practice and engineering. The dialogue of the second day can be defined as a treatise on the laws of the lever and its applications.

In the opening pages of the dialogue of the fourth day, which contain what are perhaps Galileo's most original contributions, three Greek mathematicians are introduced in this order: Apollonius, Euclid, and Archimedes. If we analyze the text philologically, we see that Euclid, with whom even those of the old school were familiar, is mentioned by

[6] For more detailed discussion, the reader should consult the works of P. Duhem, M. Clagett, and C. Truesdell.

Simplicio [Galileo 1638 in *Opere*, VIII: 270], and Apollonius, the discoverer of the properties of the parabola, which is the principle subject of the fourth dialogue, is mentioned by Sagredo [Galileo 1638 in *Opere*, VIII: 269], who is knowledgeable in modern science; but that it is Salviati, Galileo's spokesman, who mentions Archimedes [Galileo 1638 in *Opere*, VIII: 270], the founder of mechanics, which is precisely the science that Galileo sets forth in this book.

4. The Renaissance

Galileo was fully a man of the Renaissance. By the term "Renaissance" I mean a complex of correlated developments that arose in different places and at different times. Here I will mention only two: philosophy and the arts.

The Medieval knew only a part of only one of Plato's dialogues, *Timaeus*; with the arrival of the Byzantine court in Ferrara, the entire body of Platonic dialogues was translated into Latin and spread throughout the western world. It is difficult to deny the great influence of Plato on Galileo, as Koyré has made particularly evident in his *Etudes galiléennes* [1939]. The mere fact that, as in the Platonic works, two of the Galileo's books were written as a succession of dialogues – a form which, after Galileo, was widely used by others – demonstrates this influence; it is further shown by the fact that Plato is explicitly cited in the *Dialogo*. But exactly what was Plato's influence on a scientist who worked some two thousand years later? The first answer lies in the fundamental importance that both Plato and Galileo attribute to mathematics, and above all, to geometry. The laws of nature can be formulated mathematically, that is to say, by means of ratios between lengths; these are based above all on properties of curves and other geometric figures which suggest shapes and permit their quantitative formulation. Galileo was certainly guided by his geometric imagination in searching for these new laws: that he was inspired by accurate observations is clear, while the question of whether he was also guided by actual experiments, or whether these served only to confirm his conjectures is perhaps less important that often believed. In any case, it was surely the mathematical method, the results of which Plato had praised, that allowed Galileo to make his discoveries, and his enthusiasm for geometry recalls the enthusiasm expressed in many of Plato's dialogues, an enthusiasm which, after Galileo, would live on in his successors.

But there is another aspect that shows how much Galileo was a son of the Renaissance: his love for the arts and his interest in philosophical disputes about beauty. Art historian Erwin Panofsky analyzed a letter from Galileo to his Tuscan friend Lodovico Cardo (Cigoli) [Galilei 1612], a letter which was excluded from the edition of Galileo's complete works [1890-1908] because it was judged unauthentic.[7] Panofsky, not aware that M. Màrgani had proven the authenticity for stylistic reasons [Màrgani 1921-22], reached the same conclusion for intrinsic reasons. It is not so much the principal argument of the letter, that is, the superiority of painting over sculpture, which interests us, as the light that it sheds on the *hantise de la circularité* of Galileo. For the Renaissance, the circle and the sphere were ideal figures, and Galileo had admired many circles in Pisa as he strolled in

---

[7] [Panofsky 1954]. An abbreviated version can be found in [Panofsky 1956]. I wish to thank Mrs. Gerda Panofsky who kindly sent me the French translation of this work by Nathalie Heini.

the Piazza dei Miracoli. [8] Panofsky says, "If Galileo's scientific attitude is held to have influenced his aesthetic judgment, his aesthetic attitude may just as well be held to have influenced his scientific convictions; to be more precise: both as a scientist and as a critic of the arts he may be said to have obeyed the same controlling tendencies" [1954: 20]. Should we perhaps agree with Panofsky that Galileo's most serious error, that is, his tenacious refusal of Keplerian ellipses, was due to his Renaissance admiration for the perfection of in the circle, while the ellipse, the child of Mannerism, was despised? Or should we agree with Einstein, who admitted that he was inspired by Galileo, that the fact that no trace of Kepler's decisive step is found in the work of Galileo is a limited example of the fact that creative people often lack receptivity?

Readers of the *Discorsi* cannot help but be struck by Galileo's admiration for beauty and the importance that he attributes to it as a criteria for the search for the laws of nature. This attitude can still be found in scientists in our own times, such as Einstein and Dirac.

The dialogue of the third day

Before analyzing the dialogues of the third and fourth days of the *Discorsi*, I want to warn the reader that only by considering the book as a whole is it possible to appreciate its significance. Some results are already well known, but it is in the new vision of mechanics that the greatness of this work lies.

The third dialogue begins with these complacent words in Latin: *De subiecto vetustissimo novissimam promovemus scientiam* (From the most ancient subject we initiate the newest of sciences): "the most ancient" because nothing is older than motion; "the newest" because even if others had studied it, making considerable contributions, here it will be given a new formulation.

The treatise on dynamics – so we can call the third and fourth dialogues – consists in three parts: a brief one on "uniform motion"; a longer one on "naturally accelerated motion"; and a third one on "violent motion", to which the fourth day is dedicated. Uniform motion is given not even seven full pages, while naturally accelerated motion fills about eighty, which immediately shows that this is where the novelty lies.

The first part contains a definition, four axioms, and six theorems, all with their proofs. Thus it is a dry but systematic formulation in perfect Euclidean style. Here I will limit myself to an examination of only the most interesting parts.

The definition says, "By equal or uniform motion, I mean that in which the distances travelled by a moving object during equal intervals of time, however considered, are equal to each other."[9] This is followed by a caution: "It seems to us opportune to add to the old definition … the expression "however considered," meaning by this, all equal intervals of time",[10] that is to say, no matter how small. This clause is necessary because otherwise the definition would include motion that was not uniform "on average", for example 2 m/s on

[8] The circle is the dominant aesthetic element in the architecture of the Piazza dei Miracoli: careful observation can identify at least nine complete circles and more than two hundred semi-circles.

[9] *Aequalem, seu uniformem, motum intelligo eum, cuius partes quibuscunque temporibus aequalibus a mobili peractae, sunt inter se aequales* [Galileo 1638 in *Opere*, VIII: 191].

[10] *Visum est addere veteri definizioni ... particulam* quibuscunque, *hoc est omnibus temporibus aequalibus...* [Galileo 1638 in *Opere*, VIII: 191].

average, but such that in the first half-second the body moves with a speed of 3 m/s and in the second half-second it moves with a speed of 1 m/s. This is not pure pedantry: here Galileo takes the first step towards the mathematics of infinitesimals. To tell the truth, it is only a small step, but that which seemed a paradox to all the Greeks would only be understood and resolved much later by Fermat, Barrow, Newton, Leibniz, the Bernoullis and Euler. In any case, Galileo, as we shall see, will take advantage of this intuition: uniform motion means much more than average uniform motion.

A similar difficulty is reflected in the four axioms, all of which are formulated only under the form of inequalities, following a tradition that goes back to the Greek mathematician Eudoxus. For example, the third axiom states, "Given equal intervals of time, a distance traversed at a greater speed is larger than one traversed at a lesser speed".[11] This is perhaps the first time that the concept of speed has been set forth axiomatically.

This axiomatic formulation is used in the six theorems, of which I will cite only the last, which is formulated as are those that precede it, in the form of a ratio, a procedure that is characteristic of Galileo's period: "If two objects move uniformly, the ratio of their respective speeds is the product of the ratio of the distances traversed and of the inverse ratio of the times".[12] In modern language, this is equivalent to the banal relationship $V_1 : V_2 = (S_1/T_1) : (S_2/T_2)$, where *V*, *S*, *T* respectively denote indicate speed, space, and time.[iv] This gives an idea of the conceptual difficulty and of the lack of an easily-used mathematical formalism – a difficulty and a lack that it will take a century and a half of hard work by Galileo's successors to overcome. But Galileo, as the reader will see, will get around these obstacles.

For a correct comprehension of Galileo's mechanics it is necessary to underline that while today after Huygens, and above all, after Newton and Euler, we are used to considering rectilinear uniform motion as fundamental, when Galileo speaks of uniform motion he doesn't ever require that this take place in a straight line. To be sure, the figures in the text show straight lines to indicate lengths, but nothing in what Galileo says requires this. What he says is therefore equally valid for a rotating body suspended by a cable, a carriage moving on tracks, or a body that moves along a plane: motions that are today called "constrained". The constraints, as we now know, are particular forces – though Galileo didn't have that concept – that are due to a mechanism, that is, a device or a machine, or in Galileo's case, to the support of an inclined plane. It is important that in many cases the friction exerted on the support is negligible so that, as we would say today, there is no dissipation of energy. Galileo, as we shall see, does not distinguish between unconstrained and constrained motions. This explains many statements that would otherwise appear contradictory.

Salviati concludes this part by saying:

---

[11] *Spatium a maiori velocitate confectum tempore eodem, maius est spatio confecto a minori velocitate* [Galileo 1638 in *Opere*, VIII: 191].

[12] *Si duo mobilia aequabili motu ferantur, ratio velocitatum ipsorum composita erit ex ratione spatiorum peractorum et ex ratione temporum contrarie sumptorum* [Galileo 1638 in *Opere*, VIII: 195].

> What we have seen is what our Author has written about uniform motion. We will go on then to a more subtle and new consideration about naturally accelerated motion, such as that generally effected by heavy falling bodies: and here is the title and the introduction.[13]

On naturally accelerated motion

The reader now arrives at the first summit of the second part of the *Discorsi*, the crucial passage in Galileo's mechanics. We will examine the point where Galileo closes the gap between kinematics, the science of pure motions in space and time, and mechanics, the science of bodies that move in space and time, interacting with other bodies.

Galileo proceeds in the treatment of "naturally accelerated" motion in exactly the same way as he did with "uniform motion". He begins with a definition, then formulates a principle, and from there goes on to a series of theorems and problems. But this time the definition requires a much more complete and precise examination, because he is moving in unexplored territory. Galileo's search for a correct definition for "naturally accelerated" motion is described and discussed by Koyré in the *Etudes galiléenes* [1939], so here I will limit my discussion to the most important points. "Naturally" means "what nature does", rather than "what naturally comes to mind" At the beginning we find a brief description, in Latin, of the problems that must be faced. This part, which is not in the form of a dialogue, might at first appear a little dry, until we find this exclamation:

> Finally, in the investigation of naturally accelerated motion we were led, almost by hand, by the observation of the habit and rule followed by nature herself in all her other processes, of using only the first, simplest and easiest means. Indeed, I think that there is no one who believes that swimming or flying can be accomplished in a manner simpler or easier than that instinctively employed by fishes and birds.
>
> When, therefore, I observe that a stone initially at rest falls from an elevated position and begins to acquire new increments of speed, why should I not believe that such increments take place according to the simplest and most obvious rule?[14]

---

[13] *Questo che abbiamo veduto, è quanto il nostro Autore ha scritto del moto equabile. Passeremo dunque a più sottile e nuova contemplazione intorno al moto naturalmente accelerato, quale è quello che generalmente è esercitato da i mobili gravi discendenti: ed ecco il titolo e l'introduzione* [Galileo 1638 in *Opere*, VIII: 196].

[14] *Postremo, ad investigationem motus naturaliter accelerati nos quasi manu duxit animadversio consuetudinis atque instituti ipsiumsmet naturae in ceteris suis operibus omnibus, in quibus exercendis uti consuevit mediis primis, simplicissimis, facillimis. Neminem enim esse arbitror qui credat, natatum aut volatum simpliciori aut faciliori modo exerceri posse, quam eo ipso, quo pisces et aves instintctu naturali utuntur.*

Dum igitur lapidem, ex sublimi a quiete descendentem, nova deinceps velocitatis acquifere incrementa animadverto, cur talia additamenta, semplicissima atque omnibus magis obvia ratione, fieri non credam? [Galileo 1638 in *Opere*, VIII: 197].

This is a declaration of faith that the laws of nature are distinguished by a great simplicity. Effectively, from Galileo on, in their search for new natural laws, scientists have always stayed true to this creed, which is without a doubt the greatest legacy of Galileo's way of thinking. Its validity has been demonstrated by the scientists who held to it.

Then we arrive at the following definition, proposed by Salviati:

> I say that equally, that is, uniformly, accelerated motion is that in which, starting from rest, in equal intervals of time equal moments of speed are added.[15]

But how do we decide what is "simple"? This is one of the principle questions debated, in Italian, in this fascinating dialogue, during which Sagredo and Simplicio both express serious doubts about Salviati's definition. Questions raised during the discussion are scientific, philosophic, logical, and – only slightly – mathematical, all combined and intertwined, compared and discussed. It is a true melting pot – absolutely ingenious! – but difficult to summarize adequately. Here are some examples.

In one instance, Sagredo asks how can a body that departs from rest, that is, from speed zero, acquire a finite speed? In another, Simplicio, referring to the notion of slowness, *tardità*, asks how a body in motion can stop, given that its *tardività* has become infinite. This is a purely kinematic question; today the introduction of "slowness" $1/v$ is a well-known and often used mathematical artifice.

In this diatribe, Salviati suddenly introduces a purely dynamical subject, saying that the greater the fall (of a projectile) the higher its *momento***.** Thus it is reasonable to believe that the increase in speed is proportional to the time of fall. Sagredo, however, firmly holds to his definition: the speed acquired is proportional to the distance covered. Surely this law appears as simple as that of Salviati, and Sagredo maintains that it is first necessary to identify the "cause" of the fall. At this point Salviati interrupts him with the following words:

> This doesn't seem to me to be the right time to investigate the cause of the acceleration of natural motion, about which various philosophers have expressed various opinions, some explaining it by attraction to the center, others to the fact that there are increasingly fewer parts in between to be crossed, while still others attribute it to extrusion through the surrounding medium which closes in behind the falling body and press it and continually impel it. Now, these fantasies and others ought to be examined but that would be of little value. For now it is sufficient for our Author that we understand that he wants to investigate and to demonstrate some of the properties of accelerated motion (whatever the cause of this acceleration may be) such that the moments[v] of its speed continue to increase after

[15] *Motum aequabiliter, seu uniformiter, acceleratum dico illum, qui, a quiete recedens, temporibus aequabilis aequalia celeritatis momenta sibi superaddit* [Galileo 1638 in *Opere*, VIII: 198].

> departure from rest in simple proportion to time, which is the same as saying that in equal intervals of time there are equal increments of speed.[16]

Here we see the separation between science and tradition philosophy that Galileo clearly perceives: science cannot be explained by identifying causes. The explanation is obtained only after having erected a grand intellectual construction composed of mathematical relationships between concepts that, in their turn, can be verified through experiment, following a clearly specified rule.

Sagredo persists with his hypothesis that the variation in speed is proportional to distance. Salviati counters by saying that first of all he is happy to see that Sagredo has fallen into the same trap that he did. Koyré has shown this admission's autobiographical significance for Galileo, and notes that this error finds a parallel in the life of Descartes [1939]. Salviati then refutes Sagredo's hypothesis with an argument which is admittedly flimsy. He says that if speed is proportional to distance travelled, then these distances are travelled in equal times, which isn't true: the reason for Galileo's error doesn't lie in philosophical prejudice, as has been said, but simply reflects the lack of mathematical competence of his times. Rather, the hypothesis would lead to an exponential increase in speed, and therefore the time would be proportional to the logarithm of the distance; but this mathematical function was not known in that period.[vi] In any case, all of the protagonists finally agree on the definition proposed by Salviati. From this point on, *dico*, "I say", is substituted by *dicemus*, "we say".

Two further considerations can be made. The first is that in spite of the confusion in the discussion, the correct conclusion is reached, which is that the variation in speed is proportional to the variation in time. Perhaps part of this discussion's fascination is due precisely to this confusion: Galileo, with only modest – indeed, insufficient – mathematical tools and, if you will, swimming in a quagmire of apparent paradoxes and contradictions, finally finds the right path, and finds it by a purely theoretical means. This should give pause to those who hold that Galileo's discoveries were based first and foremost on experiment and observation.[vii]

The definition of naturally accelerated motion has always been rightly held to be significant, and what follows from it is essentially the crucial point that lead to dynamics, the "new science". Salviati can now proceed to bridge the gap between kinematics and dynamics that I mentioned earlier.

---

[16] *Non mi par tempo opportuno d'entrare al presente nell'investigazione della causa dell'accelerazione del moto naturale, intorno alla quale da varii filosofi varie sentenzie sono state prodotte, riducendola alcuni all'avvicinamento al centro, altri al restar successivamente manco parti del mezo da fendersi, altri a certa estrusione del mezo ambiente, il quale, nel ricongiugnersi a tergo del mobile, lo va premendo e continuamente scacciando; le quali fantasie, con altre appresso, converrebbe andare esaminando e con poco guadagno risolvendo. Per ora basta al nostro Autore che noi intendiamo che egli ci vuole investigare e dimostrare alcune passioni di un moto accelerato (qualunque si sia la causa della sua accelerazione) talmente, che i momenti della sua velocità vadano accrescendosi, dopo la sua partita dalla quiete, con quella semplicissima proporzione con la quale cresce la continuazion del tempo, che è quanto dire che in tempi eguali si faccian eguali additamenti di velocità* [Galileo 1638 in *Opere*, VIII: 202].

The basis for Galileo's dynamics: the "Principle"

At the end of the discussion, Salviati says, "That definition [of uniformly accelerated motion] established, the Author requires and holds true one single principle".[17]

Galileo's *principio* (principle), set forth by Salviati in Latin, is: "I assume that the degree of speed acquired by the same body on variously inclined planes is always the same provided that the height of the planes is the same."[18]

In order that the reader can appreciate the novelty and the significance of this principle, it is necessary to clarify a few points.

First, why does this mark the beginning of a new science that goes beyond geometry and kinematics? Is this truly something more than a pure observation regarding "speed" and "height"? Why do we speak here of "dynamics"? The crucial point is that the observation regards bodies and their interactions: the "moving body", the plane along which it descends, and of course, the earth towards which it falls, or as we would say now, to which it is attracted. But Galileo did not have the notion of a "force of attraction", a notion that would be introduced only later by Newton. How then is Galileo able to formulate it? From the modern point of view it is clear that we have here a proposition that concerns the energy of a body; in precise terms, the kinematic energy of a falling body depends only on the height of its fall. This is essentially the idea that Galileo had in mind (the kinetic energy is produced by the mass times the square of the speed divided by two). But Galileo did not know this relationship, nor did he know of that between mass, velocity and momentum, nor of that between kinetic energy and momentum. Finally, he did not know that the sum of kinetic energy and potential energy remains constant.

These are all relationships that seem obvious to us today, but which were discovered only after Galileo's time by Descartes, Huygens, Newton and Leibniz. The miracle – and it is truly an extraordinary one – is that beginning with a modest assumption Galileo was able to find all he did. In any case, Galileo now had at his disposal a precise law of nature in a new field. Having understood that it was necessary to have such a law as an axiom or "principle" is one of his greatest achievements. That his reasoning is sometimes defective is ultimately of little importance: all that he will say in what follows establishes the basis for the second new science, which we today call "dynamics".

Second, the reader who is accustomed to Newton's way of presenting mechanics must rid himself of many ideas and notions learned there. For example, only vertical fall is what we today call "free fall". Motions supported by a plane are now called "constrained". To be more precise, in our case we are dealing with a fall that is subject to "constraints".

To be truthful, the Newtonian concept of force is more powerful than the principle of energy that allows Galileo to deal with problems of only one degree of freedom, in which the trajectory is given and it is necessary only to determine the speed starting from the energy and the height from which the object falls. Galileo does not therefore need a detailed analysis of the forces, as the following discussion will show.

---

[17] *Fermata cotal definizione un solo principio domanda e suppone per vero l'Autore* [Galileo 1638 in *Opere*, VIII: 205].

[18] *Accipio, gradus velocitatis eiusdem molbilis super diversas planorum inclinationes acquisitos tunc esse aequales cum eorumdem planorum elevationes aequales sint* [Galileo 1638 in *Opere*, VIII: 205].

Third, how does Galileo arrive at his principle? Through observation? Or by making experiments? The greatest experimental difficulty is that the body can roll or slide on the plane. If it slides, friction is almost inevitable (especially if the angle of inclination of the plane is small) and the principle is therefore not valid. It would seem then preferable to use a body that rolls, but if the angle of inclination is very large, and obviously when the fall is vertical, a sphere or cylinder does not actually roll.

Further, for bodies that descend by rolling there is an additional term in the principle of energy which Galileo could not have known about. It thus appears impossible to verify the formula experimentally for all values of the angle of inclination in the case of translational motion alone as well as in the case of rolling motion.

Galileo's way of getting around these difficulties is brilliant. Let us consider a pendulum with a weight suspended from a cord. Released, the pendulum will oscillate downwards and will then go upwards again to exactly the same height, since, as we would say, it conserves energy. But what happens if we put a nail in the wall, say at point *F*? (fig. 2). The pendulum will still go upwards again to exactly the same height. The nail can be placed at any position whatsoever, as long as it is neither above the point of suspension nor too far below, since otherwise the cord would begin to wind around the nail.

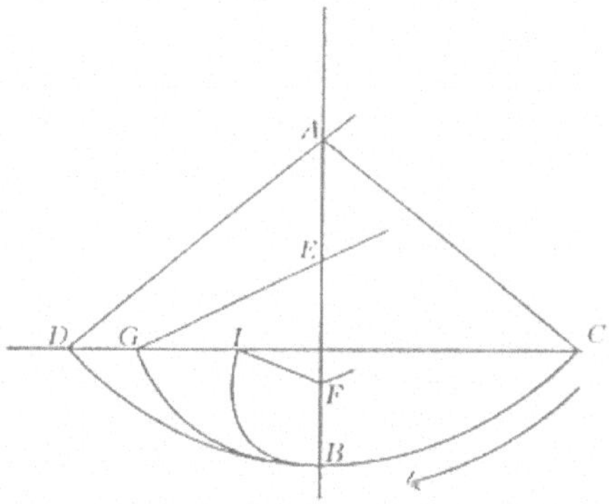

Fig. 2

This is an experimental fact that can be verified with a high degree of precision. In contrast, verifying the relationship between the height of the fall and speed of the body at its lowest point is much more difficult, even though by using two pendulums that collide after swinging from different heights, it is possible to obtain qualitative information. In any case, the relationship cannot be easily established through observation, and certainly not with a high degree of precision. Galileo is aware of this and has Salviati say only "... but beyond what is reasonable, I want to increase greatly the probability with an experiment, so that it comes very close to being a rigorous proof.[19]

Indeed, in his long final discourse before the systematic presentation of the new science Salviati explicitly states:

---

19 *...ma oltre al verisimile, voglio con una esperienza accrescer tanto la probabilità, che poco gli manchi all'uguagliarsi ad una ben necessaria dimostrazione* [Galileo 1638 in *Opere*, VIII: 205].

> So that, even if the experiment undertaken shows us that the descent along arc CB confers a *momento* to the moving body such that it carries it up to the same height on any arc BD, BG, BI, we cannot show with the same clear evidence that the same thing happens when a perfectly round ball is made to descend along straight planes, inclined in accordance with the chords of these same arcs; rather, it is to be believed that, since these planes form angles at point B, the ball descends along the incline following chord CB, and encounters an obstacle in the ascending planes that follow chords BD, BG, BI; in colliding with them it would lose its *impeto*, it could, rising, carry it to the height of line CD: but removing the obstacle, which prejudices the experiment, it seems conceivable to me that the momentum (which effectively is augmented by the amount of the descent) would be powerful enough to carry the moving body to the same height. Let us therefore take this as a postulate, the absolute truth of which will be established for us from seeing other results, built on such hypotheses, as correspond and are comparable to experiment.[20]

Fourth, the last sentence of Salviati would seem to indicate that Galileo thought that his principle, even if certainly valid, was not the most fundamental law. The search for a fundamental law of dynamics appears to have occupied him for the rest of his life, as shown by Viviani's summary of a conversation with Galileo during the last weeks of his life [1717], published by Antonio Favaro, editor of the complete works of Galileo [1890-1908].

Finally, in the same final discussion in Italian, Salviati speaks of *momentum*, which corresponds to our momentum. But Galileo did not know about the relationship between momentum and velocity since he did not have the concept of mass, nor did he have a way to measure it, for example, by means of the law of impact, as Huygens would later. He therefore could not know the relationship between momentum and kinetic energy.

Although Galileo based his mechanics on a rudimentary principle of energy, one gets the impression that his imagination brought him close to the idea of momentum. This was due principally, as the fourth dialogue will show, to his interest in ballistics. As a consequence, many of his problems and theorems that seem to us to be out of place are proven only in order to translate statements regarding momentum into his "personal" formalism based on energy. On

---

[20] *Di modo che, se ben l'esperienza addotta ci mostra che la scesa per l'arco CB conferisce al mobile momento tale, che può ricondurlo alla medesima altezza per qualsivoglia arco BD, BG, BI, noi non possiamo con simile evidenza mostrare che l'istesso accadesse quando una perfettissima palla dovesse scendere per piani retti, inclinati secondo le inclinazioni delle corde di questi medesimi archi; anzi é credibile che, formandosi angoli da essi piani retti nel termine B, la palla scesa per l'inclinato secondo la corda CB, trovando intoppo ne i piani ascendenti secondo le corde BD, BG, BI, nell'urtare in essi perderebbe del suo impeto, ne potrebbe, salendo, condursi all'altezza della linea CD: ma levato l'intoppo, che progiudica all'esperienza, mi par bene che l'intelletto resti capace, che l'impeto (che in effetto piglia vigore dalla quantità della scesa) sarebbe potente a ricondurre il mobile alla medesima altezza. Prendiamo dunque per ora questo come postulato, la verità assoluta del quale ci verrà poi stabilita dal vedere altre conclusioni, fabbricate sopra tali ipotesi, rispondere e puntualmente confrontarsi con l'esperienza* [Galileo 1638 in *Opere*, VIII: 207].

the whole, we cannot help but admire Galileo who, beginning with a vast engineering knowledge and solid traditional notions, and armed only with a slight and fragmentary mathematical knowledge, was able to place the dynamics of systems with only one degree of freedom on a solid base. With these means he was able to discover a law of nature, and postulating it as a principle, he built on it his conceptual construction. It is to this that we will now turn our attention.

The first part of the new mechanics

The remaining part of the third dialogue contains a systematic presentation of Galileo's theoretical deductions and discoveries. It is organized into twenty-two theorems, sixteen problems, nine corollaries, three lemmas, and three *scholia*, that is, into fifty-three propositions. The majority of the theorems regard dynamics or kinetics; the proofs, mostly conducted according to the style of Euclidean geometry, begin with a single axiom of dynamics, the "principle". The method is one that we still use today, that is, beginning with the simple questions and proceeding on to those of increasing complexity. Naturally, not all of them are equally important, so I will only consider a few here.

The presentation is in Latin, interspersed with some very interesting discussions in Italian. The first theorem proves the so-called "Rule of Merton" discovered by members of Merton College at Oxford University in the fourteenth century. It says that the time necessary to travel a given distance with constant speed is equal to the acquired speed of the moving body at the midpoint of the distance. Galileo's proof makes use of elementary Euclidean geometry. In order to understand his proof, the upper diagram of fig. 3 shows Galileo's figure, while the lower diagram shows the same figure rotated and reflected. It is easy to see from this that the area ABFGA is equal to the area AFBEA (time × speed = distance).

Line AE represents the speed, increasing linearly, with which the moving body falls. Line GF represents the propagation with constant speed, which is, as the figure shows, equal to that acquired by the body at the midpoint I. The distance is, in the first case, equal to the area of triangle ABE; in the second case it is equal to the area of rectangle ABFG, that is, the two areas are equal. This calculation of area is called "integration", an operation that was systematically developed and generalized only fifty years later by Newton, Leibniz, and Jacob and Johann Bernoulli. Here we see a special, elementary case in geometric form of this fundamental operation.

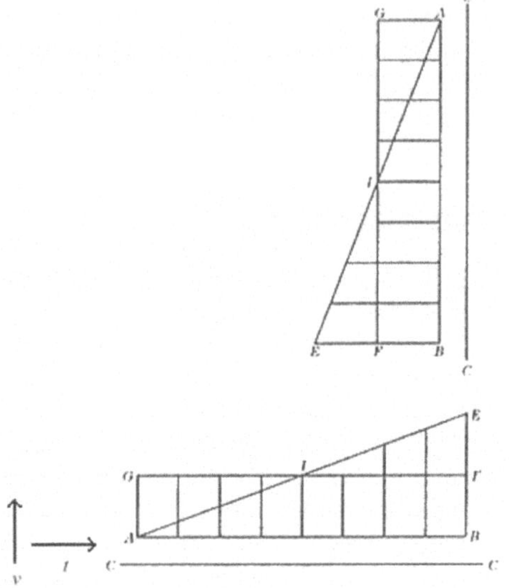

Fig. 3

But we also see something else, and it for this reason that Galileo's figure has been rotated and reflected. Here he takes a small step towards the use of coordinates, and for this reason we have added arrows that show the two coordinates t and v, and set forth the proof in the language of analytical geometry – that of coordinates – developed contemporaneously by Descartes and Fermat.

It was perhaps his musical knowledge, learned from his father, that suggested to Galileo this small step in a problem concerning time. It appears in fact that musical notation was the first coordinate system in current use: the height of the sound – its frequency, as we would say in physics – is noted as a function of time. It doesn't matter that there are only nine (or slightly more) frequencies used; what is important is the representation as a function of time of the value of a quantity, the frequency (more precisely, its logarithm) of a sound, or the speed of a body. The use of coordinates was one of the keys that opened the door to all sciences that describe events or processes that evolve over time. We will see that Galileo goes still farther in this direction.

The second theorem sets out and explains (naturally, only in words) the relationships between speed, acceleration and time ($v = gt$), and between distance travelled, acceleration and the square of time ($s = gt^2/2$), valid for any time period whatsoever, meant by Galileo as an effective variable. If, for example, we indicate by T the final instant of fall, we have $V = gT$, $S = gT^2/2$. Inserting the final speed V, given by the first equation, into the second, we obtain $S = VT/2 = gV_M T$, where S and V are respectively the total height and the final speed, and V_M the speed at the midpoint. These relationships are thus valid for the final speed as well, for the distance travelled during the fall, and for the final time. Galileo's ideas are not very far from a modern formulation, but he does not possess the necessary conceptual and mathematical apparatus.

What we have seen up to this point was in part known to others, such as the Spanish Domenican Domingo de Soto, professor at the University of Salamanca, who understood the laws of free fall of weights. But Galileo goes further, deducing from his principle results that are increasingly difficult to achieve.

As an example, I quote Galileo's Theorem VI, for which he gives *two* proofs:

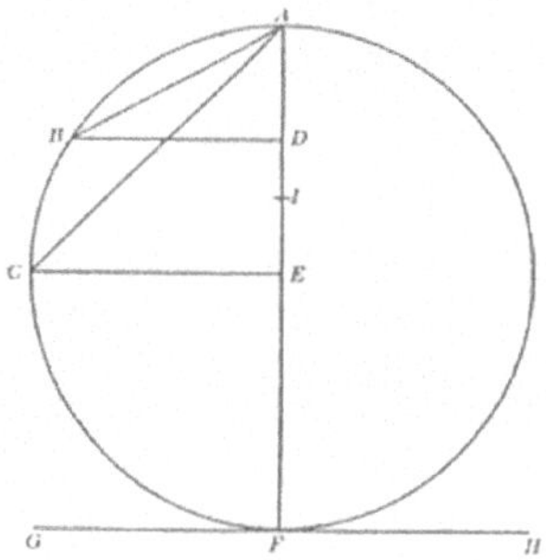

Fig. 4

In a circle that is upright with respect to the horizon, if any inclined planes whatsoever are drawn from its highest or lowest point to the circumference, the times of descent along these are equal to each other[21] (fig. 4).

In itself this theorem doesn't seem of great importance, but it is an example of Galileo's interest in rigorous and generalized results.

At almost the end of the third dialogue we find the last of the three *scholia*, which, preceded by a series of theorems and no fewer than three lemmas, answers a new question, one that in its generality and significance went beyond what had been thus far presented and discussed in consequence of the principle. As we would say today, it deals with a problem of minimization; to be more specific, it answers the question of what is the curve joining two points A and C along which a descending body requires the minimum possible time to travel? Today we call this curve the "brachistochrone", from the Greek *brachiston*, shortest. At first glance one might think that this curve is simply the shortest one geometrically, that is to say, a straight line, but upon greater reflection a curve appears more likely. After what we have understood up to this point, we can imagine the solution that Galileo gave: the curve is a lower quadrant of a circumference, or part of it (fig. 5). Indeed, Galileo is able to prove that the time required to descend along a quarter-circumference that joins two points is inferior to the time necessary to travel a straight line segment joining them.

[21] *Theorema VI, Propositio VI. Si a puncto sublimi vel imo circuli ad horizontem erecti ducantur quaelibet plana usque ad circumferentiam inclinata, tempora descensuum per ipsum erunt aequalia* [Galileo 1638 in *Opere*, VIII, p. 221].

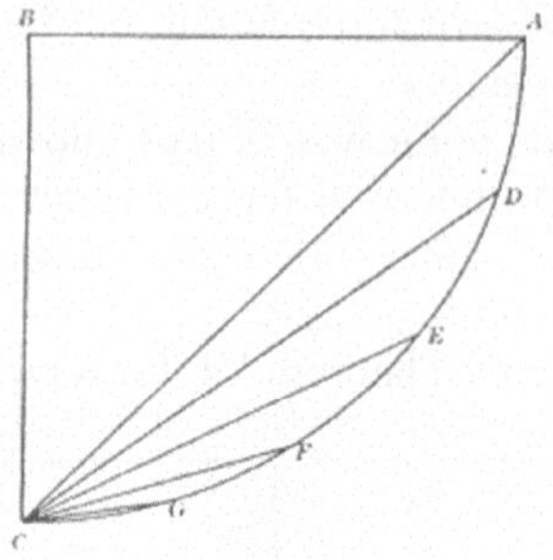

Fig. 5

The final solution is not that given by Galileo. There are in fact other curves that permit a still more rapid descent. Defining – nay, even considering – "all possible curves passing from A to C" was in itself a new idea, but finding the solution proved to be an operation that was beyond the capabilities of the mathematics of Galileo's time.

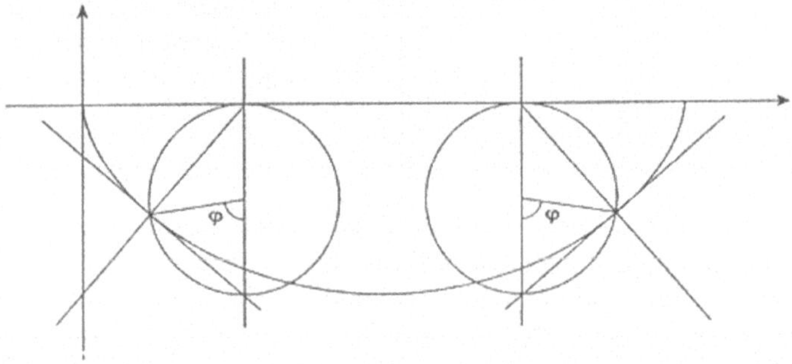

Fig. 6

In science, progress is often more about the positing of a question than its solution. The first to occupy himself with Galileo's problem was Pierre Fermat, who proved that it is the briefest time and not the shortest distance that is the essential property of the propagation of light, which would also be sustained by Huygens in his theory of light. Huygens systematically studied cycloids, the curve described by a point on the circumference of a wheel that moves without friction along a plane. In fig. 6 the curve is indicated in the lower half-plane.

In 1696, only a year after Huygens's death, Johann Bernoulli found the solution to Galileo's problem: a cycloid. Thus the circle, to which Galileo assigned such importance, indeed played a part, although not in the way that Galileo thought. Fermat then challenged other mathematicians to find the solution. Only the most illustrious responded: Newton, Leibniz, the Marquis de l'Hôpital, and the older brother of Johann I, Jacob, who had only shortly before argued that Galileo's solution was the correct one. It is worth mentioning that Johann was influenced by Fermat's theory of optics and so, although he didn't know about the questions and proposals of Galileo, he was indirectly influenced by them. In his solution, Johann Bernoulli was the first to make a mathematical analogy between mechanics and optics. It was then Jacob Bernoulli who developed a systematic method for resolving generally various "problems of minima", as we would say today. Later this method was further developed by Lagrange and by Euler, who had already set out an idea earlier suggested by Maupertuis, which hypothesizes that all laws of nature descend from a principle of minima. This idea would be made even more powerful by Hamilton. Because it is impossible here to recount this long story, it must suffice to say that today the principles of minima have become of the greatest importance in classical mechanics, and even more so in quantum mechanics.

At the beginning of this long and fascinating development is the fertile idea of Galileo. Nothing like it existed before, and at a certain point Sagredo rightly says:

> In truth it seems to me that we can concede to our Academic that he was not immodest in saying, at the beginning of this treatise of his, that he was able to give us a new science on a very old subject.[22]

The dialogue of the fourth day

In the fourth dialogue Galileo takes an even more important step. Up until this point he had only discussed problems of only one degree of freedom, that is to say, problems in which the body is constrained to travel along a given curve. In consequence, the only – albeit difficult – questions regard the speed at each instant and the amount of time used to complete the trajectory.

Here he takes on the more general problem, that of a rock thrown by hand or a projectile shot from a cannon. To be precise, Galileo considers only downwards motion: a body moving along a horizontal platform from *a* to *b* falls, describing a certain curve that must be defined. In posing this problem Galileo formulates the conjecture that the curve is always the same, even if there is no *a priori* reason to think that the fall always happens along the same curve, as occurs in the atmosphere. Galileo gives an abstract answer: a projectile subject to motion that is uniformly horizontal and naturally accelerated downwards describes a half-parabola. The parabola was already known to Apollonius, who many centuries earlier had interpreted it as a section of a cone. But the novelty introduced by Galileo was to lay in out in a grid, thus opening *ante litteram* the use of Cartesian coordinates (fig. 7). But what is a parabola? And why does this curve play a part in the science of ballistics? The answers to these two questions comprise the most fascinating part of the book.

Galileo was motivated by his boundless intellectual curiosity, and the fourth day clearly shows how much he was guided by the engineering practices of artillery. However, he was above all interested in a basic problem: what is impact? It is obvious that if we consider a body falling along a parabola, the higher the fall, the greater the impact. But how can this be described quantitatively?

---

[22] *Parmi veramente che conceder si possa al nostro Accademico, che egli senza iattanza abbia nel principio di questo suo trattato potuto attribuirsi di arrecarci una nuova scienza intorno a un soggetto antichissimo* [Galileo 1638 in *Opere*, VIII: 266].

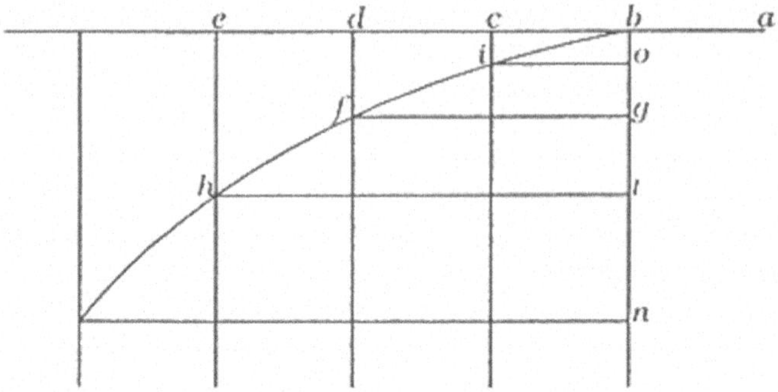

Fig. 7

Unfortunately geometry and kinematics cannot provide an answer. Today we have the conceptual tools to define it but, with the exceptions of the few notions of dynamics obtained during the third day, Galileo had at his disposition only geometrical notions and theorems. This makes his results so difficult for the modern reader to follow that – it must be admitted – we cannot be completely sure that we have correctly understood the text.

However, we must ask exactly what Galileo meant by *impeto*. Impetus was a notion of medieval science, above all in the school of Jean Buridan, usually associated with what we today call "impulse" or "momentum", that is, the product of mass times velocity. But this is the source of many a difficulty, because Galileo did not have the concept of mass. He could only say impact is proportional to speed and energy is proportional to the square of speed. To understand this *impeto* we have to ask how it can be measured. Galileo clearly understood that he was dealing with a vectorial quantity, but to measure it he used $v^2 = 2gh$, where v is the vertical velocity, h the height of the fall, and g the gravity. However, this is a measure of energy, not of momentum.[viii]

I want to conclude the description of the second part of the *Discorsi* with a few observations about a long discussion between Salviati, Sagredo, and Simplicio [Galileo 1638 in *Opere*, VIII: 273 ff] concerning the fact that the trajectory is indeed a parabola. If the axis of the parabola passes through the center of the earth, then the trajectory itself diverges from that axis; thus the body does not move towards the center of the earth (fig. 8).

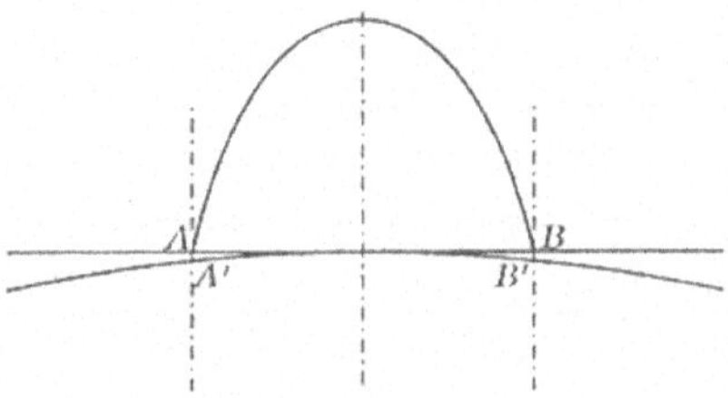

Fig. 8

Simplicio has not one but two objections. The first is that they have set forth the hypothesis that the starting point *A* and the point of arrival *B* lie on the same horizontal plane, which is that tangent to the earth's surface at the point of intersection between the earth's surface and the axis of the parabola; thus there is an ulterior displacement from the center of the earth with respect to the effective points A′ and B′ situated on the earth's crust. The second objection is that the theory neglects the resistance of the air.

Not wanting to contradict his companions, Salviati takes these objections very seriously. His defence of the theory is based on a consideration that introduces a completely new dimension into the *Discorsi*. Galileo is aware of the fact that our concept of nature is gleaned from phenomena, some of which we ignore because we consider them of slight importance; the theory that we thus formulate is what we would today call a model. Salviati is convinced that this has profound consequences.

To answer the first objection Salviati offers two examples, a theoretical one based on Archimedes' theory of the lever, and a practical one based on experience with the construction of towers. The lever has always been considered a straight line, even if we know that that can't be so. But the error can be neglected because the length of the lever is negligible in comparison to the radius of the earth, the ratio being in general a million to one. Further, all architects built towers with parallel walls, but this is only true in a rigorous sense at an infinite distance from the earth.

As for the second objection, the resistance of the air raises an infinite variety of deviations from the theory, and Salviati even wonders if there can be a good theory of friction, a problem which is still valid today.

From a historical point of view, the most interesting response is to Simplicio's first objection. This shows yet again how important the circle was for Galileo, even more important and fundamental than the straight line. Further, we see how close Galileo comes to the idea of a central force, which will become fundamental for Newton's mechanics. Perhaps of still greater interest is the response to the first objection given by Sagredo and Simplicio, that scientists propose theories that are intellectual constructions but what is important is to establish, as far as possible, their applicability. It seemed to me opportune to conclude this brief description of the *Discorsi* with this discussion because it is even today of great pertinence: it is to be hoped that those who teach physics in high school and at the universities read this penetrating discussion of the three protagonists.

Conclusion

I have attempted to shed light on some significant points of the *Discorsi*, for example, the role of Galileo's principle and the discussion regarding methodology. However, since we risk not seeing the forest because of the trees, we might ask if these points effectively account for the influence that the *Discorsi* had on Galileo's successors.

Great discoveries in mechanics were made before and after Galileo, but it was in the *Discorsi* that for the first time such discoveries were inserted in a coherent system with the aid of axioms, a principle, and many geometric theorems. This is a new occurrence in the history of the natural sciences, which only finds a precedent in the *corpus* of Pythagorean geometry and Euclid's *Elements*. I say natural science because for the Greeks, as for Kepler and Galileo, geometry *was* a natural science. To my mind, the originality of the

Discorsi lies precisely in this attempt to construct a framework of rational connections between mechanical phenomena. The way indicated by Galileo was followed by his successors, and it is thanks to their efforts that Galileo's first framework has grown into the imposing intellectual construction that is modern science.

The first of Galileo's successors was the great Christiaan Huygens.[23] By means of the analytical geometry of Descartes, Huygens's *Horologium oscillatorium* of 1673, thirty-five years after Galileo's death, contained two elements of progress. The first is an extension of Galilean mechanics through the introduction of Descartes's principle of momentum. The second is a revisited, more powerful version of Galileo's principle: the theorem of energy extended to rigid bodies.

It was through the *Horologium* that Newton came to know the mechanics contained in the *Discorsi*, which in all probability he did not read before the publication of the *Principia*. Following the example of the *Horologium*, Newton set out the foundations of point mechanics and, as Truesdell says, formulated, in the same spirit, a new program for hydrodynamics.

Just a few years later, the two subjects that Newton had omitted from the *Principia*, rigid body motions and the theory of elasticity, were redeveloped by Jacob Bernoulli, who was a keen reader of Galileo. Later, Jacob's nephew Daniel broadened Newton's program for hydrodynamics using ideas of Galileo, Huygens, and his father Johann. Later still, Euler, a great inventor of new concepts, and Lagrange would attempt a synthesis of the whole of mechanics, completed finally by Cauchy.

By means of the hypothesis of a fluid ether, Euler attempted to extend Galileo's synthesis to optics, electricity, and magnetism. This intuition was carried forth with great success in the nineteenth century with the introduction of an elastic ether by Faraday, Fresnel, Maxwell and Lorentz.

It was only with Einstein that the mechanics of Galileo, Newton, Jacob Bernoulli, Euler and Cauchy had to be modified. Thus the construction that began with Galileo continued to expand, becoming increasingly refined and abstract. When classical mechanics was finally substituted by quantum mechanics, the framework could be extended to chemistry, which today is closely linked to physics. And today, at the beginning of a new millennium, this framework of knowledge has begun to extend a step at a time to biology, which by the middle of the twentieth century had become the leading branch of the natural sciences.

Each phase of growth in the framework has been accompanied by lively discussions between scientists, as it was between Salviati, Sagredo, and Simplicio. But at the beginning of this development there was the small framework presented by Galileo in his *Discorsi*.

Translated from the Italian by Kim Williams

---

[23] See in this volume "Huygens's *Horologium Oscillatorium* and Newton's *Principia*", pp. 41-49 and "The Discovery of Rigid Body Dynamics", especially pp. 85-103.

Christiaan Huygens

Huygens's *Horologium Oscillatorium* and Newton's *Principia*[1]

While everyone recognizes and acknowledges the importance of Newton's *Principia* [1687] and the enormous achievement realized by its author, the answers diverge, sometimes considerably, on questions such as exactly what progress is made by Newton, where he went beyond his predecessors, what the precise significance of this progress is, which among the numerous achievements in this book are the most important, and which are the most difficult. Such divergences are not so surprising if one considers the very restricted number of people who have actually read this exceedingly difficult book, let alone the even smaller number of those who have fully understood it.

Much attention has been concentrated on the third and final part, entitled is *De mundi systemate*, "the system of the world", which contains Newton's results regarding the planetary system, that is, on celestial mechanics. But we know that in a certain sense Newton gave less importance to this part, which comprises only about a fifth of the whole book, and has a very different character from the rest. We also know that at one time Newton, to the horror of poor Halley, who had undertaken to publish the *Principia* at his own financial risk, wanted to publish the book without the third part!

This third book evidently contains the most spectacular parts of the *Principia*, but these are not always the most original ones, and surely not those where he treats the most difficult questions. These are probably found in chapter 11 (*De motu corporum sphaericorum viribus centripetis se mutuo petentium*) of the first of the two *Libri de motu*, books on motion, and in chapters 7 (*De motu fluidorum et resistentia projectilium*) and 8 (*De motu per fluido propagato*) of the second. These are his studies of the three-body problem on the one hand, and of his investigations on hydrodynamics and acoustics on the other. His investigations into hydrodynamics did not go very far; d'Alembert's correct assessment of this is well-known.[i] However, as Clifford Truesdell noted, these works of Newton provided a program for these fields that would be valid for the next half century.

In the present paper, I will do as everybody does and concentrate above all on the introductory chapters (*Definitions* and *Axioms or laws of motion*) to the *Principia*. I would like to compare the basis which Newton gave to his mechanics with the one that we find in the most advanced work published up to that time, that is, with Christiaan Huygens's *Horologium oscillatorium, sive de motu pendulorum ad horologia aptato demonstrationes geometricae* [1673]. This book, written in Versailles, where Huygens was working in the service of Louis XIV, was published in Paris only fourteen years before the *Principia*.[ii]

---

[1] Originally published as "Le *Horologium Oscillatorium* de Huygens et les *Principia*" in *Revue philosophique de Louvain* **86** (November 1988), pp. 485-504 (in French).

Fig. 1. Frontispiece of the first edition of the *Horologium oscillatorium*

The assessment of the influence that one scientist exerted on another is always a difficult task for the historian, and the answer is often less certain than one might think.

Thus, for instance, today we believe that when Newton wrote the *Principia* he did not know the *Discorsi* [1638], Galileo's last book in which he presents his discoveries in the field of mechanics. We also know that of Kepler's works he knew only the *Tabulae Rudolphinae* [1627], a work on position astronomy.[2] Therefore he knew the important discoveries made by his two predecessors, important especially for his own work, only through the intermediary work of others. Huygens is quoted again and again in the *Principia*, and it is clear that its redaction was influenced in part by the *Horologium*. It therefore seems that nothing would permit a better grasp of exactly what is new in the *Principia* than to compare it with this other masterpiece, Huygens's *Horologium oscillatorium*.

Christiaan Huygens and the *Horologium Oscillatorium*

The aim of Huygens's book – as the full title, *Horologium oscillatorium, sive de motu pendulorum ad horologia aptato demonstrationes geometricae* indicates – is the construction of a high-precision clock *based on the laws of mechanics*. The idea of a clock is to divide the flow of time into equal parts. But in this context, what does "equal" mean? In what does this equality consist, since we cannot directly compare two different intervals of time, and how can we verify such a measurement? We need a mechanism (or perhaps a chemical reaction) that translates a time interval into an ordinary length or angle. If the translation is effected by a mechanism, then only the universal validity of the laws of mechanics can guarantee the validity of our measurement of time, and thus of our clock. The mechanism chosen by Huygens was the pendulum. Its motion is ruled by what we today call the force of gravity, and by the "force" which Newton would later call the inertial force (*vis inertiae*) or innate force (*vis insita*). Thus, the interval of time will be the period of the pendulum. But a real pendulum always loses energy because of friction. Thus the pendulum will never return to the same height, the amplitude of its oscillation decreases, and the period changes. If, however, its mass could move along a curve such that the period was independent of the height from which it falls, this difficulty would disappear.

The question is whether such a curve, called *isochrone* or *tautochrone*, which has this property exists. Galileo, a disciple of Plato and son of the Renaissance, for whom the circle was the curve *par excellence*, had asserted that the circle possesses this property.[iii] Huygens, who knew the *Discorsi*, took it upon himself to correct Galileo and find the true isochrone, in order to construct with its help a high-precision clock. The result of his investigation is presented in the five chapters of his book.

The first chapter contains the description of a pendulum clock. The second chapter, which I will discuss below, deals with the descent of heavy bodies and their motion on a cycloid. The third treats the evolute and the length of curves. The fourth treats the center of oscillation. The fifth presents the construction of a second clock where the pendulum moves on a circle. Finally, an Appendix contains his famous theorems on the centrifugal force. Thus, the book reveals its author at the same time as a mathematician, a physicist, an engineer and a craftsman.

[2] I am indebted to I. B. Cohen for this information.

The fourth chapter will not find a continuation in the *Principia*. In effect, rigid body mechanics would be taken up directly by Jacob Bernoulli and, through him, by Daniel Bernoulli and especially by Euler.[3]

Huygens's hypotheses

In his second chapter Huygens proves rigorously that the isochrone is a cycloid, the curve described by a point on the periphery of a wheel that rolls on a plane. For the isochrone we need, we must either overturn this curve or let the wheel roll on a plane above it.

What does "to prove rigorously" mean in this context? Of course I am speaking of the theoretical proof, not of the confirmation by experiment. Since the pendulum not only concerns geometry, but time and gravity as well, a purely geometric proof based only on the axioms of Euclidean geometry would be insufficient. In any case, like Galileo before him and Newton after, Huygens insists on a proof *more mathematico*, that is, on mathematical deduction from a small number of axioms.

The four axioms formulated by Galileo in the *Discorsi* are of a purely kinematic nature, because they involve length, duration, and velocity. In contrast, Huygens's axioms refer to that which today we could call a force, that is, to a concept of dynamics.

This is the enormous innovation contained in the *Horologium*; it is one of the greatest steps accomplished by Huygens. Although the great construction called "dynamics" – that is, the science of forces that determine motion – does not begin with this book, and although here it is founded only for some special cases, for the first time the foundation is adequate and completely transparent. In other words, science had now learned to discuss questions of dynamics as rigorously as Archimedes had previously discussed problems of statics.

These are the axioms or "hypotheses", as Huygens calls them, on which his theory is based:

I. If there were no weight, nor were the air to impede the motion of bodies, each of them, having once acquired a motion, would continue with constant speed, along a straight line.[4]

II. But now, because of the action of gravity, whatever its origin, it happens that [the bodies] move with a composite motion, one part of which is uniform (*aequabilis*) in one direction or another, and another part of which is the downwards motion that comes from gravity.[5]

III. And these two movements, each can be considered separately, and neither of them is impeded by the other.[6]

---

[3] See "The Discovery of Rigid Body Dynamics" in this volume, pp. 85-103.

[4] *Si gravitas non esset, neque aër motui corporum officeret, unumquodque eorum, acceptum semel motum continuaturum velocitate aequabili, secundum lineam rectam* [Huygens 1673 : Pars II].

[5] *Nunc vero fieri gravitatis actione, undecunque illa oriatur, ut moveantur motu composito, ex aequabili quem habent in hanc vel illam partem, & ex motu deorsum à gravitate profecto* [Huygens 1673 : Pars II].

[6] *Et horum utrumque seorsim considerari posse, neque alterum ab altero impediri* [Huygens 1673 : Pars II].

Starting from these three hypotheses, Huygens deduces a series of propositions. The first to the ninth are the laws describing the fall of a body, more or less following Galileo. Then, from the twelfth to the twenty-sixth, he corrects Galileo's error and shows that the true isochrone is the inverted cycloid. I will not follow him through his calculations and proofs, but instead I shall restrict myself to a few remarks on the three hypotheses themselves.

These three hypotheses refer to what we today call forces. In effect, he is dealing with gravity, the resistance of air, and the relation between the two: since their effects are simply superposed, it is possible to study them separately.

The first hypothesis expresses what we today call the conservation of linear momentum in the case of a single body. This assertion can be dated to late antiquity; it was stated by John Philoponus (ca. 490 – 570, also known as John the Grammarian or John of Alexandria). Descartes, whose work strongly influenced both Huygens and Newton, had already based his mechanics on this principle. Many say that Descartes had in mind the absolute value of the momentum only, and not its direction; however, Koyré said that Descartes had thoroughly understood the vectorial character of linear momentum [1939]. However that may be, Huygens has clearly formulated this law as the conservation of linear momentum, and more importantly, he used it to deduce two significant results. He was the first to formulate correctly the laws of collision, which are of fundamental importance for Descartes's physics and the physics of the Cartesians, and he also obtained the form of the centrifugal force.

With the second hypothesis, he characterizes the constituents of the fall, free motion on one hand, and descent on the other. In this he follows Galileo. He also follows Galileo with his third hypothesis, but he underlines explicitly that the two motions do not interfere with each other, thus making clear the dynamical aspect of the problem.

We may sum this up by saying that Huygens provides the basis of the mechanics of falling bodies more completely, and above all more rigorously, than Galileo, and that his more powerful methods take him much further.

From the *Horologium* to the *Principia*

To come from the *Horologium* to the *Principia* is a bit like coming from Mansart's *Dome des Invalides* in Paris, built in the French classic style, clear and transparent, enchanting in its elegance (if, for a moment, we ignore Napoleon's tomb) to the great and complex English buildings such as Trinity College and King's College. But such a comparison fails to do justice to one characteristic of Newton's book, and perhaps indeed to its most impressive one, to which I shall return. So, the best comparison may be to think of Blenheim Palace (though close to Oxford and not to Cambridge), where, as we all know, another great Englishman, who left us a marvelous description of it, was born. Blenheim Palace, as gigantic as it is, was constructed according to a rigorous plan made up of an enormous number of spaces where one can marvel as well as get lost, its windows giving on to immense views of the carefully planned park that surrounds the castle, and one senses that beneath it are huge caves, where, if possible, one might become even more lost. For, like Blenheim Palace, the *Principia* is a typical product of the Baroque period, as its richness, its abundances, its multitudes of annexes and separate reflections testify; in short, the ideals of both works of art are fullness and plenitude. What are the foundations upon which Newton built his grandiose edifice?

With respect to both the matter dealt with and the sheer volume, the *Principia* is vaster than the *Horologium*; its author is a more powerful mathematician, his philosophical penetration probes deeper, and his vision of mechanics is broader. The rigour of Newton's deductions, his ingenuity as an experimenter (not less visible in the *Principia* than in the *Opticks*) and his imagination are equal to those of Huygens; but he does not attain his clarity and his transparency. This is one reason to select Huygens's much more accessible book as a guide to the *Principia*.

Let us examine Newton's axioms and compare them with Huygens's hypotheses. This comparison will show what is new in what Newton did, as well as the ideas that guided him.

Newton's three axioms are:

I. Every body continues in its state of rest, or of uniform motion either along a straight line, unless it is compelled to change that state by a force impressed upon it.[7]

II. The change in motion is proportional to the motive force impressed, and occurs along the straight line in which the force is impressed.[8]

III. The reaction is always equal and opposite to the action; that is, the actions of two bodies are always equal and directed in opposite directions.[9]

We see that Newton assumed Huygens's first hypothesis integrally into his first axiom. Huygens's second and third hypotheses are included in his famous second axiom. Then he adds a third one, for which Huygens has nothing analogous. Thus, a first, superficial reading gives us a greater impression of what the two groups have in common than of what is new in Newton's formulation. Both state three axioms and both leave statics, the science of equilibrium, aside. Also noteworthy is the close relationship between Huygens's first hypothesis and Newton's first axiom. Nor is there is a difference between the two words "hypothesis" and "axiom", except, possibly, with respect to the underlying philosophies of the two authors, but of this they say nothing.

Let us now follow the two scientists more carefully, beginning with the first statement where they seem to be closest. Before doing this I must say a word about rectilinear translation. Both assert with quite similar words that under certain conditions a body continues his motion on a straight line with the same speed. Newton alone adds "or remains at rest". What interests us here is not the addition of the latter possibility, though it is revealing and I shall come back to it later, but rather the answer to the question of what exactly are the conditions under which this rectilinear uniform motion holds. Anyone who teaches a course of mechanics for first year students knows that this is a difficult and

---

[7] *Corpus omne perseverare in statu suo quiescenti vel movendi uniformiter in directum, nisi quatenus a viribus impressis cogitur statum illum mutare* [1687: Axiomata, Lex. I].

[8] *Mutationem motus proportionalem esse vi motrici impressae, & fieri secundum lineam rectam qua vis illa imprimitur [1687: Axiomata, Lex. II].*

[9] *Actioni contrariam semper & aequalem esse reactionem: sive corporum duorum actiones in se mutuo semper esse aequales & in partes contrarias dirigi* [1687: Axiomata, Lex. III].

painful corner to turn, one that causes the teacher to sweat a lot. Huygens says that the conditions are that there is no gravity and that the air does not resist the motion of the body. In other words, Huygens refers to exactly the conditions under which the process which he studies – his experiment, if you wish – takes place. He proposes the study of a concrete problem, and he prepares the theoretical ground for studying that and nothing more.

Newton, on the other hand, says simply that the condition is that it is not made to change its state by any external force.

Today we are so familiar with the concept of "force" that perhaps we can no longer see the enormous step taken here. It is as if one goes from a beautiful but completely surrounded lake to the open sea with no limits on the horizon, or as if an immense curtain had risen. No physical context is indicated by Newton, and no specific problem which could help the reader to understand this sentence is added. Newton does not envision one specific problem, or better, he envisions all possible problems. He wants to establish nothing more or less than the foundations of the whole of dynamics, that is, of the science that tells us the laws according to which bodies move through space and time under whatever conditions. For constructing the bridge from geometry to this new science, a new element was needed, that is, a new concept or notion, as I shall call it here. He needed a notion of an extreme degree of abstraction. Newton created this notion just as it is still used today, and he made it the cornerstone of an immense intellectual construction.

Another painful question for the teacher of mechanics is what force is according to Newton. A part of his answer can be found in the eight definitions which precede the axioms, no less than six of which are devoted to this notion. The two most important ones are the third and the fourth which introduce inertia, as we say today, or inertial force (*vis inertiae*, *vis insita*), as Newton said, and external or impressed force (*vis impressa*). However, these definitions explain and teach us little. Newton merely illustrates what he has in mind by examples. Only concerning inertial force is he more explicit; he says that it is the force that maintains the body on a straight line with constant speed and that it is proportional to the mass of the body. But the reader who is desirous of understanding is kept waiting. We do find inertial force later on in the *Principia*, but strangely enough Newton does not use it in formulating his first axiom, although he could just as well, and more succinctly, have said *Corpus omne vi inertiae in stato suo perseverare cogitum esse*, that is, "each body is forced by the inertial force.... Thus later, after Euler, Newton's inertial force disappeared from the literature. What we today call an "inertial force" is something altogether different; it is not a Newtonian force at all.

I think that this lack of consequence betrays the influence which the *Horologium* exerted on Newton's formulation of his axioms. He formulated them by following the example before him. I suspect that it also might indicate that he added the introductory chapters *after* having written the two books on motion.

When we go to the second axiom, we find an analogous step taken in the same direction. Huygens mentions gravitation and its action on the motion of the body. Newton does not indicate any particular force at all, be it gravity, friction, electricity, magnetism, or any other. This second axiom is probably the most famous sentence ever formulated in the history of physics, although usually only an abridged version is quoted. For this reason it is even more important to underline the enormous degree of abstraction it contains. After three hundred years of intellectual digestion by many active scholars as well as writers of textbooks, many students still wonder what he is talking about. But this abstract formulation is no caprice of its author. On the contrary, how else could the notion of force have passed into all of physics? One might ask if there were any more powerful or successful abstract notion, a concept or ein *Begriff*, as Hegel would later call it, in the entire history of science.

The famous third axiom – the action is equal to the reaction – is Newton's alone, as far as its extension to dynamics is concerned. In statics it had been formulated by Roberval and Galileo. Only friction, the force of Biot and Savart in electromagnetism and others, seems at first to escape it. It is this law which extends the principle of relativity to the whole of Newtonian mechanics. Using this law one can show, for instance, that the center of mass of the whole solar system validates Newton's first axiom. However, it must be noted that here Huygens saw more clearly than Newton. In his first axiom Newton maintained the distinction between uniform motion and rest. In the third book, where he decides against the Ptolemaic in favor of the Copernican system, he states that the center of the solar system is at rest, a statement that is meaningless from the point of view of the principle of relativity, that is, from Newtonian mechanics as we have known it since Euler. Newton himself had not yet foreseen the final consequence here; this is one of the many paradoxes with which history, including the history of science, is filled.

One may ask if he was indeed led to this axiom only by the discoveries of his predecessors in mechanics proper, as his comments suggest. I suspect that the discovery made by Kepler, among others, that the moon excites the tides guided him here: the earth attracts the moon, but the moon also apparently attracts the earth, since it deforms its shell of water.

But the question of what force really is according to Newton still looms large. This question is asked always anxiously by students, and has received many different, often contradictory, and sometimes even incomprehensible, replies. Some answered that the concept was wrong, or at least useless. The first to say so was d'Alembert in his famous *Traité* [1743].

What does Newton himself say? On the one hand he simply shows how the force *functions* mathematically, that is, how we can compute using it. For by deriving it immediately after having stated the axioms, he derives as the first corollary the law of the parallelogram of forces; in modern parlance, he simply shows that *force is a vector*. The French mathematician André Weil said once that the idea of the group was already present in the geometry of Euclid, and in the same sense we may say that it is present also in Newton's mechanics, and with it the ideas of a vector space and of vector calculus.

The law of the parallelogram guides Newton, and through him the reader, first to the second part of the question, namely, how force is to be measured, and then to the answer to this question. In other words, how is the force distinguished from other vectors, such as velocity, momentum, etc.? The answer is, by comparing it with a standard force. Usually the standard force is gravitation, thus a force is "a vector that one can put on a scale". In this way, this abstract notion is assigned through Newton's second axiom the role which it plays in an experiment: the axiom, in a way in which Huygens's second hypothesis could not have done, shows us how we can set up the correct equations of motion for an infinity of special problems. All the forces that enter into these problems are equivalent, that is, they can be compared with each other (i.e., measured), and they can be put into the same equation. The symmetry of old scales with two cups hanging each on arms of equal length show this equivalence.

Translated from the French by Sandro Caparrini and Kim Williams

Acknowledgments

I wish to thank Patricia Radelet-de Grave for having helped me find the references, and later having corrected the manuscript. I also thank Mrs Champagne for having typed this work. Thanks are also due to Patricia Radelet-de Grave and to J. Steyaert, M. Vandyck and M. Ghins for their linguistic suggestions.

Isaac Newton

Newton's *Principia*[1]

Introduction

Few books were less understood than Newton's *Principia* [1687], but few books had a greater influence. How were Newton's ideas propagated and developed?

At the beginning of the twentieth century, the theory of relativity and quantum theory established restrictions on the validity of Newton's mechanics. What is its significance for today's physics and how do we now judge Newton's accomplishment?

No scientist was ever as famous and as honoured as Newton. Copernicus's book appeared only a few days before its author's death.[2] Kepler, the Imperial Mathematician, knew fame, but until his death he remained a poor man persecuted by worries, and he died exhausted by the many material struggles which he had to go through during his entire life. Galileo, already spoiled by glory as a young man, ended his days with his freedom severely restricted. Newton, however, held several high public offices after his period of scientific activity, and he remained president of the all-powerful Royal Society for twenty-five years. No stranger of rank who came to London neglected to pay him reverence. At his funeral, four Dukes and Marquises were his pallbearers, and his tomb occupies the most conspicuous place in Westminster Abbey.

His name, as no one else's, became synonymous with science itself. Euler refers to him often as *Summus Neutonus*. In his poem "Fantasie an Laura", the German poet Schiller calls all scientists briefly "Newtons".[i] And when the philosopher Kant, whose ideas were deeply influenced by Newton, wants to state that a mathematical biology is impossible he says simply that there will never be a Newton for a blade of grass.[ii] Only two scientists after Newton seem to have enjoyed such legendary fame: Darwin in the nineteenth century and Einstein in the twentieth, the latter precisely because he overthrew Newton's mechanics.

We must now ask in what respect we are particularly prepared *today* to understand Newton the scientist, and for appreciating his work. As an answer, one may point out the following. Firstly, while Newton's personality remained hidden for a long time behind a veil of venerability (to the point that, for example, David Brewster's biography [1831] sometimes reads rather like a hagiography), today we know it better than ever before. The American Louis Trenchard More's impressive book [1934], the article of John Maynard Keynes [1946], the well-known economist who had saved a great part of Newton's manuscripts, the analyses by Frank Manuel [1963; 1968], and others have, shown us the great scientist as a man.[3] We now

[1] Originally published in *CERN Reports* 80-02 (25 February 1980), Geneva: CERN.

[2] For Copernicus, Kepler, and Galileo, see [Koestler 1959].

[3] I am indebted to David Finkelstein, who directed my attention to the books by Manuel.

see not only the teacher, highly venerated by his disciples, but also a secretive and diffident personality filled with passions, whose behaviour towards his colleagues was often motivated by rancour and who was not above unworthy means.[iii]

Secondly, the history of science has taught us a lot during recent decades. Recent works by many scholars have made Newton's work much more accessible.[4]

With respect to the later development of mechanics, C.A. Truesdell has removed many prejudices that had become entrenched over the course of time [1954; 1955; 1960a]. He corrected the opinion, formulated, among others, by the physicist Ernst Mach that,

> Newton discovered universal gravitation and completed the formal enunciation of the mechanical principles now generally accepted. Since his time, no essentially new principle has been stated. All that has been accomplished in mechanics since his day has been a deductive, formal, and mathematical development of mechanics on the basis of Newton's laws [1883: chap. 2.3, § 1].[5]

Today, we know that this is erroneous. Much of what we now call Newton's mechanics was already known before him, and much was found only after him. Since in what follows I shall often talk of the accomplishments of Newton's successors, I hasten to state that nothing is further from my intention than to diminish Newton's importance. Rather, I do this in order to show the vastness of science itself, with respect to which even such a grandiose work as Newton's may appear small.

Thirdly, although for more than 200 years Newton's mechanics was thought to be unchangeable, in the twentieth century its limitations were discovered. The special theory of relativity, the general theory of relativity and, finally, quantum mechanics, stated limits for the applicability of classical mechanics, and today we are perhaps for the first time at a sufficient distance to correctly appreciate the significance of Newton's discoveries.

---

[4] A. Koyré and I.B. Cohen, with the assistance of A. Whitman, have collated and compared the three editions of the *Principia* printed while Newton was alive [Newton 1972]; Cohen and A. Whitman have faithfully translated the *Principia* into English [Newton 1999]. Among the numerous books and articles of I.B. Cohen, we mention especially: *Introduction to Newton*'s *Principia* [1971], "Hypotheses in Newton's philosophy" [1969], "Galileo and Newton" [1972], "History and the philosopher of science" [1974]. Newton's mathematical papers were edited by D. Whiteside in collaboration with M.A. Hoskin and A. Prag [Newton 1967-80], his correspondence by H. W. Turnbull, J.F. Scott, A. Rupert Hall and Laura Tilling [Newton 1959-77]; A.R. Hall and M.B. Hall have published some of Newton's scientific manuscripts [Newton 1962]. See also [Gray 1888], [Fierz 1954], [Koyré 1965], [Herivel 1965], [Rosenfeld 1964-65; 1969], [Wagner 1976] and the collection of essays edited by J.Z. Buchwald and I.B. Cohen [2001].

[5] *Newton hat sich in Bezug auf unsern Gegenstand zweierlei Verdienste erworben. Erstens hat er den Gesichtskreis der mechanischen Physik sehr erweitert durch seine Entdeckung der allgemeinen Gravitation. Dann hat er auch die Aufstellung der heute angenommenen Principien der Mechanik zu einem Abschluss gebracht. Nach ihm ist ein wesentlich neues Princip nicht mehr augesprochen worden. Was nach ihm in der Mechanik geleistet worden ist, bezog sich durchaus auf die deductive, formelle und mathematische Entwickelung der Mechanik auf Grund der Newton'schen* Principien [Mach 1883: Chap. 2, § 3, n. 1].

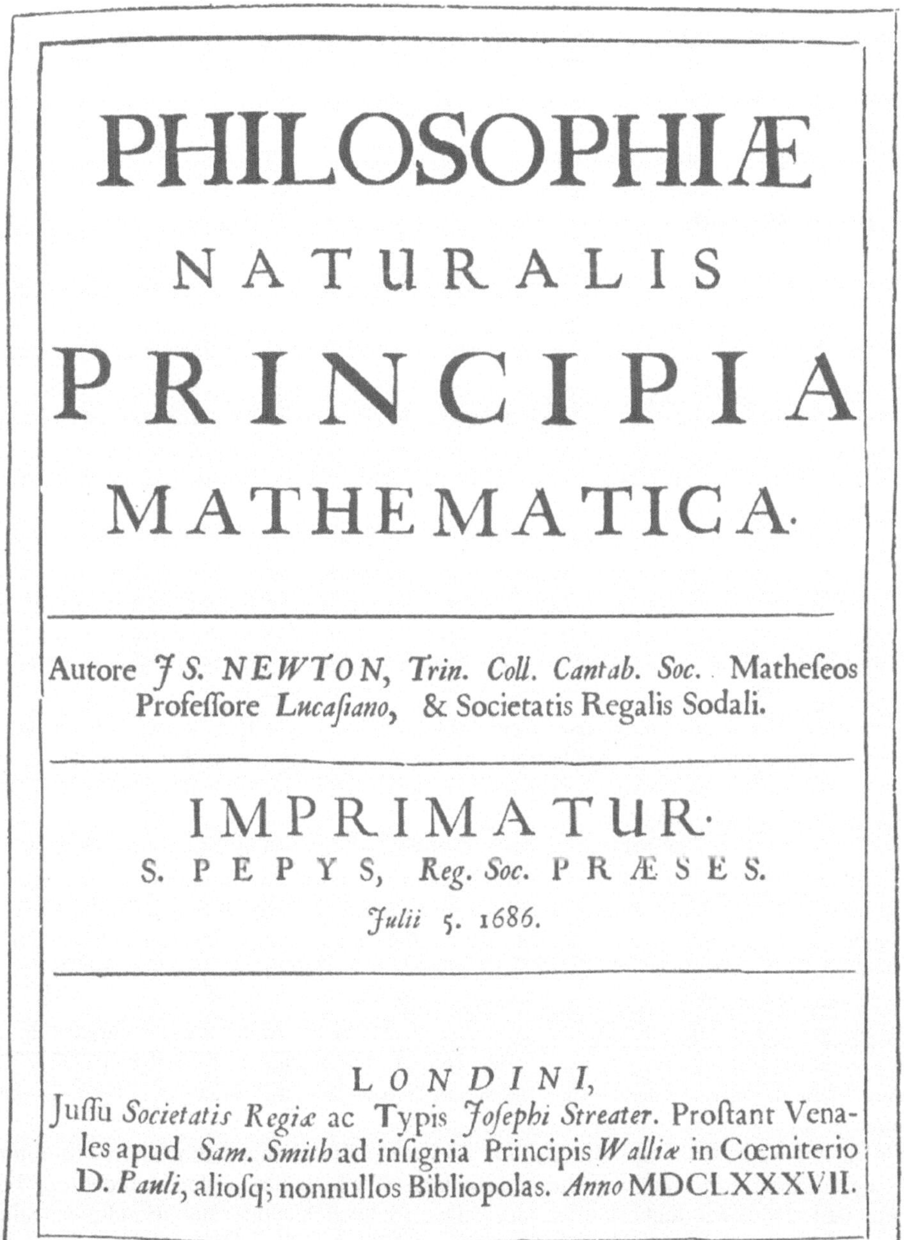

PHILOSOPHIÆ
NATURALIS
PRINCIPIA
MATHEMATICA.

Autore *J. S.* *NEWTON*, *Trin. Coll. Cantab. Soc.* Matheſeos Profeſſore *Lucaſiano*, & Societatis Regalis Sodali.

IMPRIMATUR.
S. PEPYS, *Reg. Soc.* PRÆSES.
Julii 5. 1686.

LONDINI,
Juſſu *Societatis Regiæ* ac Typis *Joſephi Streater*. Proſtant Venales apud *Sam. Smith* ad inſignia Principis *Walliæ* in Cœmiterio D. *Pauli*, alioſq; nonnullos Bibliopolas. *Anno* MDCLXXXVII.

Fig. 1. Frontispiece of the first edition of the *Principia*

Finally, one must also be aware of the fact that, perhaps for the first time, physics and chemistry are no longer at the forefront of scientific progress. Instead, molecular biology has taken over this role.

Therefore, I shall not speak as a historian of Newton, which I am not, but merely as a physicist trying to understand his book.

The life of Newton

A few words must suffice as a sketch of Newton's life. Isaac Newton was born on Christmas day, 1642 (old calendar) in Woolsthorpe. He was the posthumous son of an independent farmer, brought up first by his mother and, after her second marriage, by his grandmother. Because of his talents, and also because people did not think that he would be strong enough for the life of a farmer, he was sent to Cambridge, presumably in order to study theology. Indeed, for his whole life, theology remained the subject of his principal interest, the one to which he devoted the greater part of his time and energy. But once at Cambridge, his extraordinary gift for mathematics was soon discovered, and there he attended, among others, the courses of the famous Isaac Barrow, whose successor he would later become.

At the end of his undergraduate years (as we would say today), during the great plague of 1665-66 he was forced to return to Woolsthorpe, and there he took the first steps towards the three great discoveries which made him immortal: his calculus of fluxions, the theory of light and colours, and the explanation of the attraction of the moon. But what he discovered he kept to himself. In truth, in none of the three cases do we know exactly how much Newton had already discovered by this date. With respect to gravitation, one thing is certain: he must have grasped intuitively something decisive, but he probably thought that he did not yet have a systematic and solid enough foundation for his theory. It was at this time too that the famous incident of the apple in his mother's garden is supposed to have occurred, which he later recounted to his friend Stukeley.

However, back at Cambridge, he became involved again with other subjects. Only in 1679 and at the beginning of the 1680s did he begin to work once again on the problems of mechanics, which he had left untouched for almost fifteen years. It appears that it was an exchange of letters with his rival Robert Hooke that made him return to the still unsolved problem of gravity. And *then,* in a few years, he composed the *Principia,* or more fully, the *Philosophiae Naturalis Principia Mathematica* (fig. 1).

It was printed in 1687 thanks mainly to the efforts of his friend and colleague Edmund Halley, with the imprimatur, given on 5 July 1686, of the president of the Royal Society, the famous Samuel Pepys, whose often indiscreet diaries are still today a delight to many readers.

But a few years after this publication, Newton's life changed drastically. In autumn 1693 he suffered a breakdown, the significance of which has perhaps sometimes been exaggerated. However that may be, according to Lord Keynes, his friends, especially Charles Montague, Lord Halifax, decided that nothing less than a complete change in his life was needed.

[12]

AXIOMATA SIVE LEGES MOTUS

Lex. I.

Corpus omne perſeverare in ſtatu ſuo quieſcendi vel movendi uniformiter in directum, niſi quatenus a viribus impreſſis cogitur ſtatum illum mutare.

PRojectilia perſeverant in motibus ſuis niſi quatenus a reſiſtentia aeris retardantur & vi gravitatis impelluntur deorſum. Trochus, cujus partes cohærendo perpetuo retrahunt ſeſe a motibus rectilineis, non ceſſat rotari niſi quatenus ab aere retardatur. Majora autem Planetarum & Cometarum corpora motus ſuos & progreſſivos & circulares in ſpatiis minus reſiſtentibus factos conſervant diutius.

Lex. II.

Mutationem motus proportionalem eſſe vi motrici impreſſæ, & fieri ſecundum lineam rectam qua vis illa imprimitur.

Si vis aliqua motum quemvis generet, dupla duplum, tripla triplum generabit, ſive ſimul & ſemel, ſive gradatim & ſucceſſive impreſſa fuerit. Et hic motus quoniam in eandem ſemper plagam cum vi generatrice determinatur, ſi corpus antea movebatur, motui ejus vel conſpiranti additur, vel contrario ſubducitur, vel oblique oblique adjicitur, & cum eo ſecundum utriuſq; determinationem componitur.

Lex. III.

Fig. 2

[13]

Lex. III.

Actioni contrariam ſemper & æqualem eſſe reactionem : ſive corporum duorum actiones in ſe mutuo ſemper eſſe æquales & in partes contrarias dirigi.

Quicquid premit vel trahit alterum, tantundem ab eo premitur vel trahitur. Siquis lapidem digito premit, premitur & hujus digitus a lapide. Si equus lapidem funi allegatum trahit, retrahetur etiam & equus æqualiter in lapidem: nam funis utrinq; diſtentus eodem relaxandi ſe conatu urgebit Equum verſus lapidem, ac lapidem verſus equum, tantumq; impediet progreſſum unius quantum promovet progreſſum alterius. Si corpus aliquod in corpus aliud impingens, motum ejus vi ſua quomodocunq: mutaverit, idem quoque viciſſim in motu proprio eandem mutationem in partem contrariam vi alterius (ob æqualitatem preſſionis mutuæ) ſubibit. His actionibus æquales fiunt mutationes non velocitatum ſed motuum, (ſcilicet in corporibus non aliunde impeditis :) Mutationes enim velocitatum, in contrarias itidem partes factæ, quia motus æqualiter mutantur, ſunt corporibus reciproce proportionales.

Corol. I.

Corpus viribus conjunctis diagonalem parallelogrammi eodem tempore deſcribere, quo latera ſeparatis.

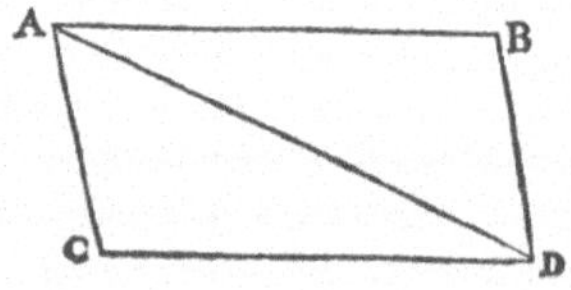

Si corpus dato tempore, vi ſola *M*, ferretur ab *A* ad *B*, & vi ſola *N*, ab *A* ad *C*, compleatur parallelogrammum *ABDC*, & vi utraq; feretur id eodem tempore ab *A* ad *D*. Nam quoniam vis *N* agit ſecundum lineam *AC* ipſi *BD* parallelam, hæc vis nihil mutabit velocitatem accedendi ad lineam illam *B D* a vi altera genitam. Accedet igitur corpus eodem tempore ad lineam *BD* ſive vis *N* imprimatur, ſive non, atq; adeo in fine illius temporis reperietur alicubi in linea illa *BD*. Eodem argumento in fine temporis ejuſdem reperietur alicubi in linea *CD*, & idcirco in utriuſq; lineæ concurſu *D* reperiri neceſſe eſt.

Fig. 3

Thanks to Halifax's influence, in 1696 he was appointed first "Warden" and then "Master of the Mint", in which capacity he was to supervise the coining of the new English money that was introduced by decision of Halifax and the Whig government. His predecessors had mostly considered the office as a sinecure, but not he: Newton, it seems, was an extremely able administrator; in fact he had already shown himself to be a very good one at Cambridge. Keynes calls him one of the greatest and most efficient of civil servants, and he performed his task splendidly; indeed, his chemical studies had prepared him well. False money was common at this time, and coins were often devalued by cutting or clipping little bits. Dr. Fierz has told me that we owe to Newton the invention of the "milled edge" for protecting money against such assaults, an invention which all of us carry in our pockets.[iv] Thanks to Newton's energy, the output of new money was increased by a factor of eight. He also defended money in a more active way, namely by organizing his own police force and a network of informers. According to Manuel, at least nineteen counterfeiters were hanged as a result of Newton's efforts, for counterfeiting was considered high treason (see [Manuel 1968]; [Craig 1963]). However we may feel about this today, England certainly profited greatly from his activities.

In London, he published his second great work, the *Opticks* [1704], which is based on work done in the 1660s and on earlier publications, and towards the end of his life, he worked on two more books, *Observations on the Prophecies of Daniel and the Apocalypse of St. John* [1733] and *The Chronology of the Ancient Kingdoms Amended* [1728].

During these last years he was universally respected, and became almost the centre of a cult. He was now the well-established protector and often generous friend of his many disciples: Edmund Halley, Roger Cotes, Henry Pemberton, and for a time, the unfortunate Nicolas Fatio de Duilliers from Geneva.

But at the same time he was the all-powerful president of the Royal Society, punishing and persecuting with all available means anyone who dared to oppose him or who he felt was attacking him. These dark, sombre traits appear in the last portraits by the painter John Vanderbank. There are light and dark features in his personality, both in almost superhuman measure;[v] Arthur Koestler calls him a "phenomenal mixture of monster and saint" [1976: 685]. But it is tragic that *both* sides should have had a stifling, almost paralysing, effect on the continued development of science in England.

During the last six years of his life he was in poor health but remained active. He died in his eighty-fifth year, on 20 March 1727, after a short illness.

An introduction to the *Principia*

I must now give an idea of the content of the *Principia*, of Newton's purpose, and of what he accomplished.

At the beginning, we find a short introduction in which Newton states his goals, and an ode by Edmund Halley, who edited the *Principia* against Newton's resistance, and at his own financial risk.

The work then begins with a series of definitions, which explain some fundamental notions such as "quantity of motion", "inner" or "inertial force" and "central force", and state how these quantities are to be measured. There follows in a scholium his famous doctrine of absolute space and absolute time, about which more will be said later.

Newton then formulates the three important "axioms" or "laws of motion" (see figs. 2 and 3), which all girls and boys learn still in high school today. These axioms, in modern terms, are:

1. in the absence of an external force, every body remains at rest or in the state of uniform motion;
2. the change of the quantity of motion is proportional to the external force;
3. the reaction is always equal and opposite to the action, or the actions of two bodies on each other are always equal and in opposite directions.

These laws are followed by a number of corollaries, including, among others, Corollary 1, the law of the parallelogram of forces (see fig. 4).

And *only then* follow the three main parts of the work: first, there are two books on the motion of bodies (*I, II Liber de motu corporum*), followed by the third book on the system of the world (*De mundi systemate*), which crowns the whole work.

In order to follow Newton's reasoning and to penetrate into his philosophy of nature and of science, one must clearly grasp the significance of this arrangement. The first two books, which comprise about 4/5 of the work, contain what we would call today "general mechanics". There, starting from theorems of geometry and from his own new axioms of mechanics, he derives and proves a great number of theorems. However, with a few exceptions they are not applied to the phenomena of nature. This he does only in the third part, the "celestial mechanics".

Since dynamics describes motions, i.e., changes, its formulation required a new mathematical language. Newton had discovered this language himself: his *calculus of fluxions,* which, however, he had never published before in a systematic form. Therefore, the first book opens with a chapter containing eleven mathematical lemmas, which in the sequel are used constantly. But for this very reason – and it seems that this was Newton's intention – this first chapter presents for the reader an almost insurmountable obstacle. Even today, knowing what Newton is driving at and being familiar with the theorems he proves, we have great difficulty following his reasoning; and this holds true for the whole work.

I Liber de motu corporum

The content of the following thirteen chapters is organized into fifty theorems, forty-eight problems, some lemmas and a few general scholia. Chapters 2 to 7 contain the investigation of the orbits of a body that moves under the influence of a central force; first there is the description of the curve itself, then the way in which this curve is traversed in time.

The presentation – and in order to understand Newton's purpose this must be kept in mind – is always conceived in view of two opposite goals. One is practical, directed towards applications, the other systematic, aimed at a rigorous deduction. On the one hand, it is clear that the needs of practical astronomy suggested many of the problems he treats, such as the fourteenth problem of Section V (how to construct the conic section that passes through five given points); this is obviously a problem of astronomy, namely, the determination of the orbit of a comet. But – and this is important – this motivation is never stated explicitly. Discussion of the planetary system is postponed to the third book, since *the systematic presentation of the whole theory must not be interrupted or obscured.*

What at once catches the eye of the present-day reader in these first chapters is the direction which Newton's investigation takes. Unlike us, he does not start from a differential equation. In fact, in the entire work, there are no differential equations at all, not even that which today we call "Newton's equation". Rather he proceeds in the opposite direction and solves the inverse problem: given a curve, for instance an ellipse, he shows that if the center of attraction is at the focus, the centripetal force must decrease with the square of the distance. Thus he finds that all conic sections are solutions of his universal law of gravitation. But only Johann Bernoulli, armed with all his analytic equipment, succeeded in proving that the conic sections are the only solutions to the problem.[6] It would be wrong to see only a completion of Newton's results in Bernoulli's theorem. Rather, we must say that Johann Bernoulli succeeded in attacking the problem from a new point of view, and, as we all know, this is one of the most difficult accomplishments in science.

There follow the sections which today we would call the "advanced" ones. In these, he treats general forces and the motion of the apsides, and Section X contains a theory of the pendulum, based on the work of Huygens.[7] Surely, the appearance of the *Horologium oscillatorium* [1673] was one of the decisive events before the composition of the *Principia.* While in these first ten sections Newton considers (as we would say today) the motion of a single body in an external field, beginning in Section XI, he considers systems of *several* bodies, simple bodies first and then extended ones. In particular, his theorem XXVI and the twenty-two corollaries that follow are the mathematical basis for his attack on the famous three-body problem. His theory on the motion of the moon in the third book is based on these theorems and later, in the second edition, it was developed further. The same holds for his theory of the tides. Sections XII and XIII treat the mutual attraction of extended bodies. It seems that the solutions of these problems kept Newton occupied for a long time and that these difficulties were one of the main reasons for the delay of the *Principia.*

The last section of the first book contains the mechanical foundation of his corpuscular optics. There he discusses the passage of a particle through parallel sheets and he derives Snell's law. This is the *first example* of how the new mechanics transcends itself and becomes the basis of another domain of physics. Of all legacies that Newton has left to physicists, this probably proved to be the most successful: to conceive of a phenomenon of nature as the propagation of a system through space and time.

However, this short summary may show the extent to which the first book is organized systematically, and therefore how much of its spirit has remained modern. Indeed, its sequence of presentation is essentially the same as we still find in textbooks today. The enormous difference from a modern book is due mostly to Newton's mathematics, and also partly, but to a lesser degree, to differences in the point of view from which he considers the problems.

---

6 See "The Kepler Problem from Newton to Johann Bernoulli" in this volume, pp. 123-134.
7 See "Huygens's *Horologium Oscillatorium* and Newton's *Principia*" in this volume, pp. 41-49.

This brief discussion cannot do justice to the enormous riches that we find in this book. In all sections, wherever we look, we find new theorems, and quite often they are theorems which still today present great difficulties to mathematicians and physicists, in spite of the powerful tools with which the mathematicians have provided us over the past 300 years.

II Liber de motu corporum

The second book does not exhibit the same unity in its structure. It culminates in the refutation of the theory of vortices by which Descartes had tried to explain the solar system.[8] But even though this book was conceived primarily in view of this goal, it nevertheless contains much more.

In the first four sections Newton treats the motion of bodies in a viscous fluid in which friction is proportional to the speed or to the square of the speed. This concept is one of Newton's most important discoveries, but strangely enough, as Truesdell has noted, it was overlooked by his successors for more than a century.[vi]

Section V is devoted to hydrostatics, and Section VI to the motion of a pendulum subject to friction, a problem that had been recognized by Huygens as fundamental for the construction of clocks.

But the most important sections with respect to both size and content, are Sections VII and VIII. Section VII contains his hydrodynamics and Section VIII his acoustics, in particular his investigations on the speed of sound. His accomplishment in this domain was summed up by Truesdell in the following words: "Newton had the genius to propose the problem, Euler to solve it" [Truesdell 1955: LXXII].

Section IX, the last, presents the refutation of Descartes's theory and Newton's triumph.

The text of the second book testifies – and this is worth mentioning – that Newton worked not only as a mathematician and theoretical physicist, but that he performed original experiments in the domain of mechanics as well. (See, for example, in the first edition of the *Principia*, Bk. II, Prop. X, Reg. 7; Bk. II, Prop. XXXI, Cor. 5; Bk. II, Prop. XXXVII, Prob. IX; and especially Bk. II, Prop. XL, scholium gen.[vii])

The structure of the first two books is dominated by his intention to systematize the whole doctrine of mechanics by trying to derive it from a few axioms stated at the beginning of the books, and thus to realize mechanics as one system. In this respect, Newton is the follower of Euclid, and his immediate predecessor is Huygens. It is true, however, that with respect to this endeavour he did not completely succeed in the second book.

De mundi systemate

The third book is the most admired of all. It contains the application of his theoretical mechanics to the solar system, that is, to the motion of planets, moons (at the time of Newton ten were known), comets and, last but not least, the sun itself. For, according to his third law of motion, the sun too moves around the solar system's center of inertia.

---

[8] See "Johann Bernoulli's Work on the Theory of Gravitation and the Weight of the Atmosphere" in this volume, pp. 105-121.

Newton made significant and fundamental contributions to the understanding of all these phenomena.

Having shown in the first book that motion on an ellipse can be understood as a consequence of the force law $1/r^2$, he now concludes in the third book that the inverse-square law of gravity is valid in the solar system without exception. This conclusion is based *both* on Kepler's laws, which rest on observation, and on the theorem from the first book.

It is worthwhile following his line of reasoning in detail. In the first edition, the third book begins with hypotheses, of which the first are general rules of correct procedure in science. The last are statements of observed facts, for example that the five planets Mercury, Venus, Mars, Jupiter and Saturn move according to Kepler's laws around the sun and that the same laws are also valid for their moons. But with respect to the motion of the earth itself, which was in dispute, he initially leaves the question open. Thus he first demonstrates the validity of the law of gravity for planets, and moons, and only after having established that Kepler's laws hold for the planets – not with respect to their motion relative to the earth but relative to the *sun* – does he decide in favour of the Copernican doctrine.

One sees that he perceived the connection between theory and experiment in a way which even Galileo and Kepler had not grasped.

But he does not leave off there. After having deduced Kepler's laws from the principles of his mechanics he goes on far beyond anything that had been achieved before him, especially in two directions.

First, we owe to Newton the first quantitative description of the orbit of a comet, who said it was either a parabola or a hyperbola.[viii] He also established the following theorem: the proportion between the instantaneous speed of a comet on a parabola and the speed of a planet on a circle is proportional to the square root of the proportion of the diameter of a planetary orbit to the instantaneous distance of the comet from the sun:

$$\frac{v_{com}}{v_{pl}} = \sqrt{\frac{2r_{pl}}{r_{com}}} \text{ or } r_{com}v_{com} = 2r_{pl}v_{pl}^2 \text{ (Book III, Prop. XL, corr. 2),}$$

which may be contrasted with Kepler's third law for circular orbits:

$$r_1v_1^2 = r_2v_2^2 .$$

This law is an extension of our astronomical knowledge far beyond what we owe to Kepler. Through this, the comets, which were still an enigma for Descartes and for Jacob Bernoulli, became accessible to observational and computational astronomy.[9] Thus was broken what has sometimes been called "the last bulwark of superstition".[ix]

The Italian Francesco Algarotti said mockingly, in a book to which we will return, "I think that the comets are the greatest enemies of heavenly vortices. They seem created on purpose to ruin [philosophical] systems".[10]

---

[9] The German amateur astronomer Peter Doerfel had proposed in 1680 that comets move in parabolas; see [Fleckenstein 1969: 120].

[10] *...io credo bastar le Comete i più dichiarati nemici, che abbiano i Vortici in Cielo. Elleno pajono fatte apposta per rovinar sistemi* [1737: 241].

Second, another major achievement is the foundation of a lunar theory. Newton's mechanics permits one to solve the two-body problem easily and, since the sun is so much heavier than the planets, these solutions are excellent approximations to reality.

However, in order to have an exact result one must solve a many-body problem. For example, in order to predict the moon's orbit accurately, it must be computed under the influence of the earth *and* the sun. Still today we do not have an exact solution and must use approximations. Newton took the first enormously ingenious steps in this direction, especially in the second edition of the *Principia*, although following these steps is quite difficult.

A comparison of different editions

Twice Newton was to reprint the *Principia* in revised form: in 1713 with Roger Cotes and in 1726 with Henry Pemberton. The comparison of the first edition with the second one is revealing. Besides an expansion of the lunar theory we find other changes. Thus the "hypotheses" at the beginning disappear, or more precisely they change their name. Between 1686 and 1713, Newton again pondered intensively and systematically the philosophical principles of science, and this must have led him to a revision of his concepts.

The attentive reader of the *Principia* will notice that the observations of John Flamsteed, the Astronomer Royal, have in part disappeared in the later editions. This was a counterblow developed by Newton in one of these battles in which he fought his enemies with all means, those that were permitted and, sometimes, those that were not. He never hesitated to attack an enemy publicly and, if possible, to destroy him. As I said, Newton's luminous genius was surrounded by dark traits of character.

At the end of the second book he adds the famous "Scholium Generale" which brings the totality of what he had achieved in science into contact with theology. There he also explains his "inductive method" and states proudly the often-quoted sentence, *Hypotheses non fingo* ("I do not invent hypotheses").

In fact, however, the hypotheses have *not* completely disappeared. We still find two of them in the second edition. First, the third lemma has now become a hypothesis. Probably this means nothing more than that Newton could not prove it. But of the original nine hypotheses, the fourth still remains. It says that the center of the system of the world is at rest. As I said, in the first edition he had left open the question of whether the world or, as we would say today, the center of inertia of the solar system, is at rest. But it *is* clear that the statement "the center of the system of the world is at rest" is neither a "rule for proceeding correctly" nor a "phenomenon", but a true hypothesis. But it appears strange, and I am not sure that the fact that this hypothesis is not only superfluous, but that it directly contradicts Newton's mechanics, has been sufficiently emphasized.[11] For, according to Newton, physical processes evolve in a homogeneous space and homogeneous time. His mechanics postulates what we today call the principle of relativity.[x] This is the meaning of his first and his third axioms. And in such a space, on which (as we say today) a group acts, there is no center: the center is everywhere and

[11] Hermann Weyl insisted on this contradiction [Weyl 1949: Part 2, chap. 1].

nowhere. It is surprising to see that here even Newton felt a great difficulty and that even he was not clearly aware of the consequences of his own ideas. This is another example of Wolfgang Pauli's observation that "a fundamental scientific discovery, sometimes even against the resistance of its creator, gives birth to further fruitful developments following its own autonomous course" [1921]. Only much later did physicists grasp this point clearly.

Let us now compare the first two books with the third one. The *truth* of the theory springs from its principles: this is pursued and explained in the two books *De motu corporum*. But its *validity*, that is, its *applicability*, is proved by its agreement with the observations: this is the purpose of the third book, *De mundi systemate*. It is precisely in the accomplishment of having recognized and achieved this double goal that we see what is sometimes a bit vaguely called the "Newtonian synthesis".

It is from this point of view that one should judge Newton's achievement. His first law goes back to late antiquity, and Galileo's successors, as well as Descartes, Wallis and especially Huygens, had grasped it clearly.[xi] The famous second law was understood by Galileo, and after him, by Beeckman. The law of gravity goes back to Kepler, and others, for example Boulliaud, Borelli and Hooke, stated it before or independently of Newton. What is Newton's most important single discovery? If I were to put my finger on one special point, my choice would be his third axiom: *action* = *reaction* (for once I agree with Mach [1883: chap. 2.3, § 11]). The law in this form, and indeed the whole idea, belongs to Newton. Perhaps it was suggested to him by Kepler's discovery that the tides are caused by the moon.[xii] This law, difficult to grasp even today, made possible the principle of relativity, which is fundamental for all of physics, including classical physics. After Newton, this axiom became a heuristic principle of great value. But to ask what Newton's most important single discovery was in this way means that one doesn't see the forest for of the trees! Newton's specific achievement is to have *put together* what his predecessors found into *one coherent system*.[12] The motions of the planets and comets could now be calculated starting from only four fundamental principles, and beyond this, as we shall see, his system is capable of an extension to the whole of physics.

Divulgation of Newton's principles

We may pause for a moment and reflect on the sobering fact that the book which more than any other supported the claim of the Enlightenment, that all questions could be solved with the help of rational thought, was understood only by a mere handful of scientists!

To understand how Newton's work spread, it is revealing to learn how large the edition of such an important work was. Dr. Whiteside of Cambridge, the editor of Newton's mathematical works, informs me that in Newton's time 250 copies constituted a large edition, and that 1000 copies was a very large enterprise indeed. His estimate is that the

[12] Whatever one may think of Newton's behaviour towards Hooke, it is from this point of view that one must understand the following sentence in his famous letter to Halley, 20 June 1686: "Now is not this very fine? Mathematicians that find out, settle and do all the business must content themselves with being nothing but dry calculators and drudges and another that does nothing but pretend and grasp at all things must carry away all the inventions as well of those that were to follow him as of those that went before" [Newton 1686: 438].

first edition of the *Principia* printed in London numbered something between 330 and 350 copies.[13] For the second edition printed in Cambridge we are fortunate to possess exact data: 750 copies were printed, of which 711 were distributed. The others were used by Newton for corrections, and he sent a few to his friends. The third edition, again printed in London, comprised 1200 copies. These numbers testify to the great success of the book.[xiii] But, in spite of its success, one must ask how many readers really understood it. Together with Dr. Fellmann in Basel, who knows Newton's mathematics and who rediscovered in Geneva the copy of the *Principia* owned by Leibniz [Fellmann 1971; 1975-76], we tried to enumerate the possible candidates. Our lists were not identical, but we agreed that during Newton's lifetime at best six *savants* understood the book and, up to the present, probably not more than one dozen or, if one is generous, two; moreover, Dr. Fellmann and I agree that *neither* of us is on this list!

A fourth edition, still in Latin, with a commentary, was edited in Rome after Newton's death by the Rev. Fathers Jacquier and Le Seur [Newton 1739-42]. This edition is noteworthy in several respects. First because it was prepared in Rome under the watchful eye of the Santo Officio, Jacquier and Le Seur had to declare that, with respect to certain assertions, they did "only wear a mask", that is, that they did not express their own opinion; this was not much more than a formality. Afterwards this book was printed in Calvinist Geneva, of all places, so that one might say that this edition was an ecumenical collaboration *avant la lettre*.[xiv]

But what interests us most here is that the two reverend fathers added to Newton's book three papers which had been submitted in 1740 to the French Academy of Sciences. For its biennial prize, the Academy had asked for an investigation on the tides, a subject about which the theory of Descartes and the more recent one of Newton were in competition. The prize was divided and attributed equally to Colin Maclaurin, one of Newton's disciples [1741], Daniel Bernoulli [1741], and Leonhard Euler [1741], all of whose papers took Newton's *Principia* as their starting point. Thus, one may see in the attribution of these prizes a sort of official recognition of Newton's physics in France and on the Continent in general.[14]

At one point in his paper, Daniel Bernoulli writes, "I see that this corresponds to what Newton says in the *Principia,* where he compares the length of the earth's axis with the radius of the equator. As far as his reasoning goes, he is probably the only one to understand it, for this great man sees even through a veil what another can hardly see with a microscope".[15] In other words, when one of the best mathematicians and physicists of his time read the *Principia*, he could see that Newton was right in his contentions, but he could not always follow his proofs.

---

[13] I am indebted to D. Whiteside for this information and I should like to thank him very much.

[14] See "Daniel Bernoulli" in this volume, pp. 137-154.

[15] *Je vois que cela est conforme à ce que M. Newton dit à la page 380 des* Princip. Math. Phil. Nat. *Edit. 2. pour déterminer la Proportion de l'Axe de la Terre au rayon de son Equateur. Quant à son raisonnement, il n'y a peut-être que lui, qui pût y voir clair; car ce grand Homme voyoit à travers d'un voile, ce qu'un autre ne distingue qu'a peine avec un Microscope* [D. Bernoulli, 1741: Ch. II, Section VIII].

The difficulty met by his contemporaries is also illustrated by the following: we have books written by Newton's disciples Henry Pemberton [1728] and Colin Maclaurin [1748] in which, as they themselves say, they explain the philosophy of Newton. Both books contain few formulae or none at all, for those were understood only by a handful of mathematicians who shared the secret of the "New Math". And this still in the middle of the eighteenth century, at the time when Voltaire was popularizing Newton's physics on the Continent with the help of his friend, Mme du Châtelet, who quite probably, like Voltaire, did not understand it either.[xv]

Count Algarotti, one of the foreign luminaries who Frederick the Great attracted to his Court, whom I have already quoted, wrote in his youth a volume entitled *Il Newtonianismo per le Dame* [1737], which was translated into French in 1738.[xvi] The major part of the work is devoted to Newton's *Opticks.* Only in the sixth and final part does the author approach the *Principia.* The *Principia* was praised, but the *Opticks* was read.

The propagation of Newtonian mechanics

If we want to investigate how a new science is disseminated and how it penetrates into research itself, we must ask where it stimulated new creative work. Thanks to the authority of its creator, Newton's doctrine became popular in British universities comparatively early,[xvii] but we find little creative response there. The first such response came from the Continent, at first in a critical, even polemic way, which I shall mention only briefly.

Science became embroiled in a sterile dispute between Newton and Leibniz and their supporters concerning the priority of the discovery of calculus, or more precisely, the independence of Leibniz's discovery. This quarrel has often been related more or less exactly; perhaps even now we do not know all the relevant facts, and therefore I shall not enter into this episode.

An essential and important step was, however, taken by Jacob Herrmann, the disciple of Jacob Bernoulli who published the first systematic presentation of Newton's mechanics after the *Principia* itself, indeed, the first book of all on mechanics after Newton, entitled *Phoronomia, sive de viribus et motibus corporum solidorum et fluidorum* [1716]. In this book we find, perhaps for the first time, Newton's laws in differential form.[16] Unfortunately, in spite of its historical importance, the *Phoronomia* has been neglected by historians, but we can hope that this situation will change in the years to come.

But one may justly say that the first important creative response to the *Principia* came from Daniel Bernoulli. Daniel Bernoulli had learned mathematics and physics from his father Johann and his elder brother Nicolaus. In physics he was first guided by the ideas of Descartes, then by those of Huygens and Leibniz, whom his father venerated and had defended in the priority dispute with Newton. But one day – we do not know exactly when and how – he began to read and study the *Principia* carefully. His father owned a copy of the second edition, which is still today at the University Library in Basel. By studying Daniel's papers we can see how he penetrates, step by step, into Newton's ideas; how, with the help of the mathematics of Leibniz, of his uncle, and of his father, he begins to reformulate them; and how in his hands these different currents begin to merge. So far,

[16] Following a communication for which I am indebted to I. B. Cohen.

these currents had been in competition on the ideal plane; on the plane of daily life they had simply quarrelled. Now, a new synthesis was to emerge.

But all progress has its price, and in this case the price was a family tragedy. We know how old Johann behaved towards his son: in 1738, when Daniel was finally able to publish in Strasbourg his *opus,* the *Hydrodynamica* [1738], the father published at the same time his much shorter *Hydraulica* [1743], backdating it by several years and implying thereby that his son's discoveries had followed his own.[17] This has often been attributed to jealousy, which certainly played an important part, but a more profound motive was the great bitterness of the father on seeing his son take over the enemy's cause and lead it to triumph. For, as we saw, it was Daniel's paper on the tides [1741], just then in preparation and later to become his most printed work, which sealed the victory of "Newtonianism" (as some would say today) on the Continent.

At about the same time, or even a bit earlier, the final and irreversible step was taken. In 1736, the young Leonhard Euler (not yet 30 years old) published in two volumes his *Mechanica sive motus scientia analytice exposita.* This synthesis of Newton's and (we must insist) Huygens's physics with the mathematics of Descartes, Leibniz and the Bernoullis, is a *second* turning point. While Newton's book remained forever closed to almost everybody, here for the first time we find a work that provides easy access to mechanics, that is, as easy as this science permits. For it was in these books that the language in which we understand and present mechanics until this very day was coined and fixed. Newton's mechanics was developed into a method which *will and must* attack all phenomena of the physical world: it has become a program for the whole of the natural sciences. The book's success was enormous. In terms of beauty and originality, the *Mechanica* does not equal some of Euler's later work, but it made the new science accessible for the first time. When Laplace later called Euler *notre maitre à tous*, he was thinking not only of the mathematician, but also of the author of the *Mechanica.* Euler's accomplishment is the more astonishing as his book was only the third systematic presentation of mechanics, and yet its appearance established mechanics in the academies and universities as a domain of knowledge. In fact, it made possible the modern way of studying. A few years later in his *Anleitung zur Naturlehre* [1862], which unfortunately appeared only after his death, Euler again discussed in detail the fundamentals of mechanics.[18] The mathematician and physicist Hermann Weyl says of this work: "Euler exposes in wonderful clarity the sum of the whole knowledge in the natural sciences of his age" [Weyl 1949: 42].

In the *Anleitung* Euler especially discussed inertia and the principle of relativity. Later these questions were neglected, even forgotten. Had these problems been understood, the quarrel about Einstein's theory of relativity would never have become so silly. For, as everyone who receives anti-relativist articles from time to time can testify, the opponents of the theory of relativity not only do not understand Einstein's mechanics, they do not understand Newton's either, which they try to defend.

[17] For a different point of view, see "The Kepler Problem from Newton to Johann Bernoulli" in this volume, pp. 123-134.

[18] See "Euler's *Anleitung zur Naturlehre*" in this volume, pp. 185-196.

Euler's *Mechanica* and his later works, especially those in hydrodynamics, have given physics its modern character. They have also opened the way to mathematical technology, a connection so characteristic of modern science.

During the eighteenth century, Newton's computations on celestial mechanics were extended, especially by Clairaut and by Euler, the true founders of analytical lunar theory.[19] For quite a time mathematicians and astronomers were not sure whether or not Newton's law of gravity would suffice to explain all observations made in the planetary system, or whether additional terms would be needed. The situation was not unlike that in quantum electrodynamics today. Only after having performed long and complicated computations did Clairaut, Euler and their successors become convinced that no discrepancy between observation and theory existed. After the enormous efforts of many mathematicians and astronomers, by the middle of the nineteenth century only a few small effects remained unexplained. And of all those, eventually *only one* remained: the perihelion of the planet Mercury advances too fast by a minimal amount of 43 seconds of arc per century, which is a tenth of a second per revolution. This discrepancy, as we know, was only explained by Einstein in 1915.[xviii] But then, Einstein showed that the discrepancy is due not only to the law of gravity but to a limitation in Newton's mechanics itself.

It is also worth recalling that some years ago, Kenneth Nordvedt in Montana showed how a precise measurement of the moon's orbit permits one to confirm the Newtonian principle of equivalence to the order of the gravitational interactions themselves [Nordtvedt Jr. 1968a, 1968b, 1996]. Subsequently those measurements were made independently by I.I. Shapiro and his colleagues [1976] and by R.H. Dicke and his colleagues [1976], with the help of laser techniques which permit a determination of the earth-moon distance to an accuracy of 10 cm, and they were able to confirm the principle.

Beyond astronomy, through hydrodynamics and the theory of elasticity – due mainly to Euler and Cauchy respectively but based on ideas of Newton and Jacob Bernoulli – the triumph of Newton's mechanics have been extended to all domains of physics. Also, through Euler it found its way to Faraday, and through him definitively into magnetism and electricity.[20]

If, in modern times, we have been forced to replace Newton's mechanics, which we call "classical", with Einstein's mechanics on the one hand, and by quantum mechanics on the other, as the foundation of physics, it is precisely this replacement that permits us to appreciate more justly what Newton accomplished with his own theory. During the entire nineteenth century, and still at the beginning of the twentieth, Newton's formulation of the principles of mechanics were subject to continual criticism from Mach among others [1883].[xix] The first and third of his axioms were often degraded to the rank of mere definitions, and some (d'Alembert already) even went even so far as to see nothing more than a definition in his second axiom (the change in the quantity of motion is proportional to the external force). Today it has become clear to us that Newton was right. The second axiom was modified by special relativity, *in order to take account of experiments;* thus, *it cannot be* a definition. All of the three are true axioms of classical mechanics.

---

[19] Cf. for example [Taton 1976].

[20] See "Euler's Writings on Optics, Electricity and Magnetism" in this volume, pp. 211-223.

However, as Truesdell [1964] and others have shown, the list is not complete. Euler recognized in 1775 as a fourth independent axiom for continuum mechanics, that the change of angular momentum is equal to the moment of the force. The independence of this axiom was also forgotten for a long time, and according to Truesdell, Boltzmann was the first to recognize again its true role (according to a written communication from C.A. Truesdell to B.L. van der Waerden, who kindly showed it to me).[xx] The same may be said concerning Newton's doctrine of absolute space and absolute time.[21] That he saw in them attributes of God may astonish us today.[xxi] But what is important is that, like Euler after him, he saw here a fundamental problem, namely, in which system must physical quantities be measured? His doctrine was derided; only Einstein recognized the importance of the question. Therefore, even though today Newton's answer has been slightly modified, the glory remains his and we can see today more clearly than ever that he understood the character of his own answers better and more exactly than did most of his successors. In this respect, Newton triumphed over his critics.

Conclusion

Permit me now to conclude with a few words about Newton's personality. In no less than three domains – mathematics, optics and mechanics – this extraordinary man achieved accomplishments of unique importance and left us incomparable works. And in stating this, one must be conscious of the fact that these occupations were only a small part of his activities. In addition, he was a great chemist and alchemist; for example, his famous law in thermodynamics grew out of these occupations.[xxii] But in fact, as we know today through the enormous mass of papers which he left, his principal interests were theology and history, to which he devoted probably the greater part of his time. And here too, as in the exact sciences, Newton did not see himself as a scientist among others, but as a prophet, to whom it was given to reveal the great secrets of nature and of history.

With whom can we compare this infinitely gifted man, of such a complex character, generous towards his pupils, but also diffident, defiant, sensitive, easily offended, rancorous, and filled with strong passions, especially an insatiable desire for power? Galileo, Kepler, Euler, Faraday, Riemann, Maxwell and Einstein do not show traits of similar passions to the same degree, and none was filled with such a sense of a prophetic mission. Kepler, it is true, felt such a mission, but his ideas lay always open and unconcealed, and he was not given to being the judge of others. Besides his splendid discoveries, Newton, certainly more than any other great scientist, has also left science with an uncomfortable legacy. Science and research can become the victims of ambitions and of power, and certainly in many places fell foul of entrenched power. I think it is fair to say that, in this respect, Newton's grasp for power was accepted with the acclamation of the majority of his fellows and successors, and thus he has, more than any other, infused European science with this often deadly danger. We should not overlook this constant menace, because it is our work, and with it our dignity, that are at stake.

If we want to find a kindred spirit in Europe's history, I think we have to look for it outside science. What other man was so imbued with a quasi-religious mission to reveal to

[21] For Newton's ideas on space and time, see [Fierz 1954].

the world the secrets of the universe and of its history, and to be judge of all previous periods? I think that we may find Newton's closest spiritual relative in the Italian poet and scientist Dante. Each saw himself as a prophet sent to reveal what had been secret (see [A. Speiser 1932] and [Callahan 1976]). Each accomplished this mission in a work whose composition must have required an almost unimaginable degree of intellectual concentration and physical strain. Of both we can see the boundless dedication to their work and feel the enormous hidden passions, and of both we can admire their sense of responsibility and dedication to the political community in which they lived.

Neither work is easily accessible and both present to the reader great difficulties. Like Dante's *Divina Commedia,* the *Principia* too has a majestic beauty which fascinates and captivates the reader. The impression one receives from the *Principia* can be compared to an immense dark cavern, full of fascinating byways revealed by occasional flashes of light. And at the far end burns a brilliant fire whose dazzling reflections are seen from time to time. I hope to have given you a little reflection of this fire and of the fascination it radiates. But as with Dante's work, the full impression is reserved for the reader.

Acknowledgments

Prof. J. O. Fleckenstein and Dr. H. Stähelin, former president of the *Basler Naturforschende Gesellschaft*, invited me to talk at the Newton celebration in Basel. A few weeks later, Prof. R. Hagedorn invited me to give a talk at a CERN colloquium.

Mrs. P. Radelet-De Grave helped me with the documentation and Dr. J. L. Pietenpol smoothed out my English. Miss E. Finger typed the German, Mrs. N. Van Olmen the French, and the CERN Text Processing section looked after the English version.

Stimulus, encouragement and help from all of them are most gratefully acknowledged.

Jacob Bernoulli

The Science of Mechanics in the Eighteenth Century[1]

Introduction: The predecessors

Allow me to begin with a few words evoking seven great predecessors of the scientists of the eighteenth century: Galileo, Kepler, Huygens, Newton, Jacob Bernoulli, Descartes and Leibniz. Five of them are known mainly as scientists, and two mainly as philosophers.

I begin with Galileo. On the third day of the *Discorsi* [1638], after the definition of "uniformly accelerated motion" Salviati says:

> our author postulates only one principle and supposes it to hold: "I accept that the degree of speed acquired by the same body on variously inclined planes is always the same provided that the height of the planes is the same".[2]

From this rudimentary energy principle Galileo claims to derive each of his theorems. This idea that the whole of science ought to be connected through the deduction from one principle, or as we say today less ambitiously, from one set of consistent axioms, was time and again the unspoken ambition of precisely the greatest scientists of our own times. And in what follows, we shall encounter here and there Galileo's energy principle.

Kepler's three laws showed like nothing else, even to non-scientists, what is meant by a "law of nature", even if still today a deeper understanding of this idea may elude us.

Huygens, by refining Galileo's energy principle, became, indirectly at least, the founder of the domain of *rigid body dynamics*. Also, his writings set a standard of clarity surpassed only by few.[3]

Newton, in his celebrated book, the *Principia* [1687], bestowed above all else four gifts on mechanics. His concept of force is still today the best door for entering into it. Through the concept of force he gave a lasting axiomatic foundation to *point mechanics*, accepted by all but a few. He also opened the door to the domain of *hydrodynamics*. Finally, of

---

[1] Lecture given at the symposium "Il fiorire della scienza nel XVIII secolo", 18 October 2003, on the occasion of the opening of the C.A Truesdell library at the Scuola Normale Superiore, Pisa. Previously unpublished.

[2] [Galileo 1638 in *Opere* VIII: 196]. See "The Second New Science in Galileo's *Discorsi*" in this volume, pp. 17-38.

[3] See "The Discovery of Rigid Body Dynamics" in this volume, pp. 85-103.

course, he gave us his doctrine of universal gravitation, which made its author famous and mechanics prestigious to an unprecedented degree.

Jacob Bernoulli, the oldest of the Bernoulli dynasty, who worked on the transition from the old century to the new, put Huygens's idea concerning the pendulum on its true basis and opened the door to the domain of *elasticity*.

But I must add to this list two men who are better known as philosophers than as scientists, for it should be recalled that the eighteenth century was also *le siècle des lumières*, and the guarantor of what was claimed to be philosophical enlightenment was, of course, the new science: mechanics.

Descartes's discoveries in mechanics are not always highly valued, but besides creating the indispensable tool of coordinates, he expressed the first clear ideas of what is meant by a "mechanism". Today we know that he was very important for the formation of Newton's ideas, as Truesdell always stressed.

Leibniz is on this list not so much for his discoveries in our field, although, as Truesdell showed, they are respectable, but for two other reasons. First, infinitesimal mathematics is essential for mechanics, and Leibniz's calculus of differentials was its ideal formulation, used throughout the Continent. To Leibniz is due also the concept of the *functio continua*, or, as we say today, the analytic function, which became the mathematical expression for causality and determinism.[i] As we shall see, this idea, the breeding ground of so many philosophical speculations, underwent a sharp crisis in the middle of the eighteenth century.

Point mechanics

The first decades of the new century were still overshadowed by Newton's *Principia*.[ii] With the single exception of Jacob Bernoulli, all scientists now worked following Newton's discoveries, which were overwhelming. The influence of his model of the solar system, which was based on his theory of universal gravitation and Kepler's three laws, on scientists as well as on popular opinion had can hardly be overrated,. But it is wrong to say, as some have, that after Newton, for solving a problem, it was only necessary to apply fundamental principles.[iii] Not only was his *Principia* extremely hard to read, but the problems of mechanics are still attacked from the side of geometry. For example, if the orbit of a planet is an ellipse and if it is attracted by a force towards a focus, then this force is proportional to the inverse square of the distance. At the time, this statement was called the solution to the direct problem. Later in the *Principia* Newton shows that conic sections are indeed the only solutions.[4] It soon became obvious that only a few problems could be solved this way. Only in 1710 was the problem attacked by Jacob Hermann taking Newton's force and the force-law as the starting point, just as we do today [1712]. Johann Bernoulli simplified Hermann's deduction and carefully determined the constants of integration [1712]. The perception that the laws of nature must be completed by initial data in order to take the empirically given circumstances into account, was an essential step which enriched Newton's mechanical model of the solar system. Newton had already attacked three-body problems, but for doing this he had to use complicated, ad hoc

[4] See "The Kepler Problem from Newton to Johann Bernoulli" in this volume, pp. 123-134.

hypotheses, which were difficult to handle.[iv] Now at least differential equations for a system not only for two or three but of any number of planets could be written without difficulty. Thus Newton's model for the solar system was now mathematically realized as an n-body problem, and the way to analytical dynamics was open. This required, however, a long development that lasted through the whole century.

It was Clairaut who achieved the first breakthrough here. He was followed by d'Alembert and then especially by Euler.[v] The most important goal of their investigations was to discover the exact orbit of the moon under the attraction of the earth and of the sun. This, and the more general problems of planetary perturbation theory, were pursued throughout the century.[vi]

I must now say a few words on the development of the principles of mechanics themselves. Already in 1727 Euler, who computed the aberration which had been discovered by Bradley shortly before, noted that this effect depended, if only minimally, upon whether one assumed, following Newton, that light consisted of corpuscles or whether it was a wave motion in an ether, as Euler himself believed.[5] This difference is due to the fact, that, while in a corpuscular theory the principle of relativity holds, in an ether wave theory it does not, since the ether defines an absolute frame. This result showed Euler the importance of the principle of relativity and of the Galilean transformation.[6]

In a paper in written in 1750 Euler announced *un nouveau principe universel* [1752b]. It said that Newton's law, if written in a space-fixed Cartesian coordinate system, is the universal law for mechanics.[vii] One may ask, what is new? One answer is that Euler had formulated Newton's law in what is today called an inertial system. This provided the basis for a great part of mechanics, as Euler claimed, and it lead him to one of his greatest triumphs.[viii] However, failing to recognize the limits of his principle, it delayed him for twenty-five years, before he fully reached another goal towards which he had struggled again and again. By this formulation in space-fixed, Cartesian coordinates, the closest surrogate for vector calculus, Newton's intuition of an absolute space and time was now mathematically formulated as the totality of all inertial frames. These are connected by a group, named today for Galileo, since he stands at the beginning of this endeavour to formulate mechanics in a relativistically invariant way, as we say today. This group then governed mechanics for more than 150 years.

The second half of the century saw the tremendous rise of analytical mechanics, due to those I mentioned at the beginning of this paper and others, especially Lagrange and later to Laplace.[ix]

To Lagrange we also owe the generally covariant formulation of mechanics, covariant under the general linear group GL(3), leading to the well-known equations named after him, which became, like Euler's, part of the regular language of physicists.[x] I should like to mention also the famous works of Lagrange and Laplace on celestial mechanics. Laplace's idea of writing Newton's law as a field equation proved to be of great value in other fields of physics as well.[xi] And then, of course, Lagrange's *Méchanique analitique* [1788] exerted a profound influence during the century that followed.

---

[5] See "Euler's Work in Physical Optics" in this volume, pp. 197-209.

[6] See "The Principle of Relativity in Euler's Work" in this volume, pp. 167-178.

Already with Poisson we face ideas that lead to Hamilton's formulation of mechanics, and Dirac's use, 150 years later, of the Poisson brackets for quantum-mechanics is well known.[xii]

It must also be stressed that the research in pure mathematics, and especially in analysis, caused by celestial mechanics was considerable, for in this period mathematics and mechanics were in closer and more fruitful contact than they would ever be again.

Hydrodynamics

While hydrostatics goes back to Archimedes, whose work was continued in modern times by Simon Stevin and by Pascal, the systematic beginning of hydrodynamics, after the first discoveries by Torricelli and Riccati, is due to Newton. In the second book of the *Principia* Newton first presents his results on the motions of bodies in a viscous fluid and then, in the seventh section, he presents his attempts at a formulation of hydrodynamics.

Newton's attempts are based, as d'Alembert observed, on two models that are used side by side: an atomistic model, and a model based on the idea of a material continuum. As Truesdell said, this use testifies to Newton's versatility and it would be a mistake to hold it against him. But the importance of his work in this domain lies not so much in his results as in his having formulated new problems, and as Truesdell said, Newton had the genius to conceive the program. I shall say later who, according to Truesdell, had the genius to solve it.

The next comprehensive presentation of this field, the first in the eighteenth century, was Daniel Bernoulli's *Hydrodynamica* of 1738.[7] The bases for his theory are the energy principle in the form Galileo and Huygens had given it, and the postulate of incompressibility.[xiii] As for the content, I must restrict myself to three of the most spectacular accomplishments, two of which are still well known today. The first one is the so-called Bernoulli equation, which says that the sum of the kinetic energy, the pressure and the potential of the outer force is constant. I say "so-called", since in the *Hydrodynamica* we look in vain for the equation in the form in which we know it today. In fact, Daniel did not formulate a general equation at all, but rather he computed in an example the connection between Stevin's outer pressure against the tube and the speed of the fluid, and the reader learns how to compute the connection through other examples as well. Daniel checked the prediction of all examples by measurement. It is surprising, though, that he does not choose an example in which he could have brought to bear the full power of the calculus. I say these things not to belittle what Daniel achieved, but to make clear the importance of the various fundamental discoveries that had to be made in order to formulate modern hydrodynamics. Daniel's great achievement is to have united two domains of science which so far had remained separate: hydrostatics, the science of fluids at rest and Newton's hydraulics, the science of moving fluids. To underline this unification, he created his title word, *hydrodynamica*.

While this achievement stands in the mainstream of eighteenth-century science, the other achievement of his that remains famous to this day is quite singular. I am speaking, of course, of his derivation of the law of Townley-Boyle-Mariotte based on an atomistic

[7] See "Daniel Bernoulli" in this volume, pp. 137-154.

model using his uncle Jacob Bernoulli's theory of probability [1738: sect. X, §§ 1-7]. This success points to the future kinetic theory of gases, which, however, would have to wait for another century.

Clearly we still see here the same dichotomy between two models that we saw in the work of Newton. While, again, this dichotomy is witness to its authors ingenuity, it also indicates that something is missing.

A third high point of the book, but one often overlooked, is found in the many applications to technical and engineering questions, an interest that Daniel shared with his friend Euler. In the ninth section he discusses quite a few of them, evaluating their efficiency. It is in this section that we find, perhaps for the first time, the concept of the "power" of a force, which is so characteristic of modern technology. Later in his life he would return again and again to technical and engineering problems. The Académie des Sciences in Paris regularly offered prices for the best solution of a problem, which were mostly technical. Daniel Bernoulli received the prize no less than ten times, either alone or *ex aequo* together with others. I shall return to Daniel Bernoulli, but now after having shown how he prepared the stage for the final construction of modern hydrodynamics we must leave him.

Four great discoveries that were crucial for this construction must now be mentioned. The first came from Daniel's own father, Johann I Bernoulli, who, after Daniel's *Hydrodynamica* appeared, published his much shorter *Hydraulica* [1742b]. One must only look at the preface of the book to see what is fundamentally new here: Johann introduced Newton's concept of force into hydrodynamics. The first one to see this was Euler, who stated that while Daniel attacked the subject with the indirect method – the principle of energy – Johann did so with the direct method – Newton's concept of force. Even if Johann only attacked the problem in a one-dimensional current, he clearly asked the question of precisely what was needed to do it in three dimensions; the answers were not trivial.

The second discovery was due to d'Alembert, who was the first to realize that what was needed here was a formulation of the laws in which both the time and space coordinates enter as independent variables. In other words, he developed a field description, at least in two dimensions [1752].[xiv] It was in this context that he discovered the Cauchy-Riemann equations [1752: §§ 59-60]. Both Johann Bernoulli's and d'Alembert's discoveries must be considered turning points in the history of hydrodynamics. And just as point mechanics needed a new language – the differential equations – so did hydrodynamics – partial differential equations. D'Alembert, Euler, Clairaut and, later, Lagrange would develop this new branch of mathematics.

But two more discoveries were needed. The first was a new concept: Euler's internal pressure. Stevin's outer pressure is a force divided by a skew tensor, the surface element, defined on a given surface. For this reason Daniel Bernoulli's *Hydrodynamica* was restricted to fluids moving in a tube, and possibly to liquids moving in a channel. The internal pressure introduced by Euler is a much simpler concept mathematically: it's a scalar, or more precisely, a scalar density. It is more complicated as a mechanical quantity. Like the potential, it is not fully measurable as the outer pressure is; only its gradient can be measured.[xv]

I already mentioned Euler's other discovery, the inertial system. It is only in an inertial system that the form of the laws of nature governing a complicated phenomenon is so simple that we can freely guess and imagine them by observing the symmetries of space and time. It allows the use of Cartesian coordinates, the closest substitute for vector calculus. The Euler equations of fluid dynamics, as they are called today, appear for the first time in "Principia motus fluidorum", a memoir written in 1752 but published much later [1761].

The 3 + 1 equations are:

$$\frac{d\mathbf{u}}{dt} + grad\, p = \mathbf{f}\ , \qquad \frac{\partial \rho}{\partial t} + div\, u = 0\ .$$

The vector triplet is the equations of motion; the singlet is the continuity equation which yields the conservation of mass; u, ρ, p, f, for the velocity, the mass-density, the pressure and the force-density respectively.[xvi] The vector notation is, of course, modern, but since Euler used Cartesian coordinates in these papers, they can directly be transcribed into the vector formalism.

Thus we have entered into a new world, that of vector fields and field theory. This statement, alas, requires a comment. "Field" is primarily a purely mathematical concept, yet it entered into the language of physicists as the electromagnetic field and became reserved for Lorentz invariant fields. But this first field theory is Galileo-invariant, as Euler showed explicitly in a later publication, to which I shall come back.

However, while there are five unknown variables, there are only four equations. This brings us back to a discovery made by Jacob Bernoulli towards the end of the seventeenth century, namely, the difference between dynamical laws and constitutive equations. The latter describe the constitution of bodies, which in classical physics cannot be derived from dynamical principles, but must be assumed beforehand. To the Euler equations we must add the constitutive equation $p = p(\rho)$ that characterizes the material, that is,

for an incompressible liquid: ρ = const.
or for an ideal gas: p = const. ρ.

Thus hydrodynamics proper and aerodynamics now followed the same dynamical laws. But the problem concerning the constitutive equations is open-ended in classical physics, and on the other hand today is also a part of technology. Viscosity had to wait for the nineteenth century, and the twentieth century saw the development of rheology, which became an immense branch of mechanics in its own right.

Through these field equations the language of hydrodynamics now became, once and for all, partial differential equations, and the historian observes that quite a few who did not bother to follow this development, among them eminent scientists, dropped out of this field.[8]

Thus, for instance, to the initial values, which can be freely chosen, must be added freely chosen boundary values too, and as we saw, one constitutive equation as well, whence results an enormous increase of the structure of mechanics. Thus models now became much richer and more refined. For all these reasons I am inclined to consider the completion of hydrodynamics – that is, the creation of modern hydrodynamics and of field

[8] See "Daniel Bernoulli" in this volume, pp. 137-154.

theory – the supreme achievement of eighteenth-century mechanics. Euler first published his theory in three memoirs in French [1757a; 1757b; 1757c], and later, in partial fulfillment of the program of research set out in the *Mechanica* [1736], in a treatise in four parts in Latin [1761; 1770a; 1771b; 1772].

These two writings, especially the French versions, still deserve to be read today by every physicist and all those interested in the history of science. I believe that it is impossible, with perhaps only the exception of Lagrange, to find in the eighteenth century a more easily accessible writer than Euler. In clarity and transparency he surpassed everyone before him; even Huygens and his own teacher, Johann I Bernoulli, are surpassed.

Of course, research in hydrodynamics did not end here. In particular, it was greatly advanced by Lagrange's new mathematical formulations. However, as I said, the main, immediate task, a theory of viscosity, could not yet be attacked.

The dynamics of rigid bodies

I now leave the domains where the scientists followed the path laid out by Newton, and go over to rigid body dynamics and to the theory of elasticity, which Newton left untouched in the *Principia*.

Jacob Bernoulli, who, while he dealt with neither point mechanics nor hydrodynamics, achieved the decisive breakthrough in rigid body dynamics and elasticity. We do not know whether Jacob ever studied the *Principia*; there is no good evidence that he did, but there are many indications to the contrary. The study of rigid body dynamics begins with Huygens's formulation of a rigorous theory of the pendulum, based on Galileo's energy principle. Huygens's goal was, of course, practical, namely, to construct a high-precision clock. The theory is presented in his *Horologium oscillatorium* [1673], one of the most beautiful books on mechanics, which remained a fundamental guide for all of the older Bernoullis.

But Jacob, though an admirer of Huygens, did not trust the energy principle as a general basis, since he knew from his experiments that in a mechanical process energy is not always conserved. Thus he looked for another basis, and he found it in Archimedes' law of the lever.

In a first note [1686], he had presented his ideas, but without a tangible result. However, after a few years [1691] he could, with the help of an important hint from the Marquis de l'Hôpital [1690], who had defended him against criticism from Huygens [1690b], solve his problem, at least for an arbitrary mass distribution along a rigid line. But in one of his last papers [1703a], he stated his idea in the following form: suppose a number of weights can turn around an axis, each one at the end of a rigid arm of given length, such that the whole system can oscillate around the axis. Clearly, he says, this whole system of levers is in equilibrium, if and only if the sum of all lever arms (that is, of all products of weight times length of the arm of the weight) is zero; or, as we would say today, if the sum of the moments of the forces is zero. Thus, for obtaining the law of the oscillation of a pendulum, that is, a rigid system of masses, one must, in today's parlance, add to the resultant of the moments of the weights the time derivative of the sum of all angular momenta of the masses. From this axiom he could indeed reproduce all the results that Huygens had derived from his energy principle, and solve a few more problems as well.

Here the interesting question for us is if Jacob saw that his principle, contrary to Huygens's, could be generalized to three dimensions, since in this lies the superiority of the principle of moments over the one Huygens had used. There are indications in this direction, but so far I have not found any clear evidence.

For while the generalization to three dimensions is straightforward for Newton's equation, for the angular momentum principle it is not, since here we are dealing not with vectors but with skew tensors, and at that time there was not even a symbol for angular velocity. This is where Daniel Bernoulli and then Euler came in. Both solved the problem of the collision of a mass point with a rigid bar and the collision of two rigid bars [Daniel Bernoulli 1744; Euler 1744b]. Daniel Bernoulli still used the old formalism based on Huygens's oscillation length, but Euler wrote down equations of motion. Thereby he developed a new concept: the two-dimensional moment of inertia, which does for rotational motion what the mass does for the linear one: it connects the angular velocity with the angular momentum. But, of course, for three dimensions the problem is much more complicated.

When Euler published his *nouveau principe*, which I mentioned earlier, his first ambition, before attacking hydrodynamics, had been to build on it the theory of the general motion of a rigid body in three-dimensional space. In a first mémoire [1765c] he constructed the three-dimensional tensor of inertia, the first symmetric tensor in mechanics, and perhaps in geometry as well.

He then set out in a series of papers to find the equations of motion, just as he had proceeded for hydrodynamics. For formulating the complicated kinematics he introduced what we today call the three Euler angles, and for the derivation of the dynamic equations he used his new principle.

His deduction of the result was criticized for lack of rigour. Euler accepted this criticism and set out to correct the shortcomings. This time he met with approval, except from one scientist: Euler himself. To him the deduction still seemed too complicated. So he presented a third deduction in his great work *Theoria motus corporum* [1765]. But it was only in 1775 – he was then 68 – that, guided by his work in elastic bodies, he recognized that Newton's equations alone were not sufficient for attaining his goal [1776]. Here, contrary to nonviscous hydrodynamics, both Newton's second axiom and Jacob Bernoulli's principle are needed:

$$\dot{\mathbf{P}} = \mathbf{F} \text{ and } \dot{\mathbf{J}} = \mathbf{M}\,.$$

That is to say, the rate of change of the total momentum **P** is equal to the sum of the forces **F**, and the rate of change of the total moment of momentum **J** is equal to the sum of the moments of the forces **M**. Truesdell calls the combined system of the Newton- and the Euler-equations "the true fundament of mechanics".[xvii]

If both are used, the derivation of the Euler equations for the rigid body becomes a matter of only a few lines; it is essentially this deduction which one still finds today in all textbooks. And, as a student of mine once pointed out to me, this deduction contains the first use of a *repère mobile*.

This story of Euler's repeated deductions recalls a remark by André Weil, who wrote in his book *Number Theory. An Approach Through History*:

> Hardly less striking is the fact that Euler never abandoned a problem after it had once aroused his insatiable curiosity. ... Once he had taken up a question, not only did he come back to it again and again, ... but he also loved to cast his net wider and wider with never failing enthusiasm, always expecting to uncover more and more mysteries, more and more, more and more 'herrliche proprietates' [Weil 1984: 284].

It may be that the history of rigid body dynamics, even with its splendid conclusion, appears less rich and overwhelming than the history of hydrodynamics. But we should not overlook the fact that mechanics had now entered – through the pursuit of entirely its own motives – into tensor algebra and into the theory of compact non-commutative Lie groups. Witness to this is the first and successful exploration of the space group SO(3), with the help of the Euler angles. Physicists may take pride in the fact that here it took pure mathematics a long time to catch up with them here.

Elasticity

The domain of elasticity is the generic case of mechanics,[xviii] hence it is the largest and also the most complex one. In what follows, I shall distinguish from elasticity proper and linear flexibility, a special case, which fills a considerable part of the story, and then I shall also speak on acoustics. Elasticity is to rigid body dynamics what hydrodynamics is to point mechanics, thus it needs Newton's law and the moment of force principle and, as in hydrodynamics, the laws are localized. Thus a new, more complicated concept and more complicated partial differential equations are needed. Small wonder then, that the construction of a complete basis eluded the whole century. Jacob Bernoulli stands at the beginning of the modern theory of elasticity. He had tested by experiment Hooke's law, which says that the elongation of a hanging cord is proportional to the weight attached to it, and found it to be in general invalid. He wrote about his results to Leibniz, who had been engaged in a discussion with Mersenne about Galileo's formula concerning the breaking of a beam. Stimulated by this investigation he suggested to Bernoulli, as he said "a new and not inelegant problem", namely, to compute the form of a beam that is bent by a weight.[9]

Jacob recognized that the weight acts through the mechanism of a lever whose fulcrum is on the *neutral fiber*, that is, the fiber that is neither compressed nor elongated when the beam is curved [1694]. The weight acts on the longer arm and the force of resistance of the fiber acts on the shorter arm; it resists the force induced by the shorter lever arm. Thus, the dynamical principle is Archimedes' law of the lever, but then how can its effect be calculated?

Assuming that a straight section of the beam remains straight after the bending, Jacob found a proportion between any fiber, its elongation, the section through the beam and the

---

9 See "The Beginning of the Theory of Elasticity from Galileo to Jacob Bernoulli" in this volume, pp. 1-14.

radius of curvature of the beam. For its length of the radius of curvaturehe had discovered a formula that he called the *theorema aureum*. (This, incidentally, was the first nontrivial differential expression of second order.) But this does not yet solve the problem: we still must know the relation between the elongation and the force. The simplest answer, always correct to first order, would be the one given by Hooke's law. But Jacob, as I mentioned, knew that in general this law was not valid, and his experiments had convinced him that there was no such universal law. So he had to insert into the solution, which he had in the form of an integral, an arbitrary function, which then had to be determined by measurement. This is the first example of a "constitutive equation", as distinguished from a genuine dynamical principle. In the modern axiomatic formulation of mechanics created by W. Noll in the 1950s and developed by Truesdell and his school, this distinction is fundamental.

But if one accepts Hooke's law, one finds an elliptic integral, and for the arc length even one of the first kind. Bernoulli computed both integrals numerically through a series expansion. The theory of elliptic integrals later became a central branch of mathematics, which exerted its powerful influence far into the twentieth century.[xix]

Jacob Bernoulli's first paper makes for difficult reading even today, and at the time neither Leibniz nor Huygens could follow it. In particular it seems that neither understood that the law of the lever was the basic principle upon which the rest of the deduction depended. So, the next year Jacob wrote a second, clearer paper [1695]. Even so the reader is grateful for the help of the very careful notes of Gabriel Cramer, the faithful editor of Jacob Bernoulli's *Opera omnia*!

Bernoulli worked on these problems until his death. His ambition was to determine the position of the neutral fiber. The error he committed was shown by Antoine Parent [1713], a forgotten French physicist whom Truesdell resuscitated from oblivion.[xx] This problem remains unsolved still today, but all the same, the theory of elasticity had now found a first basis: the principle of moment of momentum. Truesdell states, "The form of the elastic band, the deepest and most difficult problem yet to be solved in mechanics, is his alone".

The famous exchange of letters between Daniel Bernoulli and Euler on problems of elastic curves took place during the 1730s.[xxi] One of the first ideas communicated by Daniel to his friend was an expression for the elastic energy stored in a beam. The main mathematical instrument used by Euler was the calculus of variations.

While in the eighteenth century, unlike the nineteenth, theory moved faster ahead than did experimental research, thanks mainly to the enormous progress of mathematics, it would be wrong to neglect the progress made through experiments as well as through the engineering sciences and the perfection of artisanship. I mentioned Jacob and Daniel Bernoulli's observations and measurements; I should like to mention also, for the first half of the century, van Musschenbroek and, for the second half, Chladni.

To Musschenbroek [1729] is due the experimental discovery that in general work done on an elastic body produces heat, thus incidentally also the discovery of a limitation of pure mechanics. Also, according to Truesdell, to Musschenbroek is due the first systematic experimental program on the strength of materials [1960a: § 22]; Truesdell calls his discovery of buckling "the climax of his work".

Concerning engineering it suffices to mention Euler's many papers and his textbook for naval engineers[xxii]

Concerning artisans, laurels must go to John Harrison for his construction of high-precision clocks, which indeed changed the art of seafaring by, for example, providing the best, and for a long time final, answer to the terrible "problem of the longitudes".[xxiii] Since this problem was the cause of many naval disasters, the English parliament had offered a large prize, which after many tribulations was finally awarded to Harrison. Many, including Daniel Bernoulli, had worked for the solution to this problem; it was the single-minded Harrison who solved it.

Accounts of the history of experiments, where we are not only told that *A* discovered *X* and *B* discovered *Y*, but where we learn the exact difficulties that had to be overcome and how, are quite rare.[xxiv] But on John Harrison and his clocks we have now a series of papers by M. Vandyck, professor of physics in Cork, Ireland, who with an assistant studied Harrison's achievements, both from the theoretical and from the experimental side.[xxv]

Flexibility and acoustics

The mechanical theory of flexibility was opened by Brook Taylor's computation of the fundamental frequency of a string [1714]. This computation is significant also for the fact that Taylor performed it by using Newton's axioms, and this also determined what followed, for from then on all theories of oscillations were derived from the general principles of mechanics.[xxvi]

In 1732-33 Daniel Bernoulli investigated the frequencies of the multiple pendulum. His results were that a pendulum with n masses has always exactly n stationary solutions whose frequencies depend on the masses and on the distances of these masses from the point of suspension [1740a]. Taking all masses equal and equidistant and going to the limit of a uniform, finite, hanging cord he could derive an infinite series, whose infinitely many zeros yield the frequencies of the cord's stationary oscillations. This series is what we call now the Bessel function $J_o(x)$.[xxvii] In the following year [1740b], he derived his results from a principle which is practically the same as that later stated by d'Alembert in his famous *Traité de Dynamique* as the basis of all of mechanics (and as Truesdell remarks, d'Alembert stated it as obscurely as Bernoulli).[xxviii]

Then history repeated itself. We saw how Daniel Bernoulli in the *Hydrodynamica* had investigated the stationary states of a fluid, and how his father Johann then introduced Newton's equation into this field. Now we see the son analyzing with great success stationary motions, and again the father, by now 75 years old, working on the same subject. This time, the older Bernoulli is the first to publish a set of differential equations for a system of linked bodies, the double pendulum, that serves as a model for the cord [1742b: art. LVI]. It is an equation of motion, albeit in a veiled form.[xxix]

And now in 1743 appears d'Alembert's *Traité de Dynamique*.[xxx] That d'Alembert was impressed by the work of Daniel Bernoulli is attested by the fact that of the twelve applications which he makes in the second part of the *Traité*, the one on the oscillating cord fills no less than a quarter of the second part. However, he goes beyond Daniel and formulates the partial differential equation of the hanging cord [1743: § 110], which is the first general statement of the law of motion of a continuous medium. This is a brilliant achievement indeed, but it produced no echo.[xxxi] One reason for this may be that after the presentation of this discovery d'Alembert turns to much more trivial questions, so that it is very well buried in the text. In our days, it took Truesdell to unearth it [1960a: 192].

A few years later d'Alembert published his discovery of the wave equation for a string, together with a solution [1749].[xxxii] This time it produced not only an echo, but, as you might say, a mighty resonance, because this discovery gave rise to another – luckily the last – of the great fights and polemics for which this century is still famous today. The polemic did not concern the solution itself, which no one contested, but its generality. D'Alembert, surprisingly, maintained that a solution ought to be an analytic function, while Euler, who also presented a solution, maintained that continuity and piecewise smoothness is all that was required here. His fundamental example was the triangular solution.[xxxiii]

Daniel Bernoulli took a stand apart, to which his investigations on the pendulum had led him. He solved the problem by a trigonometric series [1755a; 1755b] and claimed full generality for his solution, but neither of the other two would agree, and Bernoulli could not present a general, non-analytic example. But here he discovered his general theorem, that all solutions can be found as superpositions of the eigenfunctions, in this case the trigonometric ones.[xxxiv] In retrospect we see that on this point Daniel was right, and it is surprising to see how long it took the mathematicians to realize what an enormous field, namely, the theory of eigenfunctions, lay open before them. Thus when we look at the important role that Daniel's discoveries on eigenfunctions would come to assume in quantum mechanics, I think that his greatest achievement is precisely his theory of oscillations.

Yet at the time, research followed d'Alembert and Euler, that is, the line indicated by partial differential equations. There Euler's assertion that non-analytic solutions are admissible is shown explicitly by his well-known linear transformation into an equation with a mixed second order differential operator.[xxxv] Commenting on this dispute Truesdell says that this was the century's most important development in the methodology of science, and the historian must add, a blow for the philosophy of Leibniz too.

But only one two-dimensional problem was solved – the drum-equation found by Euler – while another was missed – the vibrating plate. The young Jacob II Bernoulli, the last of the dynasty, had thrown himself into this problem. He followed a wrong approach, and thus missed a term, without which his equation was not locally isotropic. When experiments did not confirm his predictions, he set out to correct the mistake. But he died shortly later in an accident in the Newa. Recently Diarmund Ó. Mathúna [1995] showed that it was a very narrow miss, but the problem would be solved only forty years later by Lagrange [1788].

Here is the place to mention Chladni, a Hungarian who had systematically studied the nodal lines of all kinds of bodies, comparing them with what theories were available [1787]. He himself had a good understanding of theories and was a careful judge. Here we see the dawn of the empirically-oriented nineteenth century.[xxxvi] I could have discussed acoustics in the section on hydrodynamics, since sound waves are solutions of the linearized Euler equations. The historic development shows that research followed the theory of vibrations, since brass and wood instruments are cousins of string instruments.

The 24-year old Lagrange first became famous through his papers on acoustics, since at the time sound waves were considered the most difficult problem of science. In an exchange of letters, Lagrange and Euler solved in a short time most of the basic questions;[xxxvii] only the speed of sound would remain a deep puzzle for a long time to come.

Before concluding, I must return to elasticity proper. Two more discoveries must be mentioned, in order to show how close the century came to the final foundations of elasticity. One was Euler's discovery of the shear force [1771b; 1776b]. The other one, coming even closer, was made by Coulomb, who, studying torsion, and constructing his pendulum to confirm the $1/r^2$ law of electrostatics, discovered the shear stress [1776].

Curiously enough, shear stress had almost been discovered a hundred years before by the same Antoine Parent, mentioned earlier as having been unearthed by Truesdell. Was there really no one between Parent and Coulomb who made some progress here, or it is that there are not enough Truesdells?

The legacy

What was the main legacy that the eighteenth century passed onto the nineteenth? I must begin with elasticity, precisely because in this field the foundation of a dynamical basis eluded the scientists of the eighteenth century. This had to wait for Cauchy, who in 1826 discovered the stress tensor, which does for elasticity what the internal pressure does for hydrodynamics.[xxxviii] The stress-strain relations could now be investigated systematically in linear and nonlinear three-dimensional models. Through this discovery the whole of continuum mechanics became a unity as a field theory formulated through partial differential equations. This united field theory found an immense range of applications in modern technology.

Hydrodynamics also profited from Cauchy's discovery, since viscosity, a shear effect, could now be properly formulated. And only partial differential-equations permitted a general attack on problems of turbulency.

The rigid body now remained a singular limit-model.

Point mechanics had a different destiny. After all, the mass point, the smallest model of them all, cannot be incorporated into continuum mechanics. From Jacob Hermann and Johann I Bernoulli it grew into analytical dynamics, and in the nineteenth century it found its classical formulation through Lagrange. Then, in the hands of Jacobi, it was formulated in a way such that partial differential equations were of the essence. Hamilton saw the mathematical affinity between mechanics and optics, which became of central importance in the nineteenth century. But that is not all: mechanics also influenced other fields of physics.

Euler had always maintained that electricity, magnetism and optics ought to be understood as phenomena due to the mechanisms of a fluid ether.[10] In his first contacts with Lagrange, he wrote a letter, published in Turin [1762], in which he communicates him the 3+1 dimensional differential equation containing what is today called the *d'Alembert operator*, writing that this must be the true equation not only of sound but of light as well.

After Cauchy, Euler's model was enriched to incorporate polarization from a fluid to an elastic ether as well. And if I may be permitted to give a look at the twentieth century, the ether disappeared, but the electromagnetic field took over its heritage. Finally the Galilean model of space and time was abandoned for a dynamical structure of space-time, formulated through partial differential equations.

---

[10] See "Euler's writings on Optics, Electricity and Magnetism" in this volume, pp. 211-223.

I chose these examples on purpose, for throughout the nineteenth century mechanics operated in a totally different intellectual climate. Research now became primarily empirically minded and therefore its results often appeared disconnected. But in the nineteenth century mechanics remained the backbone of physics, around which the other fields were eventually organized, thanks to the discoveries made in the eighteenth century.

At the beginning of the twentieth century theory and mathematics became once again the fundamental and leading influences. This fact means that we today are much closer to the eighteenth century than scientists of the nineteenth century were. Especially, as in the eighteenth century, we once again appreciate that physics, indeed the science of nature, forms a logical and mathematical connected unity, and I am especially happy to say this in honor of Clifford Ambrose Truesdell, who did probably more than anyone else for our understanding of this century!

The Discovery of Rigid Body Dynamics[1]

In memory of Fr. Ciaran Ryan

Introduction

Historic research has taught us the important lesson that the development of all branches of mechanics proceeded simultaneously. Work on the most complicated ones, hydrodynamics and elasticity, did not wait at all for the completion of point mechanics. Newton, building on the many and great achievements before him, formulated the foundations of point mechanics and conceived a program for the theory of fluids which for several decades afterwards guided research in hydrodynamics. On the other hand, the theories of elasticity and the theory of rigid bodies were both developed independently of the *Principia* [1687] and, in part, before it. To be sure, Newton talks in the *Principia* of a rigid body – our Earth –, but he does so in a special way that is very different from the dynamics of rigid bodies as we know it. Rigid body dynamics is a work of many hands (as is often the case in science), but it is fair to say that three among the pioneers in this domain were not only the most prominent but played decisive roles: Christiaan Huygens, Jacob Bernoulli, and Leonhard Euler. As different as the lives and the careers of the three were, they do have one thing in common: their accomplishments, especially those in mechanics, are vastly underrated today.

The history of rigid body dynamics cannot boast of a major synthesis like the one we find in the development of hydrodynamics, where in the hands of Daniel Bernoulli and Euler hydrostatics merged first with hydraulics and then with acoustics. Rather, its development proceeded around only two or three major questions. Nevertheless, it had its influence on other domains, such as elasticity, and as we shall see, it had a great influence on the understanding of one fundamental point of mechanics: the principle of relativity. Or, better, it had an influence on two men's understanding of the principle of relativity, since after them it was forgotten again. Indeed, as we shall see, a book by one of the authors on the dynamics of rigid bodies contains the clearest formulation of the principle of relativity up to 1900. This is not so surprising, since in a certain very precise sense, the rigid body constitutes the essential element of what would come to be called the Galileo group. From the geometric point of view, it defines the *repère mobile* and thus may help to define the fiber bundle; from the point of view of physics, it is the transportable yardstick

[1] 1984 Ciaran Ryan Memorial Lecture, presented in Dublin. Previously unpublished.

as well as the transmitter of instantaneous signals that define the present, that is, the famous simultaneity of classical physics, which in 1905 people learned they had to abandon, and indeed found so hard to give up. No wonder that there is no rigid body in Einstein's mechanics; it was replaced by the linear field.[i] On the other hand, it can be said that through the work of Kronig and Rabi [1927] and others, it survived into quantum mechanics.

All the same, you may ask if this is a pertinent subject; indeed, if you look at small distances in high energy physics, or at large distances using general relativity, then it would seems that you could not be further away from rigid body physics. Yet we must not forget that physicists never achieve their successes by looking only at observations and experiments, and handling them with some formalisms called "theories". Rather, they are guided by their acquired ideas, gained as a matter of course through the long and often complex elaboration of their results.

The history of our subject was outlined by Lagrange in the *Méchanique analitique* [1788], and though we are happy indeed to have his account, we should not remain content with it.[ii] It is not so much that inspection of the original sources reveals much that may be added, but that the discoveries made in physics during this century and the historical research have modified our views.[2]

Christiaan Huygens

Christiaan Huygens is one of the most fascinating scientists in history. He was the second son of the Dutch statesman and poet Constantin Huygens, who is described to us as one of the impressive men of his country at that time, one of the more glorious periods of the Netherlands. The young Huygens received the best possible education; while still in his youth, he may even have received instruction from Descartes himself, who was in Holland for a few years. Later on, Huygens spent some time in England where, on the grounds of his discoveries, he was made one of the first members of the newly-founded Royal Society. A few years after this trip, in 1666, Louis XIV called him to Versailles, where he worked in the personal library of the *Roi Soleil*. He remained there for fifteen years, even though the King was at war with Huygens's own country, and only later, in 1681, did he return to Holland.

Today he is mostly known for three great accomplishments. First, the proper explanation of the "companions" of Saturn observed by Galileo, as a *ring* and the discovery of its planet Titan. These discoveries were made possible thanks to the new lenses that he and his older brother were able to construct. Second, the laws of collision, to which Marcus Marci had paved the way, were discovered by Huygens independently and, it would seem, somewhat earlier by Wallis and by Wren (cf. [Szabó 1979: 427]). Huygens had derived them by explicitly using the "principle of relativity".[iii] Third, the *Traité de la lumière* [1690], which contains his largely intuitive but nevertheless rigorous foundation of

[2] The greatest credit for this new view must go to the American Clifford Ambrose Truesdell, whose views on this subject, and especially on the work of Jacob Bernoulli, often shaped my own. I should like to mention also the book by Istvan Szabó on the history of mechanics [1979]; it is often dependent on Truesdell's work and contains much valuable information and. sound insight.

geometric optics. The *Traité* laid the ground for this branch of science, which eventually made its biggest advance and received its definitive formulation by Hamilton in Ireland.

Apart from these major discoveries, we find in his *Oeuvres complètes* a host of minor ones.[3] Among these, I may cite the *Cosmotheoros* [1698], a speculative investigation of the possibility of life and civilisation on other planets. The work is interesting enough alone for the daring of the questions that its author raises and the thoroughness with which he pursues and discusses them.

But undoubtedly Huygens's *masterpiece*, although much less well-known, is the *Horologium oscillatorium* [1673]. The work, written in Latin, was published in 1673 in Paris. Today we are fortunate to have the anastatic copy of this beautifully printed book made in Brussels, in the series "Culture et Civilisation".

Before I discuss its content, I cannot resist making a small remark about the dedication. The book opens with a magnificent dedication to Louis XIV printed with especially beautiful letters. The first sentence says literally, "the reborn geometry, great king, and its institution in this century, we owe above all other to France". In Latin the sentence begins, "*Renatam* *rex maxime* ...". The complete edition does not say anything about this sentence, but I do not think that one must be a particular lover of puns to recognize in the very first word "Renatam = Renée", which precedes even the allocution of the greatest king himself, a special tribute paid by Huygens to his scientific mentor, René Descartes. Descartes, and even more so the Cartesians, are generally given bad press in the histories of physics. "Cartesianism" sometimes became synonymous, quite often unjustly so, with a kind of purely deductive rationalism that pretends to be superior to experience and experimental knowledge. However, we must carefully distinguish: there was Descartes, there were the Cartesians, and then there was Huygens.[iv]

The second part of the *Horologium*'s title indicates two of the book's most patent virtues: the clarity of its presentation and the rigour of its arguments. But these two qualities can induce us to underestimate some of its other specific achievements, among them many new discoveries and innovations.

The titles of the book's five chapters convey an impression of its range and scope:

1. A description of the oscillating clock;
2. The fall of weights, and the motion of these along a cycloid;
3. The size and evolution of the curve;
4. The centre of oscillation or movement;
5. The construction of another kind of clock is shown, in which the motion of the pendulum is in a circle, and theorems concerned with centrifugal force.

The theorems on centrifugal force, which became of central importance to the later development of mechanics, are contained in an appendix.

While in the first part and at the end of the fifth we see Huygens the engineer and clockmaker, the second part contains a systematic exposition of the mechanics of heavy

---

[3] Huygens's *Oeuvres complètes* [1888-1950], edited by the Dutch Academy of Science, is often praised as a model for today's editions of complete works.

bodies. To be sure, Huygens's scope is less general than Newton's, but it anticipates a great deal of the latter's accomplishments.[4]

It may be said in passing that the third chapter took an enormous step forward in the geometry and the kinematics of curves because there Huygens investigates the fundamental properties of the cycloid, establishing it as the "isochrone": at whatever altitude a body starts its oscillations, the period will always be the same. Joachim Fleckenstein, editor of one volume of the *Collected Works* of Jacob Bernoulli, called the cycloid "the curve of destiny" of the late seventeenth and early eighteenth centuries, and indeed, in the hands of Huygens and the Bernoullis, it revealed itself as the solution to some of the deepest problems that have stimulated the progress of science.

The fourth chapter, which interests us here, comprises fifteen definitions, two hypotheses and twenty-six propositions. (The last one, incidentally, is numbered incorrectly, maybe because Huygens added it only later.) His style might be called "Cartesian", though "Euclidean" may be more appropriate. As I have noted elsewhere, it should be noted that the *Principia* is organized in quite the same way.

What is the "center of oscillation" to which the fourth chapter is dedicated? Huygens, who indicated in the second chapter some of the laws of motion of "bodies whose extension can be neglected", asks what the laws are that rule the oscillations of a suspended body. As always, the methods by which one attacks a problem are largely determined by the common knowledge of the epoch one lives in. In this case, this means that Huygens knew only a few elementary laws of motion, but he possessed an *extensive* knowledge of geometry, including some powerful methods equivalent to integration, though he did not know infinitesimal calculus. Thus he attacks his problem by asking the question of how, given a physical pendulum of arbitrary form, an equivalent simple pendulum is found, that is, how does one find a simple pendulum having the same mass that oscillates with the same frequency? In other words, Huygens was looking for the position of a point on the line from the suspension point to the center of mass, whose distance from the suspension point to the center of mass is the length of the equivalent simple pendulum. This point was called the *oscillation point*.[V]

This notion seems to have already played a small role before Huygens in the work of Mersenne and Descartes. The oscillation length turns out to be proportional to the inverse of the square of the frequency (as we would say), and Huygens and his followers often say "oscillation length" where we would simply speak of the frequency.

The greater part of the fourth chapter is devoted to the systematic development of this notion through several important theorems and its computation for various homogeneous bodies, one-, two- and three-dimensional bodies, and in many cases for various different points of suspension. In other words, Huygens computes the frequency of the various oscillations. Those who are used to measuring the advancement of physics by the amount of computed numbers must consider that during three-quarters of a century, only Newton and Daniel Bernoulli produced more that could be confirmed by experiment. In these computations we may recognize some of the most refined integrations accomplished up to this time.

---

[4] See "Huygens's *Horologium Oscillatorium* and Newton's *Principia* in this volume, pp. 41-49.

In fact, for the oscillation length Huygens found the following relation (written in modern form):

$$l = \frac{\theta}{Ms},$$

where l is the length of the equivalent pendulum, M the total mass, s the distance of the center of gravity from the point of suspension, θ the moment of inertia with respect to the axis of rotation.[vi]

But here this latter concept is not introduced into mechanics in the way that we do it today, because the moment of inertia does not occur here in the place where it is most effective, namely, in an equation of motion. Its significance is not grasped, and it appears only as an intermediary tool.

From the many theorems which he proves, I may quote the following two. Theorem XVI says that the oscillation around points lying at the same distance from the center of gravity have the same frequency. This is a weaker form of Steiner's theorem, which, for reasons that I shall make clear in a moment, Huygens could not yet formulate.

Proposition XVI and theorem XX contain the basis of the reversion pendulum. They say that the point of suspension and the center of oscillation can be interchanged, that is, if we choose the center of oscillation as a point of suspension, the new center of oscillation will coincide with the old point of suspension; thus the frequency in both cases is the same. This made it possible to fix the standard length of time with an exactness hitherto totally unknown. Huygens even wanted to base the standard of length on this standard of time.

What is the physical basis on which Huygens built his edifice? What are the two hypotheses from which he rigorously deduces his results? The first says that if any system of weights begins to move, its center of gravity cannot rise more than the height it had when the motion began.[vii] The second says that if the air and all other hindrances are removed, the center of gravity of a pendulum in motion always falls and then rises again through equal arcs.[viii] In those two statements is clearly recognizable a *rudimentary form of the principle of energy*, expressed here by Huygens in two statements, an inequality and an equality (much the same way as we do today for entropy). Insofar as the physics is concerned, all his propositions are deduced from those statements.

This energy principle became the starting point for a long line of investigations, independent of Newton.[ix] Its central notion is not *force* but *energy*. Following Huygens, Leibniz become the champion of this idea, for instance, in his famous dispute with the Cartesians. But it had its greatest successes in the hands of Johann I Bernoulli, who reformulated this hypothesis in terms of the *vis viva*, that is, kinetic energy, and through him, in those of his son Daniel. Daniel Bernoulli filled an important gap in Huygens's theory, namely, if the axis of suspension goes through the bodies' center of gravity, the oscillation length becomes infinite and thereby the moment of inertia takes the form $0 \cdot \infty$! For this reason Huygens could not handle this case. But for us, this special case has become through what is called Steiner's theorem, the most important one, as Huygens's theorem which I quoted before had already indicated. Daniel Bernoulli closed this gap with the introduction of what is today called the "radius of gyration" [1729; 1744]. In fact, Daniel discovered Steiner's theorem in two dimensions and in principle he could simplify Huygens's theory considerably. I say "in principle" because, alas, his presentation lacks the clarity of Huygens's, although Huygens had been for a decisive period of Daniel's life his principal source of inspiration. I quote this paper especially because it is the next to last one written with the help of Huygens's tools and along the lines of his ideas. To Daniel's very last work I shall return later.

As perfect as Huygens's theory was in its own way, its successes were necessarily limited to rotational motion in a *plane*. The deeper reason for this limitation, not difficult to see today, must now be indicated. The energy principle provides us with only one integral part of the equation of motion and thus it can yield the solution of problems of only one degree of freedom. Such is indeed the case for the rotation of a body around an axis or its motion along a curve, but for describing the motion of a body in space, the energy theorem does not suffice. Finding the path here was left to others to whom we must now turn.

However, this shortcoming should not induce us to underrate the greatness of Huygens's achievement. In order to be fair to him and his accomplishment it must be stressed that his principal goal was *not* the formulation of the motion of a rigid body but the construction of a high precision clock. Thus he discovered that he must also consider the movement on a cycloid, which is not circular, and for this the energy theorem was indeed the appropriate basis. For this achievement Arnold Sommerfeld called him "the greatest genius among clockmakers".

But what is more, Huygens gave us a systematic, rigorous mechanical theory, the most important one before Newton's.

Jacob Bernoulli's discovery

The theory of Huygens was attacked in 1682 by the *philosophe* and *savant*, Abbé Catelan, who offered a counter-example to Huygens's theory. Imagine two weights that are fixed to a rigid bar and oscillate. First the weights descend together, but once the bar is in vertical position, the link that connects them is cut and they now rise separately. Catelan claimed that in this case Huygens's hypothesis about the center of gravity is shown to be impossible. But of course he could not support his argument by computation; neither, by the way, could Huygens.

It is at this moment that an as yet unknown man rose to Huygens's defense. Notice that we are still in the period prior to the publication of Newton's *Principia*, and before Leibniz published his paper on the infinitesimal calculus, the paper that was to be the turning point in the young man's life and indeed, through his work, a turning point in mathematics.

In terms of character, upbringing and career, Jacob Bernoulli, the oldest and greatest of the Bernoullis, contrasts sharply with Huygens. His ancestors had left Antwerp as religious refugees for Frankfurt, and from there his great-grandfather went to Basel. In the closed society of Basel, even though Jacob's father, a well-off merchant, was a member of the city government, the Bernoullis were still considered newcomers. The father allowed Jacob to study theology, although Jacob would have preferred to study astronomy and mathematics (and indeed did so behind his father's back). He chose as his motto that of the unfortunate Phaeton, *Invito patre sidera verso* (Against my father's will I am among the stars). In other words, Jacob had to fight his way to mathematics and to the sciences. This may have strengthened his determination, but may have also made him even more introverted. At any rate, these traits contributed to the quarrel with his extroverted, aggressive, one might even say loudmouthed, younger brother, Johann.

It may be noted, however, that his studies of theology left their traces in his publications. While the Latin of most of these authors is far from the classical one, Jacob Bernoulli, of all the authors I know, most indulges in complicated constructions of the type we had to unravel at high school; this is doubtless a fruit of his theological training!

In his papers "Extrait d'une Lettre du Sr. Bernoulli" [1684] and "Narratio Controversiae inter Dn. Hugenium & Abbatem Catelanum" [1686], Jacob Bernoulli first makes two statements. On one side, he shows clearly that Catelan's criticism is unfounded. But then he expresses doubts as to whether Huygens's hypotheses are well-founded. Not that he thinks that they must be wrong, but to him they do not seem well suited as a principle. Instead, he attempts a theory of his own to formulate correctly what we would now call the effect of the constraints that one body suffers from another because the two are rigidly linked. Although this theory is neither complete nor correct, in Jacob's discussion one seminal idea enters that was henceforth to become of central importance: *the idea of the lever*. The laws of the lever, you recall, had been formulated by Archimedes and mark the beginning of mechanics. As he did in differential geometry (we may remember his spiral), here too Jacob became the successor of Archimedes. Eventually this idea, as we shall see, will lead him to his goal. But it was to take him almost twenty years!

It should be noted that the discussion of these authors is made difficult for us by the fact that the notions used sometimes have a meaning that is different from today's. For instance, they speak of "velocity" where we would speak of "momentum". This is of course possible since the masses, thanks to the principle of equivalence, drop out of the discussion; all the same, the addition of velocities and the addition of moments are the same thing! The ambiguity of the notions will become even more pronounced as we now turn to the next participant in the discussion, Marquis de l'Hôpital.

L'Hôpital, the most brilliant French mathematician of his time, entered the scene. He had learned the new infinitesimal calculus from Johann Bernoulli as it was being developed by the two brothers starting from Leibniz's original paper, He now claims that Jacob's basic idea was correct and that his mistake occurred only in the application [1690]. In fact, starting from Jacob's principle, he could now correctly compute (that is, in agreement with Huygens's result) the oscillation length in the special case of a bar loaded with two, three or four mass points. Though limited, l'Hôpital's results were very helpful and he deserves credit for having reopened the case and pushed it in the right direction.

After l'Hôpital's paper, we find another one by Huygens, also in the *Histoire des Sçavants* [1690b]. There, he thanks Jacob for having defended him against Catelan (who seems to have been an irritating person indeed). But at the same time he expresses his displeasure at the fact that Jacob did not accept his hypothesis, which seems to him the only really sound one, as shown by its successes. He seems to be a bit annoyed.

Jacob's answer appeared in the *Acta Eruditorum* [1691]. The title begins with the words "Demostratio Centri oscillationis ex natura Vectis" ("Proof of the center of oscillation from the nature of the lever"). This is not his last work on the subject; for although he is able to generalize l'Hôpital's result, in fact he does not find the general solution either. I will only say that the notions here are ambiguous, and that he talks explicitly of accelerations and moments of forces, etc.

It must be stressed that at the same time Jacob recognized that the infinitesimal principle of the lever was the basis for computing the bending of an elastic beam, that is, the basis of the (rigorous) theory of elasticity! In fact, he was the only one who could solve this most difficult problem.[5]

It would be another ten years before Jacob's ideas matured. Huygens had been dead for about eight years when Jacob – almost twenty years since he had first taken up the subject and only shortly before his own death – published three papers that contain his last ideas published on the subject [1703a; 1703b; 1704]. It is the first of these three that concerns us here. A fourth was in preparation, published after his death [1744].

This short paper entitled "Démonstration générale, Du Centre de Balancement ou d'Oscillation, tirée de la Nature du Levier", about six and a half pages long, is an impressive piece of work indeed. It begins with a historical summary of the discussion, where he freely acknowledges that although earlier he had been able to progress beyond de l'Hôpital, he was somehow prevented from finding the general solution; he could not apply his idea to any curved or solid body. But now he had penetrated to the root of the matter and so, he proudly announces, he has found the solution to quite a few other problems as well. He starts from the law stated by Mariotte in his *Traitté de la percussion ou chocq des corps* [1673]: a lever or any other rigid body is in equilibrium if the sum of all moments of forces that act on it is zero. This law, he says, is generally accepted. If we now analyze the *oscillations* of a rigid body, we see that the different points are not accelerated in the same way as they would be if they were only linked to the axis or were free. If we consider two arbitrary points, one is sometimes pulled by the other, while the second one is pushed. He says that if we consider two arbitrary points, one is sometimes pulled by the other, while the second one is pushed. We may say that Jacob Bernoulli tries to respond to the question of which force gravity works against here. His answer is that this internal pulling and pushing also follows the same simple law of the lever: to the gravitational moments we must add what we would now call "the moments of inertial force".

We may say that he tries to respond to the question of which force gravity works against here. His answer is that this internal pulling and pushing also follows the same simple law of the lever: to the gravitational moments we must add what we would now call "the moments of inertial force". In other words, the motion is such that the sum of all gravitational and inertial moments is zero. From this assumption, as he now shows in a straightforward computation, Huygens's entire theory can be deduced.

A very few minor slips in the paper suggest that the otherwise very careful man wrote hastily under the impact of his discovery. Moreover, he sometimes says "velocity" where we would say "acceleration", but this does not unduly cloud the paper.

How must we look at his principle? The rotational motion is such that the sum of the moments of all external forces and all inertial moments is zero. If you replace "moment" in this statement with "force", then you get what is sometimes called the "principle of d'Alembert".[xi] This is a misnomer, since the principle stated by d'Alembert in the *Traité de Dynamique* [1743] is a different one, only loosely connected to the one I have just

[5] See "The Beginnings of the Theory of Elasticity from Galileo to Jacob Bernoulli" in this volume, pp. 1-14.

quoted. This is, as Truesdell has shown, one of the three principles cited with d'Alembert's name in the literature. Jacob's principle is also noteworthy for the fact that it was stated first for the moments and not for the forces. At least for planar motion, he found the true principle of rigid body dynamics.

Jacob stops a hair short of writing down an equation of motion, which would have to wait for another thirty years. For one thing, angular velocity had not yet become a handy tool of the geometers, and this is more than just progress of notation. But at any rate the theory of rigid bodies had been given its true foundation.

Jacob died shortly after the publication of the three 1703 papers, as I mentioned, and it is idle to speculate whether he would have gone further, as indeed one gets the impression. In order to be fair to him, we must remember that he was already 34 when his real career began, and that he was allowed to be a full-time scientist only at an advanced age.

While Huygens had erected a whole building, Jacob Bernoulli had just opened a door, but this door, as we shall hear, happens to be one of the main entrances of mechanics. The discovery made by this enigmatic but impressive figure came to fruition mainly in the second generation of his disciples, and it is to the greatest of them that we must now turn.

Leonhard Euler

Leonhard Euler's personality as well as his life are very different from Jacob Bernoulli's. While Jacob was away from the city of Basel only for a couple of short trips, Euler left it at the age of twenty and never returned. While Jacob occupied a chair at a university, Euler worked as a member of an Academy and, like Huygens, spent his life if not at a court itself, always very close to one. While Jacob was an introvert, Euler was open to the point of being, as André Weil called him, "ebullient". While Jacob was secretive and suspicious, Euler was easily accessible and looked for contact.

These differences come out strongly in their papers and books as well. While those of Jacob are short and compact, Euler's are broad and they present to us in great detail the author's reasoning, even more so than those of Huygens. Where Jacob's papers must be "looked up", sometimes with great pain, Euler's desire to instruct the reader, as well as his great clarity make his papers paradigms of pedagogical transparency, and can still serve as models today. Where Jacob takes the decisive step but then often stops – since, to quote Truesdell, "approximations were abhorrent to him" –, Euler proceeds and pursues the consequences of a discovery in all directions, often by discussing a great number of examples, as many as his incredibly fertile imagination could produce.

Two personal links connect Euler directly with Jacob Bernoulli. First, his father, the clergyman and amateur mathematician, Paul Euler, who gave the young Leonhard his first instruction in mathematics, had been a colleague and pupil of Jacob's. Second, Euler became the preferred student of Jacob's younger brother, Johann, who was acknowledged as Europe's leading teacher of mathematics during that period. Johann had himself been Jacob's student and collaborator, before they fell out in a quarrel which was both the intellectual delight as well as the scandal of the scientific community, and which filled the learned journals of the day. Johann Bernoulli led Euler to the summit of knowledge of his time. But it is perhaps due to the influence of Euler's father that, unlike Johann's son

Daniel, Euler thoroughly acquainted himself with the work of Jacob Bernoulli, not only with the calculus of variation and the theory of elasticity, but also with his work on rigid bodies. This must be stressed, for Euler is usually presented, and rightfully so, as one of the foremost Newtonians on the Continent, between Daniel Bernoulli and Clairaut. But Euler, even more so than Daniel Bernoulli, is the man in whom the Continental legacy merges with the English one. Euler inherited this legacy first from Huygens and then, to an even greater extent, from the Bernoullis.

Up to this period, the development of theories of rigid bodies had been entirely Continental. But now, with Euler, Newtonian mechanics affects its development as well. From his early days, Euler had been an ardent student of the *Principia* and it is fair to say that, after Daniel Bernoulli's investigations in hydrodynamics and his great endeavor to establish Newton's theory of gravitation beginning very early and culminating around the years 1738-40, Euler was the first to accept the principles of Newtonian mechanics on the Continent. In 1736, following Jacob Hermann's *Phoronomia* [1716] and Johann Bernoulli's paper published in the *Memoires* of the Académie des Sciences [1712], which presented in analytical form what was then called the inverse Kepler problem[6] and is today called simply, if a bit wrongly, Newton's solution of it, Euler published his *Mechanica* [1736], the first book on mechanics in purely analytical form.

Euler's occupation with the *Principia* had taught him that the true principles of dynamics are equations of motion. Henceforth through all his life he tried to establish for all branches of science such dynamical equations of motion.

The work on rigid bodies fills the greater part of four volumes of Euler's *Opera omnia*. The most important ones are "De communicatione motus in collisione corporum" [1744c], "Découverte d'un nouveau principe de Mecanique" [1752b], "Recherches sur la connoissance mécanique des corps" [1765b], "Du mouvement d'un corps solide quelconque lorsqu'il tourne autour d'un axe mobile" [1767b] and especially his great *Theoria motus corporum solidorum seu rigidorum* [1765a]. This was book two of the great series of six planned to cover the whole of mechanics.[xii] Euler had conceived this project at the age of twenty-seven, when most of the laws that he was to explain had not yet been discovered.

The first of these papers is just a short but important prelude. In a letter to Euler, Daniel Bernoulli had raised the question of what happens if two asymmetric two-dimensional bodies collide. He then worked out the solution with the help of a theorem that refers to Huygens's oscillation length: this was the very last success due to this approach, to which I have alluded earlier [1744].

Euler now took up the question and at once gave his own solution. But his approach is very different from Daniel's. Euler follows Jacob and writes down equations of motion. To Newton's:

$$\dot{\mathbf{P}} = \mathbf{F}\,.$$

he adds

---

[6] See "The Kepler Problem from Newton to Johann Bernoulli" in this volume, pp. 123-134.

$$\dot{\mathbf{J}} = \mathbf{r} \times \mathbf{F}$$

The rate of change of angular momentum is equal to the moment of force. Now the moment of inertia finds its rightful place, namely in an equation of motion. Thus he has four equations. Eliminating the force, he at once obtains three scalar linear equations. To these, with the help of a very elegant trick – albeit one which is not completely satisfactory from the physical point of view – he adds a fourth one, which is also linear. And this scalar system is easily solved.

While Jacob's solution is a bit obscure, Euler's is transparent. Of course, both solutions are restricted to planar motion: the kinematics of three-dimensional motion still remained too complicated, and not the slightest progress was made for a long time. Neither Bernoulli's nor Euler's papers gained much attention; not only the solution, but the *problem* itself was totally forgotten. Only Poisson, as Truesdell informed me, took it up and solved it in three dimensions.[xiii] Truesdell himself gave a derivation (unpublished) which is unobjectionable, and another based on a different approach was given by Yves Elskens and myself and then used in an application of the kinetic gas theory [Elskens and Speiser 1982: 539]. Euler's formulae are so simple and teach so much about angular momentum that they deserve to be part of every course on elementary physics.

As I said, thus far the three-dimensional motion of a rigid body (if we do not include Newton's treatment of the precession of the earth, which is, however, based on a completely different and very special approach) had defied everyone.

We must now turn to Euler's paper "Découverte d'un nouveau principe de Mecanique" [1752b]. What is this "New Principle"? It is nothing but the equation:

$$\mathbf{F} = m\mathbf{a}\,.$$

Two things are new. First, this is an equation (not a description in words, as in Newton), valid in a space-fixed Cartesian coordinate system and equivalent to the modern vectorial form (three scalar equations). Second, it is *general*; in other words, it is part of the foundation of every domain of mechanics. Although Euler, as we shall see, later corrected this claim, he was, for instance, able to complete with its help the foundations of hydrodynamics.

For the moment, however, he applies the principle to the motion of rigid bodies. One difficulty that had not yet been overcome was the enormous variety and diversity of the possible motions of a rigid body. Euler begins by distinguishing purely translational from purely rotational motion and from motion possibly composed of both kinds. He then says that the law (quoted above) remains the same if we pass to a new system that is in uniform motion relative to the old one. If, for instance, the body moves freely, one can attach the "space fixed" coordinate system at the center of mass of the body, which moves uniformly on a straight line. Thus the translational part of the problem is eliminated and only the purely rotational part remains. He then sets out to solve this. As he does not yet accomplish it in a totally satisfactory way here, I will not stop but pass over to his next work. But, as we shall see, the importance of his step cannot be underestimated: for one thing, in this paper, Euler made it clear that part of the difficulty comes from the fact that rotational inertia is much more complicated than linear inertia, which can be represented by the mass alone.

In the"Recherches sur la connoissance mécanique des corps" [1765b], Euler sets out to solve the problem of rotational inertia. First he says that we must distinguish the center of mass or center of inertia from the center of gravity.[xiv] The center of mass may also be defined as the point around which all moments of forces in a constant field vanish. Such a point, as he proves, always exists, and it is interesting to note that he shows explicitly that its determination does not depend on the coordinate system used.

But the rotational inertia depends on the moments of second order,

$$x^i x^k dm ,$$

that is, the moments into which the second powers of the coordinates enter. These he now discusses as well, and to do so, he develops what we today call "tensor algebra".[xv] He develops the transformation formulae and shows that there are always three mutually orthogonal axes with respect to which the mixed terms disappear. Thus the tensor of inertia was presumably the first one to be investigated systematically and for which the diagonalisation theorem was proved.

Like many of Euler's papers, this one can still be easily read, and with profit, since he discusses many questions (such those on the center of mass) which today we often find presented confusedly, or even wrongly.

But, while the correct formulation of the inertia of a rigid body is an important step, the dynamics of rigid bodies is still miles away from a solution.

Euler himself eventually found the complete solution in the important piece "Du mouvement de rotation des corps solides autour d'un axe variable" [1765c]. But, as this work has been criticized, I will pass over it too, and turn right away to the famous "Du mouvement d'un corps solide quelconquc lorsqu'il tournee autour d'un axe mobile" [1767b], written in 1760 but published only seven years later. Here indeed we do find the final solution and I should like to discuss it in detail.

In an inertial, that is, "space fixed", frame we have simply

$$\dot{\mathbf{J}} = \mathbf{M} .$$

The difficulty is that while the mass is a scalar, and thus a constant, the components θ_{mn} of the tensor of inertia are not constants in the frame, and the equation

$$J_m = \theta_{mn} \omega^n ,$$

which relates the angular momentum J and the angular velocity ω, does not help us to solve the problem in a simple way. Thus we go over to another body-fixed system, which, of course, is not an inertial one; therefore we will have additional terms. If d_t denotes the derivation in the new system we have:

$$\dot{\mathbf{J}} = d_t \mathbf{J} + \boldsymbol{\omega} \times \mathbf{J} ,$$

thus

$$d_t \mathbf{J} + \boldsymbol{\omega} \times \mathbf{J} = \mathbf{M} ,$$

and there we are!

If we now introduce the tensor of inertia in this equation and let the new system coincide with the body's main axis, which implies that $\theta_{12} = \theta_{21} = \theta_{23} = \theta_{32} = \theta_{13} = \theta_{31} = 0$, we find in component form:

$$\theta_{11}\frac{d\omega_1}{dt}-\omega_2\omega_3\left(\theta_{22}-\theta_{33}\right)=M_1,$$
$$\theta_{22}\frac{d\omega_2}{dt}-\omega_3\omega_1\left(\theta_{33}-\theta_{11}\right)=M_2,$$
$$\theta_{33}\frac{d\omega_3}{dt}-\omega_1\omega_2\left(\theta_{11}-\theta_{22}\right)=M_3.$$

But even as late as 1760, most of this was still far away; most of the ingredients had not yet been extracted from the enormous heap of notions that lay around. Angular momentum had not yet been formulated, really studied or understood. This had been done only in two dimensions, where the fundamental difficulties cannot appear. To be sure, the properties of rotational inertia had been formulated, but they had not yet been pursued thoroughly enough. The transition from one Cartesian coordinate system to another is a difficult thing to formulate.

But most of all, the idea of the angular momentum equation as a principle in its own right was not yet clearly grasped. Mathematicians found it difficult to free themselves from linear momentum and to recognize angular momentum as an independent tool for handling extended bodies. All this was gained only as the result of the studies that I am now going to sketch.

At the beginning of the paper, Euler sums up in a few words what he had done so far and adds a reflection that is quite characteristic of his own way of making investigations:

> If one excepts some simple enough cases in themselves, one can say that the rules that one must follow in this research, are again almost entirely unknown. Because although, in a piece entitled *Discovery of a new principle of Mechanics*, I have already developed the formulas that can lead to this goal, their application is in any case extremely difficult; and to overcome these difficulties, it seems that the most secure means would be to undertake the same research in several different ways, and to represent the rules that I already found in other forms, in order to make them more familiar to us, & to better understand their powers. Because one knows from experience, that when a research is very thorny, it is usual that the first efforts to solve it clarify but little; & it is only through repeated efforts, & by considering the same thing from several points of view, that one arrives at a definitive knowledge.
>
> I therefore turn myself to new attempts to search for the theory of the motion of rotation of solid bodies, which will certainly shed more light on this matter that still appears so obscure.[7]

[7] *Si l'on excepte quelques cas assez simples en eux-mêmes, on peut dire que les regles qu'on doit suivre dans cette recherche, sont encore presque entièrement inconnues. Car, quoique j'aye déjà dévelopé dans une Pièce, qui porte le titre: Découverte d'un nouveau principe de Mécanique, les formules, qui peuvent conduire à ce but, l'application en est pourtant souvent extrèmement difficile; et pour surmonter ces difficultés, il semble que le plus sur moyen sera d'entreprendre la même*

The many theorems, propositions, corollaries, etc., are organized in this paper as a series of questions, to each of which Euler gives us the solution. Since they show us the trend of his thought I shall mention some of them. Some of the solutions had, of course, already been found and published by him previously.

Proposition I. Given a body that turns arbitrarily about its center of gravity, we ask what is the best way to represent this motion and form a correct idea about it. [8] The answer is, of course, given through the angles that he had already discovered.

Problem II. Whatever motion the body has about its center of gravity O, find for every instant the axis ON, such that the elements of the body situated along this axis remain immobile during this instant. [9] The solution rests upon his own theorem, according to which any rotation, and especially an infinitesimal one, always leaves one axis invariant. Euler gives us an illustration: it is his famous explanation of the regular precession of the earth!

Problem III. For any motion of a body around its center of gravity O, whatever it may be, find the forces to which each element of the body are subject such that the body may continues its motion. [10] Thus, a bit like Newton had done for the planetary system, Euler starts from the geometric description and then computes the force.

With the next problem he prepares now the decisive step.

Problem IV. Having found the three forces in directions OP, OR *and* OC, *perpendicular to each other, to which the element Z is subject, assign to it three other directions,* OM, OS, *and* OT, *which are also perpendicular to each other, and which depend on the first Meridian* OML *of the body.* [11] This is a purely geometrical or, if you wish, algebraic problem, whose importance need not be underlined. Euler solves it by giving the parametric presentation of what we today call the group SO(3), a result which, by the way, has never been adequately appreciated in the history of group theory. It is thc first noncommuntative group investigated systematically. The importance of the result becomes evident from the next problems.

---

recherche en plusieurs manieres différentes, & de représenter les regles que j'ai dejà trouvées sous d'autres formes, afin de nous les rendre plus familieres, & d'en connoitre mieux la force. Car on sait par l'expérience, que lorsqu'une recherche est fort épineuse, les premiers efforts nous en éclaircissent ordinairement fort peu; & ce n'est que par des efforts réiterés, & en envisageant le même chose sous plusieurs points de vue, qu'on parvient a une connoissance accomplie.

Je m'en vai donc faire de seconds efforts pour rechercher la théorie du mouvement de rotation des corps solides, qui ne manqueront pas de nous fournir de plus grands éclaircissements sur cette matiere, qui paroît encore si obscure [1767b : Introduction].

[8] *Proposition I. Si un corps tourne d'un movement quelconque sur son centre de gravité, on demande de quelle maniere on peut le mieux représenter ce movement; & s'en former une juste idée* [1767b : § 2].

[9] *Probleme II. Quelque mouvement qu'ait le corps autour de son centre de gravité O, trouver pour chaque instant le rayon ON, de sorte que les élémens du corps situés dans ce rayon demeurent immobiles pendant cet instant.* [1767b : § 6].

[10] *Probleme III. Le movement du corps qui tourne autour de son centre de gravité O. étant supposé quelconque, trouver les forces dont chacun de ses élémens doit être sollicité, pour que le corps soit mis en état de poursuivre son mouvement.* [1767b : § 15].

[11] *Probleme IV. Les trois forces, dont l'élément* Z *est sollicité, étant trouvées suivant trios directions* OP, OR, OC, *perpendiculaires entr'elles, réduire les mêmes forces à trois autres directions* OM, OS, OT, *qui sont aussi perpendiculaires entr'elles & qui dependent du premier Méridien* OML *du corps* [1767b : § 20].

Problem V. *Find the moments of the forces to which the element* Z *is subject, with respect to the three axes* OM, OS, OT, *whose position always remains fixed with regards to the body.*[12]

Problem VI. *In order that the body can pursue the movement that is indicated by the quantities* p, q, r, *determine the moments of the forces to which the body must be subjected.*[13] The combination of the results yields the desired goal. But the formulae are very long and involved indeed: they fill an entire page!

It is beneficial to pause here for a moment. The formulae which Euler had found after years of labor belong to the most intricate ones that we have encountered thus far. They are, as we heard, the result of investigations done in very different fields, such as mechanics, spherical trigonometry, and algebra.

If I may, it is only after this long trip through a dark jungle, during which, however, he never lost his bearings, that Euler eventually found the simplification that permitted him to recognize the significance of his achievement. For so far, he had still used the angular velocities with respect to the space-fixed system as basic variables whose angular velocities now answer the apparent ones of angular momentum.

He now remarks,

> As these expressions are quite complicated, it is appropriate to introduce, in place of the three variables *p*, *q*, *r*, three others that are determined by them and thanks to which our expressions become simpler."[14]

In fact he now switches to the angular velocities with respect to the body-fixed system. He says:

> And thus, by the means of these substitutions, introducing the letters *P*, *Q*, *R*, instead of *p*, *q*, *r*, our formulae become not only considerably simpler, but one also notices a very beautiful uniformity, by which we see that these three new quantities also enter in the determination of our three moments.[15]

I shall come back to this sentence in a moment. First I quote his general comment at the end of the paper. There he says:

> This resolution of the problem, that I have just developed, is unquestionably simpler than the one that I gave at an earlier time, since the formulae that contain the solution are less complicated. But the biggest advantage consists

---

[12] *Probleme V. Trouver les moments des forces, don't l'élément* Z *est sollicité, par rapport aux trios axes* OM, OS, OT, *qui conservent toujours la meme situation à l'égard du corps.* [1767b : § 24].

[13] *Probleme VI. Pour que le corps puisse poursuivre le movement, qui est indiqué par les quantités* p, q, r, *determiner les momens des forces requises, dont le corps doit être sollicité* [1767b : § 25].

[14] *Ces expressions étant fort compliquées, il sera à propos d'introduire, au lieu des trois variables* p, q, r, *trois autres qui en sont déterminées, & par lesquelles nos expressions deviennent plus simples.* [1767b : § 30].

[15] *Et partant, par le moyen de ces substitutions, en introduisant les lettres* P, Q, R, *au lieu des* p, q, r, *nos formules deviennent non seulement considérablement plus simples, mais on y remarque aussi une uniformité fort belle, par laquelle nous voyons que ces trois nouvelles quantités entrent également dans la détermination de nos trois moments* [1767b : § 30].

> in that this solution is much more suitable for application to all cases that one can propose. The obvious reason is that here I reduced the calculation of the elements that depend on the configuration of the body, to axes that are fixed with respect to the body, so that these elements always stay the same, while for every different position the first solution requires a new search of these elements, since they were in related to stationary axes in absolute space, with regard to which the position of the body can change at any instant.[16]

Thus we see that what impressed Euler most was the tremendous simplification which now suddenly emerged. The final formulae are:

$$dp \sin q \cos r - dq \sin r = P\,dt,$$
$$dp \sin q \sin r - dq \cos r = Q\,dt,$$
$$dp \cos q - dr = R\,dt.$$

Comparing these short equations, each of which fills only one line, with the enormous expressions through which he had to struggle, this does indeed look like a miracle. Sommerfeld, somewhat understating the case, calls these equations "remarkably symmetric and elegant", and indeed they are both. But with our advantage of looking at them from such a distance, we can see that something fundamental appears here which Euler did not yet fully grasp, but of which he was aware. I come back now to the statement which he makes after his introduction of the new variables. There he adds: "Such regularity also serves as a proof to justify the calculation that I have just developed".[17] The wording shows that Euler was aware that after ten years of hard work he had established a new set of fundamental equations.

Remember how in his "problems" the forces are replaced step-by-step by the "moments", and also that a moment is not necessarily the moment of a force. Thus to the equations

$$\dot{\mathbf{P}} = \mathbf{F},$$

now comes independently the equations

$$\dot{\mathbf{J}} = \mathbf{M}.$$

Here we have seen how it emerged in a very roundabout way, as is often the case in science.

[16] *Cette résolution du probleme, que je viens de déveloper, est sans contredit plus simple que celle que j'en ai donnée autrefois, vu que les formules qui contiennent la solution, sont moins embarrassées. Mais le plus grand avantage consiste en ce que cette solution es beaucoup plus propre à être appliquée à tous le cas qu'on puisse proposer. La raison en est évidente, parce que j'ai réduit ici le calcul des élémens qui dépendent de la figure du corps, à des axes qui sont fixes dans le corps, de sorte que ces élémens demeurent toujours le mêmes: au lieu que la premiere solution exige pour chaque situation différente une nouvelle recherche de ces élémens, puisqu'ils étoient rapportés à des axes fixes dans l'espace absolu, à l'égard desquels la position du corps peut changer à tous momens* [1767b : § 59].

[17] *Cette régularité sert aussi de preuve pour justifier le calcul que je viens de déveloper* [1767b : § 30].

According to Truesdell, Euler realized fifteen years later that these two dynamical principles, together with the necessary constitutive assumptions about materials, are the complete basis for the whole of dynamics. But again, this step belongs to the history of the theory of elasticity.[xvi]

Of course, history did not stop here. Euler himself did not expect this. At the end of his paper he writes:

> In any case, while this solution is complete, there is still much to do in order to develop it thoroughly: undoubtedly, the surest method to bring this material to a higher degree of clarity is to apply it to cases that are as defined and as simple as possible; in that case we cannot help but discover the means for resolving the formulae, which, while they may seem particular to the case in point, will in any case lead to a greater generality.[18]

These new results, and with them, the greater generality predicted by Euler, were realized by the French school of Lagrange, Poisson, Poinsot and others. Then of course Hamilton showed us the full beauty that is hidden behind these equations. Through the quaternions he opened the path to the Cayley-Klein parameters and to the theory of the top by Klein and Sommerfeld, which then led the way to the Pauli spinors.

Conclusion

I hope that I have shown the important role which the rigid body played in our understanding of the principle of relativity. Elsewhere, I have talked about Euler's contribution to it and to discussion about it.[19] Here again, there is a point of contact with Huygens. Where Huygens uses the principle of relativity to find the laws of collision, Euler gives his most penetrating analysis precisely in his *Theoria motus corporum solidorum* [1765a], as well as in the neglected *Anleitung zur Naturlehre* [1862], where he definitively settles the old dispute between Newton and Leibniz about the nature of space.[xvii] This, I think, is no accident. True, the significance of this principle must have first become clear to him through his work as a practical astronomer, as he himself observed in his first Petersburg period, a fact we have known for a few years thanks to Russian historians.[xviii] But it is the rigid body, the *repère mobile* par excellence, as René Dugas had remarked, that is connected to his principle. Then too, the rigid body belongs to the class of objects where the dynamics is easier than the kinematics and here, therefore, the physicist has an advantage over the mathematician!

This principle is, as physicists have learned from Hamilton, related to laws of conservation. And whenever physics must live through a crisis, then we fall back on the

[18] *Cependant, quoique cette solution soit complette, il s'en faut encore beaucoup qu'elle soit déja assez développée: le plus sur moyen de porter cette matiere à un plus haut degré d'évidence sera sans doute d'en faire l'application à des cas déterminés, & aussi simples qu'il sera possible; car alors on ne manquera pas de découvrir des artifices pour la résolution de ces formules, lesquels, quoiqu'il paroissent particuliers aux cas qu'on traite, conduiront néanmoins à une plus grande généralité* [1767b : § 59].

[19] See in this volume "The Principle in Relativity in Euler's Work", pp. 167-178 and "Euler, the Principle of Relativity and the Fundamentals of Classical Mechanics", pp. 179-183.

principle of relativity. We find ourselves in just such a crisis whenever nature forces us to look for new and, hopefully, more generally valid laws. It is then that we must use the theory of invariants, in order to *faire flèche de tous bois*, as the French say. But few if any endeavors had contributed so much in one piece to our understanding of tensor covariants and invariants as the theory of angular momentum and of rigid body dynamics.

In the recent past, before we had the new electronweak theory for organizing the not inconsiderable experimental data, people again had to use what is sometimes called phenomenology, together with all its tools: coupling constants, form factors, etc.

Of course the group that guides us is a new one; still, what we had learned in the old days is present in recent methods. The greatest and most detailed discussion and summary of the results of forty years of this kind of pioneering work was written partly here in Dublin, and it bears as one of its authors the name of Ciaran Ryan.

Remembering Ciaran Ryan

It was at the Heidelberg Conference High Energy Physics in 1967 that I first met Ciaran Ryan. He had, I think, learned from Father McCrea that Louvain might offer possibilities for collaboration. Fortunately, we were able to arrange a visit, and thus began a series of exchanges that lasted for many years. Ryan's lectures were appreciated and so were the discussions which the students could have with him; for he was easily accessible, he loved to discuss and to explain, and besides his competence, of course the students felt his particular kind-heartedness and his charm. And so did, I may say in passing, we, my wife and my children, when he came to our home.

The encounter between Ryan and Louvain was also a fortunate one scientifically. The topic of his lectures was usually the weak interactions and our students were working on these questions. Of course, Ryan, mastered this subject completely, as later his book written with R.E. Marshak and F. Riazuddin was to testify [Marshak, Riazuddin and Ryan 1969]. Two of our students were so fortunate as to even follow him here to Dublin as a married couple. That both were extremely happy in Dublin was especially due to Ciaran Ryan. For Ryan was a real teacher if by this one does not just mean, a man with a good lecturing technique, but someone who has the sincere desire to help the students and to push them forward if necessary, who can understand them, who can stimulate them, excite their enthusiasm and above all, give them courage. I remember him best when on a misty December day (I was on my way back from the United States) he showed me Dublin. I remember how Dean Swift's Church appeared from the mist and he showed me Hamilton's Bridge, Trinity College, the Institute for Advanced Study and many other places. The city's charm emerged strongly under his guidance and held me captive under its spell. Invariably, from time to time, someone in the streets came to him and appealed for his help and generosity. One could always see that Ciaran saw through this game, and that he knew; one saw that the other knew and that he knew that Father Ryan knew, and that he knew that he or she knew, etc. But as invariably the appeal was successful in the end, and I think that it is this generosity which, besides his competence and his understanding friendliness, impressed me most. This is the picture of him that will always stay with me. For all these reasons, it was a great honor for me to have been chosen as the

1984 Ciaran Ryan lecturer, an honor which I appreciate greatly and for which I thank the members of the committee very deeply!

The relations between Dublin and Louvain, which Ciaran Ryan established, have luckily never broken off. We had the great pleasure of having Dubliners in Louvain regularly, Dr. L. O'Raiffertaigh, Dr. D, McCrea, Dr. P. Hogan, and at the KUL, Dr. J. Lewis. Many of us received a most warm and kind reception in Dublin.

Acknowledgment

It is a pleasure to thank Mrs P. Radelet-de Grave for her help in composing the index; my wife and Makiko and Alan Weiss for linguistic advice.

Johann I Bernoulli

Johann Bernoulli's Work on the Theory of Gravitation and the Weight of the Atmosphere[1]

Introduction

Why talk about Johann Bernoulli at a symposium on gravitation and especially one on the weight of air? One does not often find his name mentioned in the textbooks on mechanics, especially on point mechanics, but there are indeed two good and very different reasons for doing so.

One is his decisive contribution to the development of Newton's theory of central forces into modern analytical mechanics. Indeed, Bernoulli belongs to the very few who understood the *Principia* [1687], and so became the first to take a step – in fact, the first decisive step – not only towards analytical mechanics, but also towards what would become the modern way of looking at and understanding mechanics. In Newton's book this understanding, of which I will say more later, is only sketched out, and although many of the most important elements are assembled, they are not yet in the place where we have become used to seeing and using them for 250 years.

But there is another reason, although it is less straightforward. As is well known to historians, Newton's new mechanics was not received by the scientific community with the triumph bestowed on a victorious conqueror, as physicists sometimes think. Looking today at this reception, we do well to ask ourselves once more what Newton's new mechanics was. The answer is that there were two very different things, written down in the *Principia* in two different styles, as he himself said. There was the general mechanics, presented in the first two books called simply *Libri de motu corporum*, and there was what we would call today celestial mechanics and cosmology, presented in his third book, to which Newton gave the title *De systemate mundi*. The receptions of these two new sciences were very different. While the first one, understood by only a handful of specialists, was for over fifty years a way to solve mechanical problems, among many others, the content of the second one was grasped immediately by many, but fiercely rejected by most adherents of Descartes's older cosmology, which was based on his idea of the cosmic vortices. The bitter scientific contest between the two doctrines remained undecided for fifty years after the publication of the *Principia*, that is, well into the 1730s. We shall see that Johann, thanks to the "Essai d'une nouvelle Physique Céleste", his prize

[1] Conference given in Wolfenbüttel at the symposium on "Die Schwere der Luft in der Discussion des 17. Jahrhunderts". Originally published in in *Wolfenbütteler Arbeiten zur Barockforschung*, vol. 29, Wim Klever, ed., Wiesbaden: Harassowitz, 1997, pp. 187-208 [Klever 1997].

essay of the year 1734, is still a chief witness to this whole process. But before examining this testimony I shall begin with a short discussion of the three theses on the atmosphere and the weight of air written in Groningen by students under the direction of Johann Bernoulli who, following the custom of the time, reported their teacher's work. Then, for reasons that will become clear presently, I will briefly discuss Johann's most important contribution to point mechanics and only then will I turn to my main topic, Johann's "Essai", sent to the Académie des sciences in Paris, on the inclination of the planetary orbits. Thus we shall meet Johann first as a Cartesian, then as a Newtonian, and eventually as the arbiter between the two doctrines.

It may seem then that there we are very far from atmospheres; however one atmosphere will triumphantly reappear and play a central role in a problem that is for all of us, still today, of the greatest importance.

The theses of Phaebus Themmen and Casimir Hoorn

The first of the three short theses written in Groningen by students under the direction of Johann was defended publicly in 1701 by Phaebus Themmen. Of the three it is by far the most interesting scientifically, and the most valuable. In passing it may also be mentioned that the thesis is distinguished by an extraordinarily large vocabulary; rarely have I been forced to consult the dictionary so often. At first I thought that this testified to the superior acquaintance of the Dutch with the old languages, as compared with the people of Basel, but the reading of the two other papers changed my opinion. Themmen's rich vocabulary must be due to his theological studies, to which he alludes.

The subject of the thesis is the weight and the elasticity (*elasticitas*) of air. It is an experimental thesis, that is, the description of experiments, sometimes mere observations, forms the main bulk of it. Indeed, one measurement seems to me to deserve respect. On the other hand, the theory is entirely auxiliary.

Johann Bernoulli might rightly be claimed to have been the first who could penetrate into the *Principia*. Already as a young man he had found a mistake in Book II, which Newton, at least privately, acknowledged and corrected.[i] Yet here his physics is not Newtonian; in some respect it is outright anti-Newtonian. More important, at this point, is the question of how Johann became an experimentalist. In fact, he was twice initiated into the science of experiments. First, he became a doctor of medicine – *Bernoulli le médecin*, as he distinguished himself from his older brother Jacob – and it is clear that already by the time of these studies he was a well-trained observer. Second, Jacob Bernoulli had given public lectures in Basel on experimental physics, and in 1683 had published a *Dissertatio de gravitate aetheris*, about 110 pages long [1683]. It is written in the Cartesian tradition and Johann was certainly familiar with it as well as with Jacob's experiments. Indeed, much of the ground covered by the Groningen papers had been examined by Jacob as well.

Themmen's thesis opens with a definition of heavy and light (*gravis* and *levis*): heavy is what has a tendency towards the center of the earth (§ 3) light is what is not pressed by any force (§ 4): the air therefore is heavy. We find also a definition of elasticity: a body is elastic, if after compression it takes on its former shape. But later he describes the somewhat grisly experiment with a cat that is put into a glass box, in which is then created a vacuum, so that the cat is inflated because of its "elasticity" and even may burst. Clearly for us it is the pressure, not the elasticity, which caused the poor animal's death, a very typical example of the kind that would bedevil scientific thinking for a long time to come. It is only in Jacob's last work, written about the same time as Themmen's thesis, that we find the beginning of a sharp distinction between dynamical and constitutive concepts [1705].

Elasticity, according to Bernoulli-Themmen, is due to the pores of a body which contain light ether, which leaves the pores when the body is pressed and flows back when the pressure is released. Of the experiments I will mention only the most interesting ones. Themmen describes two measurements of the specific weight of air, as we would say, made by his teacher. The first is accurate to within 5%, the second to an almost unbelievable 0.5%. Though Johann must have been lucky, the result certainly shows his skills as an experimentalist. Furthermore Themmen shows that air is elastic, explains and discusses the barometer using mercury as well as other liquids, shows the change of pressure that occurs when one goes up to higher altitudes (the measurement was performed at the 97.5 m high St. Martin's tower in Groningen; according to my Baedeker the church and tower had been badly damaged by a fire not long before, but luckily for Johann and Themmen, they had been restored in time). He also reports on the study of the weather with the help of the barometer, and especially the influence of condensation of water in the air on its pressure, the burning of gunpowder, whereby much air is released. Finally he explains with gusto how tobacco smoking works.

Everybody recognizes, of course, the strange and sometimes hilarious jumble of measurements, those of great worth side-by-side with trivial, even worthless ones, which characterizes this and even later periods of experimental physics and chemistry. But, as I hope I have made clear, Themmen's thesis does honour to his University.

By comparison, the second thesis by Casimir Hoorn is considerably less interesting. It is a theoretical but non-mathematical discussion of gravity and no experiments are presented. It suffices to condense the content of the thesis into a few statements. While *gravitas* is defined as the tendency to go to the earth's center, *levitas* is the power to go freely away from it. Both properties are relative, as the reduced weight of a body immersed in water shows. Descartes's various celestial matters are *levis* and after a shock with earthly matter they can freely leave the earth, for instance, through the centrifugal force. The denser earthly matter is, the heavier it is; for the less porous it is, the more often it is shocked by the celestial matter. All this can be seen by many experiments, and best of all by the well-known experiment made by Huygens, where wax powder is put into water. If the water turns, the powder will flee to the edge of the vase, if the water stops it converges towards its center.

It is obvious that Johann's ideas here are strongly anti-Newtonian, but that he is at great pains with *levitas*. It is not clear whether his theory is consistent, but one can see that for obtaining a theory of the solar system in agreement with his principles he must supply a mechanism, and of this I shall have to speak much. This thesis was defended shortly before Johann left Groningen, and from about the same time stems the third thesis, to which I shall return later.

Johann's new derivation of Kepler's laws

In order that one can appreciate Johann's position in what follows, I must now mention with a few words his most important contribution to Newtonian point mechanics, for we shall meet here a man with a thorough grasp of the *Principia.*[2]

Recall that Newton's derivation of Kepler's laws proceeded as follows. From his axioms he had deduced that first, if the motion of a body observes the aerial law, this motion is due to a central force, and second, that if the orbit is an ellipse or another conic section and if the central mass is in one of the foci, then this force follows the inverse square law.

Whether Newton had also proved that conic sections are the only possible orbits for this law is a question that has often been disputed, and still today there are partisans on both sides of the fence. Here this question seems to me of only secondary importance.

The major step taken by Johann, one that cannot be overestimated, is that, unlike Newton, who starts from the orbit, Johann starts from the force. He considers the law of the force, that is, the inverse square law as given, and looks for the orbits. He finds them all by writing down for the first time a differential equation for this problem. Before him his fellow countryman Jacob Hermann had tackled the problem and this attempt had stimulated Johann [1712]. However, Hermann's work was unnecessarily complicated and a constant of integration is missing. Johann discusses this always a bit tricky point carefully. His work then is followed by three papers of Varignon on the same subject [1703a; 1703b; 1712].

Today we usually take an only slightly different route. Johann's work can still be read with ease and it presents radical progress over Newton's procedure. Here suffice it to say that he shows explicitly that for any central force the problem can be reduced to one integration only, whose solution will in general be a transcendental curve. All later progress in mechanics is based on this insight of Johann's, that Newton's concept of force and not the orbit must be put in the center of the formulation of any mechanical problem, whence the enormous power of analytical mechanics.

This is not the place to go into this question, but I must mention it in order to explain how the stubborn Cartesian and anti-Newtonian whom we have met in Groningen will later show himself to be one who knows Newtonian mechanics very well indeed, and who also had a firm grasp of most of its fundamental mechanical ideas.

Johann's second prize essay and the historic situation

For its annual *concours* in 1732 the Académie des Sciences had asked the following question:

> What is the physical cause of the inclination of the planes of the orbits of the planets relative to the plane of the rotation of the sun about its axis, and why do the inclinations of the orbits differ among themselves?[3]

---

[2] See "The Kepler Problem from Newton to Johann Bernoulli" in this volume, pp. 123-134.

No prize was awarded in 1732 and the question was repeated two years later, in 1734. This time the Académie awarded two first prizes, one of them to Johann, who was by then 67 years old, but still very active.

Johann's prize essay is the subject of a very interesting paper by William Shea [1988b]. I agree with most of what Shea says, but as you will hear at the end of my talk, I take a slightly different attitude towards the Newton versus Descartes dispute. Before turning to Johann's paper, it is worthwhile to point out that still today the subject of the *concours* remains essentially "a good question", since there is still no generally accepted theory of the formation of the solar system, even if a lot more is known now – sometimes with a quite high degree of precision – from observation as well as from our theories. I shall come back to this point.

But it is clear that this question did not refer to a subject among many others, where only a valid answer is added to our corpus of knowledge. I need not tell you that what was at stake was the validity of Newton's theory of gravitation (but not of the validity of his general mechanics!), nor must we forget the possible validity of Descartes's rival theory, which had still many supporters at that time.

In the third book Newton applied the theorem which I have mentioned above to the orbits of the planets, the moons, and the comets, and as far as the astronomical observations of the period were concerned, with overwhelming success. This application is probably the one single statement in the *Principia* that everybody, scientist or not, could grasp and understand. So great was its success that it was easy to forget that it was not yet possible to compute more than the motions of a two-body system.

Then there was another success: the theory of the tides, although it remained only qualitative until 1740. Finally there were the two famous expeditions sent by the Académie to Lapland under Maupertuis and to Peru under La Condamine, but Maupertuis, the first of the two, returned to Paris only in 1735. Small wonder then that on the Continent the theory met with great resistance, a resistance that should not be attributed solely to blindness, prejudice and incomprehension, nor to nationalism. Sometimes these factors contributed to the resistance, but many of the opponents were, if not Newtonians, at least men of high scientific standing. After all, today we also constantly observe resistance from scientists to new yet valid proposals. Who has never sinned in this direction himself?

Indeed, if we look at the solar system, what strikes us first is that all planetary orbits lie almost in the same plane, and this plane is, to boot, the equatorial plane of the sun. There are small deviations, the biggest one, the earth's, is 7°. Although it is easy to see, we tend to forget that this fact cannot be explained by central forces, since these cannot define a plane. On the contrary, Newtonian mechanics tells us that if there is no other force present than the gravitational one, all orbital planes are equally probable. In any case, there must be a second agent besides Newton's. Thus it must be admitted that for those who claimed, as still many do today, that the scientific knowledge of nature must and can only emerge out of observation, Descartes's old vortex theory was a much more plausible candidate for the description of the heavenly phenomena. A vortex provides a straightforward qualitative

---

[3] *Quelle est la cause physique de l'inclinaison des Plans des Orbites des Planetes par raport au plan de l'Equateur de la revolution du Soleil autour de son axe, & d'ou vient que les Inclinaisons de ces Orbites sont différentes entr'elles?* (cited by Johann I Bernoulli in [1742b: § I].

explanation of the fact that all planets together with their moons move in the same plane and indeed in the same direction.

We can then ask what we mean by "an explanation"? This word is used by physicists all the time, but still today it is present in their thinking only in a vague and ill-defined, if not outright confused and self-contradictory, way. For Descartes and his followers, a "physical explanation" meant the precise indication of a mechanism, that is, an assembly of bodies which interact through direct contact and thereby transmit effects also over long distances. In this precise and very restricted sense, Cartesian physics was a "mechanistic" one.

Johann's discussion of the two theories

The strengths and weaknesses of both theories are thoroughly discussed and summed up by Johann in the introduction and in the first part of his "Essai". To the reader it is obvious that his sympathies are on the side of Descartes and he may even be accused of playing up to the Cartesian members of the Académie, especially to Cassini, whose measurements, which supposedly confirm the earth's prolateness, are praised more than once. Yet Johann was not blind; he had a sure grasp of the mathematical tools and he understood well the strength of Newton's arguments. It is precisely the combination of both these qualities, which indeed Johann displays to an exceptional degree, that make his "Essai" a document of historic interest still today.

This is his verdict. Descartes's vortex theory explains in a qualitative way the motion of the planets, especially why they move in the same plane. This explanation is a sound mechanistic one: some matter moves the planets and the moons, and it is in agreement with philosophical requirements. On the negative side, Kepler's laws show clearly that there is an attractive force directed toward a center and not towards an axis around which a vortex turns. Moreover, and for Johann this is a serious defect, the velocity of the vortex, if we extrapolate following Kepler's third law, does not agree with the speed of the sun's rotation, which is 230 times slower. Newton's theory, on the other hand, can boast of deducing Kepler's laws mathematically from a few assumptions, which in turn are confirmed by observation to an enormous degree. Johann shows no hesitation in acknowledging this success and in stating the requirement that any theory of gravitation must incorporate a central force. But Newton's law of gravitation, as he insists, has no real foundation: there is no mechanical agent responsible for it, and thus it is incomplete. Johann could not know that Newton himself had privately expressed the same opinion. Furthermore, Newton's theory supposes a vacuum, and Johann considers an action at a distance *une chimère* [1742b: § IV].[4] It is hard to say exactly why, but I shall come back to this important point. Finally, to him Newton's assumptions seem hopelessly insufficient for explaining the coincidence of the various orbital planes as well as the rotation of the planets around their own axis.

Thus two things are clear. First, Johann is neither a Cartesian nor a Newtonian; as we shall see now, his own theory tries to combine the best of both worlds, which is the scientist's normal attitude. Second, his criticism of Newton's theory is only partly ill-

[4] See "The Kepler Problem from Newton to Johann Bernoulli" in this volume, pp. 123-134.

taken. To us, as I said, it is clear that the mechanics of central forces cannot explain the features of the solar system, and we know equally well that even today we cannot do this in a satisfactory way.

Johann's own theory of gravitation

From Johann's evaluation of Descartes's and Newton's theories we may almost deduce the line along which his imagination will run. The attraction of the planets towards the sun demands, as Newton had proved, a central force, but the fact that all orbital planes coincide defines an axis, orthogonal to these planes. And since according to him all actions must be transmitted through a medium, two media are thus needed. The *matière première*", which forms the vortex, is a continuum which he says offers no resistance – or, as we shall see, almost none – to any of the bodies it may contain. It cannot push the planets and thus, unlike the Cartesian vortex, is not responsible for their motion. As I said, Johann's vortex is 230 times slower than Descartes's, and its speed goes continuously over into the rotational speed of the sun:

> I regard the Sun and its Vortex as **all one** [my emphasis], of which one is, so to speak, the continuation of the other... [5]

and later,

> ...here is the simplest and most natural idea that one can have about the formation of the vortex: because why do those exaggerated Cartesians have to set the constraint that the first stratum of the solar vortex circulates 230 times faster than the surface of the Sun, with which the first stratum is contiguous?[6]

In a way, and I shall come back to this, we can see today that his sentence *Je regarde le Soleil et son tourbillon comme un tout* is a precise shot in the right direction, more precise than anybody at that time, with perhaps one exception, could have known. That he takes his distance from *les Cartésiens outrés* perhaps also shows the smart politician. Later we shall see the role played by this vortex.

However, the gravitational attraction must be transmitted by a mechanical agent and it must be a central force, as Newton had correctly deduced from Kepler's laws. Johann fulfils both requirements by the following scheme. From the "outer surface" of the vortex comes from all directions a torrent of tiny particles (*massules*) towards the sun.[7] This

---

[5] *Je regarde le Soleil et son Tourbillon comme un tout, dont celui-ci est, pour ainsi dire, la continuation de celui-là* [1752].

[6] *...c'est la l'idée la plus simple et la plus naturelle qu'on puisse avoir au sujet de la formation du tourbillon: car quelle contrainte ne faut-il pas se donner avec les Cartesiens outrés, que la première couche du Tourbillon solaire fasse sa circulation 230 fois plus rapidement que la surface du Soleil a laquelle la première couche est contigue?* [1752].

[7] Bernoulli's idea is similar to N. Fatio de Duillier's which probably stimulated it ; see [Zehe 1997]. Cf. also Fatio's correspondence with Jacob Bernoulli, in *Der Briefwechsel von Jacob Bernoulli* [1993: 161-200].

torrent flows with no interruption and from all sides with the same intensity. The particles that make up the torrent hit the planets not only on their surface but are absorbed by all parts of it (think, for instance, of neutrinos). Thus they produce a force that is directed everywhere towards the sun, increasing towards the sun with the square of the distance, and proportional to the volume of the planets.

Thus he obtains at once Kepler's laws. It also follows (at least if the various elements are built from the same elementary particles and differ only in their internal configuration) that the force is proportional to the mass of the body. But Johann does not elaborate on this last point.

A similar mechanism forces the moons to turn around their respective planets.

So far so good. Johann's mechanism is probably the only conceivable one which reproduces Kepler's laws, but its author must now answer two questions: how is this torrent produced, and where does it end up? To answer them he contrived the following mechanism. Because of centrifugal force the vortex expands slightly, leaving at its center a space where the down-pouring "gravitons", as I shall call them, will collide and assemble, thus forming the sun. But under the impact of the newcomers, the already assembled gravitons will break up into tinier and tinier parts. Of course, there is now an enormous amount of kinetic energy present, and eventually the very finest of these gravitons are emitted in all directions as photons, that is, as light. So with one stroke Johann has seemingly solved two problems; now in addition to having a theory of gravitation, he has explained the sun's radiation. We must admit that this transformation of gravitational into optical energy exists, even if it plays only an imperceptibly small role. But the second question of how these gravitons are produced still looms. Here his answer is adventurous indeed.

The photons propagate far outside the sun's vortex and there they encounter photons coming from other stars with the same speed. If they collide, they stop each other and stick to each other. A third photon will then push it towards either the sun or the other star. Often this will happen only after several photons have already coagulated into a heap, *un peloton* as he says. This pushing is the origin of the torrent, and it is clear that this torrent is isotropic and stationary only in the average.

It is easy, of course, to criticize and even deride this attempt at a theory. Here I mention only a few points. All collisions ought to be central and totally inelastic, a requirement not easily fulfilled. Worse still, even a torrent which is isotropic only on the average cannot be constructed without difficulty, and the central symmetry of the torrent is not easily reconciled with the axial symmetry of the vortex.[8] But the worst is that even if we succeed in constructing a torrent that fulfils all necessary requirements, how can we construct the infinity of torrents which the idea of a universal attraction of all bodies and of all parts of all bodies postulates? This is, of course, a hopeless task.

But even so, Johann's construction has the merit of showing that within the narrow frame of the push-and-pull mechanics of Descartes and Huygens, Newton's law of gravitation cannot be reproduced. We must remember that long after Johann such ambitions still existed. The most interesting and refined one is probably the theory by H.

[8] Contrary to Fatio, Bernoulli seems aware of the difficulties connected with infinities of infinite particle currents, hence my "validity in the average only", though Bernoulli does not use this word.

Hertz, sketched if not executed, in his book on mechanics [1894] written at the end of the nineteenth century, but still interesting to read for the ideas presented and discussed.[9]

The rotations of the planets

Johann thinks that the torrent affords him a special bonus, namely, that he can explain the rotation of the planets, which Newton's theory cannot. While a sphere at rest in the torrent feels only the gravitational force, according to him, a moving sphere undergoes yet another effect. We all know that if we move through the rain, our front will become wetter than our back. Moreover the difference in wetness is proportional to the speed: it is greater on the cape of the cyclist than on the coat of the pedestrian. Similarly, he thinks that more gravitons will hit the front of the planet than the back, and thereby through their momentum they will induce a rotation. Such an effect would be what we call today a "spin-orbit coupling", and as a consequence, which he could not know, Kepler's laws would no longer hold exactly.

Johann's idea reminds us of James Bradley's explanation of aberration, discovered in 1725 [1728]. Johann knew this effect, as we see in a letter written by him only a few years later. However that may be, it is precisely the aberration that shows that Johann's mechanism will not yield the expected result. The particles simply appear to come obliquely, slightly directed against us, as everybody who carries an umbrella against the rain knows. But they still appear to come in a parallel direction. Thus while they produce a force on the planet, they cannot produce a moment of force around the planet's axis. His theory fails, and in any case Kepler's laws would no longer be exactly valid, since only the total angular momentum, not the orbital angular momentum separately would be conserved.

It is interesting to note that this refutation, as Euler was to show in 1739 [1750a], is based only on the principle of relativity, and this had escaped Johann.

The inclination of the orbits

After having presented the fundamentals of his "new celestial physics", in the fourth part of his "Essai", which comprises about a third of the whole, Johann sets out to discuss the question asked by the Académie.

The vortex, he repeats, is not responsible for the motion of the planets, and can neither accelerate nor retard them. The velocities of the planets are something elementary, given at the beginning of time, and as the vortex, and *globules célestes* all turn in the same direction, so does the sun. It is at this place, he underlines, that sun and vortex are one.

Why then do the orbits of the planets not lie in the equatorial plane of the sun? His answer is ingenious, in principle at least. If the vortex cannot accelerate the planets through the *globules célestes*, which swim in it, it can change their direction. He seems to think of a force perpendicular to the motion that does not change the planet's energy. Although Johann concedes to Newton that the velocities of the planets are something given (by God?) at the beginning of time, he is evasive on this point.

[9] Cf. also the article by William Shea [1988b].

His answer would seem to be qualitatively correct, but the predicted effect, if computed, would certainly turn out to be much too small. This is how the *globules célestes* can deviate the planets: If they were perfectly spherical such a deviation would not be possible, but in reality they are all a bit prolate, like the earth – or so he thinks together with Cassini – or oblate, like Jupiter, for example. When such a planet moves through the vortex, and if its axis of rotation is not exactly perpendicular to the orbit, as is the case for example with the earth, then it is not hit by an equal number of *globules* above and below its orbit. Twice during a revolution it happens that the axes of the sun and of the planet lie in the same plane, namely, at the solstices. At this moment the direction of the planet's axis is orthogonal to the direction of the vortex, so that at those two points there is no deviating action on the planet. At all other points the *globules célestes* push the planet out of its natural plane but:

> With this difference, that the nodes of the two spheroids on the equator of the Sun will be of contrary names, which is to say that where there is the ascendant or Boreal node in the case of a flattened spheroid, this will become a descendent or Austral node, if we suppose that this is an elongated sphere, and reciprocally, the descendent not of the flattened sphere will become the ascendant for the elongated spheroid.[10]

Johann discusses the various possibilities very carefully indeed, trying to obtain qualitative agreement between his computations based on not very reliable observations of the shapes of the planets and the well-measured angles between their orbits and the sun's equator. This discussion is based on an analogy with the drift of a ship, and thereby he takes a step of the greatest importance.[11] As Shea formulates it succinctly, paraphrasing Galileo, "Bernoulli brought the argument down from heaven to earth, comparing the motion of the planet with the drift of a ship" [1988b]. I shall discuss this step later, but first I must return for a moment to Johann's proposal. I called his explanation "ingenious in principle"; since his mechanism for explaining the rotation of the planets is a spin-orbit coupling we might speak here of a tensor-force. In this case, as I said, the scheme would work, but it would also violate Kepler's law, and of course the effect would be negligibly small.

[10] *...avec cette difference, que les noeuds de ces deux sphéroïdes sur l'équateur du Soleil seront de noms contraire, je veux dire que là ou se fera le noeud ascendant, ou Boreal, dans le cas du sphéroïde aplati, il deviendra noeud descendant, ou Austral, si on supose que c'est un sphéroïde allongé, & réciproquement le noeud descendant du sphéroïde aplati se change en ascendant pour le sphéroïde allongé* [§ LXXIV]. P. Villaggio pointed out to me that Johann's idea is at variance with d'Alembert's principle, which was discovered somewhat later. However, a coupling through the anti-symmetric parts of the stress tensors could possibly do the trick. (Such a coupling would not violate the conservation of angular momentum.)

[11] It is obvious that Bernoulli is guided here by his ideas published in his book *Essay d'une Nouvelle Théorie de la Manoeuvre des Vaisseaux* [1714]. In fact the last words of the title are used in sect. LXXIV p. 66, giving thus, it would seem, his authorship away. Mrs. J. Peiffer has forcefully insisted on the importance of this book for Bernoulli's later work.

Before giving you my opinion of the paper as a whole, I wish to say a few words about the mémoire of the rival with whom Johann had to share his prize,[ii] for this rival went even one step further.

Daniel Bernoulli's mémoire

Johann's rival was none other than his second son, Daniel. Daniel's prize essay, "Recherches physiques et astronomiques..." [1735], has been republished with an introduction and commentary in volume 3 of Daniel's complete works. The first difference is that while, like his father, Daniel accepts Newton's central force, Daniel is not bothered by the lack of a mechanism in Newton's theory. The second difference is that, like his father he accepts a vortex, whose velocity at the sun's surface is the same as the sun's itself, but Daniel's is no longer a Cartesian vortex at all; his axially-symmetric agent is a modern one, as we shall see. The third difference is that Daniel starts from a very different basis than his father: he thinks that originally all orbital planes were randomly distributed. The basis for his reasoning is that, according to Newtonian mechanics, the probability that a certain plane carries a planet is the same for all planes. And then, even before setting out to explain his theory, Daniel makes clear through a calculation that the probability of finding all planets in almost the same plane is exceedingly small. (This, incidentally, is perhaps the first use of a probabilistic computation for supporting a physical hypothesis.) Therefore he affirms squarely that a very long time ago these planes were all oriented at random. Thus, unlike his father, he needs an agent which produces the departure from an original alignment, an agent which must produce the observed alignment itself. And for him this agent is the sun's atmosphere, that is, an ordinary gas governed by laws of mechanics! We must recall that at the same time Daniel was rewriting the first draft of his *Hydrodynamica*, the so-called St. Petersburg version. Of the existence of this atmosphere, and of the existence of atmospheres around all planets, he is convinced through the analogy with the earth's own atmosphere.

Clearly such an atmosphere is a more tangible object than either Descartes's or Johann's vortex. For one thing Johann's postulate to consider sun and vortex as one is automatically and plausibly fulfilled. Secondly the mechanical and thermodynamical structure can be discussed through an analogy with the earth's atmosphere or simply by discussing various laws, as indeed he does. For instance, he makes it clear that the maximum of the atmosphere's density may not lie at the surface of the sun but way outside. This atmosphere then, together with gravity, forces the planets over an enormously long time span into the sun's equatorial plane. Could Daniel support his claims through a computation? A careful examination of his arguments by Dr. J. L. Pietenpol, then at Louvain-la-Neuve, shows that his argumentation is sound throughout, and an explicit calculation on a computer made by Prof. L. Bossy at the IRM, Bruxelles, confirmed his predictions. Their results can be found in the introduction to Daniel's "Recherches" in vol. 3 of the new edition. Of course, as he makes clear, the planets reach the equatorial plane only asymptotically, and noticeably so only after a long period, so that we should not be surprised that observations made during historic times do not show this alignment. How does the atmosphere drive the planets mechanically? He cannot formulate this, of course, but he seems to think of an effect of the atmosphere's viscosity.[12]

A comparison of the two 1734 prize papers by Johann and Daniel Bernoulli

I will begin with a comparison of two of the more superficial aspects. First, while Johann is the more indefatigable calculator, Daniel displays a sharper and more acute sense for the experimental situations. Noticing, for instance, that there is disagreement concerning the indications of the sun's equator, he tries to define it through a kind of χ^2 (*chi* squared) method. What he finds may be the first appearance of what today is called "the invariable plane".[iii]

Second, there is a difference between the two papers in attitude and in style. Daniel is more reticent and cooler than his passionate and sometimes zealous father. He is also more critical with respect to the reported observations. While the father eagerly accepts all evidence that might support his cause, the son is often cautious and evaluates it carefully. While the father, as we saw, is much more ambitious, the son is more to the point.

How do we rate them today, and how would we attribute the prize? For answering this question, we must compare the proposals more closely.

The contestants who fought for the prize faced two rival theories. One was Newton's theory of universal gravitation, whose predictions were confirmed with astounding accuracy and promised to make numerical predictions better even than those found in the *Principia*. The other was Descartes's vortex theory, which was no longer tenable in its original form, but seemed to address the problems posed by the actual structure of the planetary system much better than Newton's theory possibly could.

Both papers make this quite clear, but Johann's, whose introduction is a magnificent example of a solid review paper which is still worth reading, presents the situation more clearly. And today we can see that while Newton's theory dominated for another century and a half, a vertical motion seems indispensable for the formation of the planetary system (and we may recall that today many feel even entitled to use this expression in the plural).

Both Johann and Daniel accept Newton's theory, but for the father this theory is not complete, because Newton cannot indicate a mechanism responsible for the force. Therefore, before answering the question asked by the Académie, he must present his own proposition for such a mechanism. For this he will not obtain our agreement, but I cannot help thinking of all the good – and sometimes even very good – physicists who still today adamantly refuse to be satisfied with quantum mechanics for reasons that seem to me mere prejudice. Johann's failure has at least the merit to have made clear what formidable difficulties the construction of such a mechanism in the frame of Descartes's and Huygens's mechanics had to surmount. On the other hand, there is reason to doubt that Johann was aware of the full implication of Newton's theory. At the time, Daniel's

[12] In his *Leçons sur les hypothèses cosmogoniques*, [1911: 117-129], Poincaré presents "L'hypothèse de M. See", which is very similar to Bernoulli's. Poincaré proves that because of the nebula, as Bernoulli claims, the eccentricity will diminish in time.

position, which evaded this question of a necessary mechanism, must have appeared to many as negligent; today, we simply find him more modern.

Further, both authors agree that something like a Cartesian vortex is needed for explaining the coincidence of the orbital planes of the planets as well as their rotations about their own axes. And both agree that the vortex and the sun together form an entity such that the speed of the vortex must be 230 times smaller. Finally, and this is a real achievement, both see clearly that our present situation evolved over the course of a long period of time: it is the result of a history. As we shall hear, the son goes further in this direction.

But now the differences begin. Johann's vortex is, like Descartes's, a cosmic institution, something elementary in its own right, which we must accept as a hypothesis. For him, as for Newton, the orbital planes are given, but unlike Newton, for Johann the planets were originally in the same plane. For Daniel, the planets originally moved in randomly distributed planes. Therefore he needs the vortex for a reason opposite to Johann's. To be sure, that this vortex is the sun's atmosphere is also a hypothesis, but a more tangible one which we can discuss through the analogy with our own atmosphere. Here the son leaves the father behind.

Johann too has taken a step away from the view of the world as a unchanging mechanism. It is a mechanism, yes, but one which allows the planetary system to change its shape. The divine arrangement of the planets as we see them was a changing one, and only the vortex remains immutable. But Daniel goes further: according to him, the agent of this process itself, the sun's atmosphere, also undergoes changes. For him, only the laws remain unchanged. For these reasons today's readers will find Daniel's paper more modern and my guess is that most would call it superior. But we must not forget that today it would seem that Johann's hypothesis, according to which all orbits were originally aligned, now seems closer to the truth. However, as I said, we do not know this for sure.

Why then was the prize split between father and son? I no longer think that this decision was just another jury compromise. Partisanship and friendship may have played a role; indeed, it is inconceivable that the honourable *membres de l'Académie* did not recognize Johann as the author of this pugnacious piece. But I think that the breadth of his outlook and his careful discussions added much weight to his side of the balance. And Johann's ideas are on the whole more clearly presented: it was not for nothing that he was Europe's foremost teacher of mathematics.

All the same, I will now cite a part of Daniel's "Recerches" that makes clear the philosophical resistance that had to be overcome. Speaking of the atmosphere and the laws which govern it he says:

> Is it not better to use these principles rather than invoke the immediate will of the Creator, as happens with relation to various phenomena by those who want to deduce everything from simple mutual gravitation of bodies that move in a void? & is it possible that the will of God is not perfectly effective? Could He have willed that the planetary orbits be in the same plane, without them being perfectly so; that they should be circular, without them being perfectly [circular]? & so on for many other points, to which it is necessary to relate that the Earth and all the planets move in the same way,

> & that is to say, that according to which the sun rotates on its axis; and that it is the same in the worlds of Saturn and Jupiter, which are such that even were they still hidden, our theory would dictate them to us, while Newton himself, the greatest philosopher of our century, declares in his *Opticks*, that he would not know how to provide a mechanical explanation.[13]

In other words, Daniel is well aware of the fact that his ideas considerably enlarge the domain of the problems that are accessible to science; he extends it to the very formation of the solar system! And the crucial point is indeed an atmosphere, although it is the sun's rather than the earth's.

Johann's prejudices

What is it that made Johann, who, unlike many others who refused to be satisfied with Newton's theory, understood the *Principia* on the whole very well indeed, better perhaps than anyone of his generation, refuse so stubbornly the new theory of gravitation? And why was his reaction so emotional? That he was an emotional man we know, but in a great many of his writings his emotions remained well under control. Here they are not. Sometimes people try to fight an evil in others, when in fact this evil has a safe place in their own hearts. It is indeed interesting to observe that two of his students devote a whole section of their theses to what amounts to a declaration of war against prejudices. And shortly before leaving Basel for Groningen, Johann himself had given a lecture on prejudices and the remedies against them. Whether we call his demand that a theory of gravitation must be implemented by a mechanism a prejudice is, up to a point, a matter of dispute. Newton himself thought so, as did Euler, and of the scientists immediately before Einstein I mention only Heinrich Hertz. All these later ones shared this "prejudice". Thus our criticism must be reduced to one, namely, that he had not yet realized one fundamental lesson of the new science: one law, in this case Newton's, which is experimentally confirmed by observation and experiment is a much more valid guide for one's research than all extra scientific postulates. Again, I cannot help thinking of all the adversaries of the usual interpretation of quantum mechanics; here there are striking similarities.

With respect to his second prejudice, the vacuum, there is a more subtle point involved. Why no vacuum? Obviously Newton's explanation of Kepler's laws is based on a computation *in vacuo*, and we saw that Johann understood this very well. Thus only a philosophical bias could possibly have provoked his adamant refusal of this idea. Whence

---

[13] *Ne vaut-il pas mieux employer ces principes, que de recourir à une volonté immédiate du Créateur, comme le font par rapport à plusieurs phenomenes ceux qui veulent tout déduire de la simple gravitation mutuelle des corps mûs dans un vuide? & peut-il se faire que la volonté de Dieu n'ait pas tout son effet? qu'il ait voulu que les Orbites planétaires fussent dans un même plan, sans qu'elles le soient parfaitement; qu'elles fussent circulaires sans qu'elles soient tout-à-fait telles; & ainsi de plusieurs autres points, ausquels il faut rapporter que la Terre & toutes les Planetes se meuvent d'un mêmes sens, & et nommement de celui duquel le Soleil se tourne autour de son axe; qu'il en est de même dans les Mondes de Saturne & de Jupiter, lesquelles choses sont telles, que si elles étoient encore cachées, notre theorie nous le dicteroit, pendant que M Newton même, le plus grand Philosophe de notre siecle, déclare dans son Optique, qu'on n'en sçaurot donner aucune raison mechanique*. [1734: § XXVI] .

these emotions? One point is clear. For the Bernoullis, Cartesian physics opened up the possibility of a straightforward and transparent application of mathematics to nature, while at the same time doing away with all the occult qualities which so often seemed to have been invented only in order to escape a rational explanation. For them Descartes was a liberator, while Newton's "action at a distance" seemed to bring back this hated obscurity. However that may be, we can at least follow Johann's various arguments and try to find the source of his failure to understand Newton here.

Scato Trip

Now we arrive at the third thesis over which Johann had presided over while still in Groningen: *Disputatio Philosophica de Vacuo* by Scato Trip [1705]. Trip did so well that he received from his teacher praise in form of two poems, one in Latin and one in Dutch. As the title indicates, the thesis is a philosophical one, that is, the argumentation is carried on in philosophical rather than in scientific terms: *substantia*, *essentia*, *in re*, and *modus*, to name a few.

The argumentation of the thesis itself runs as follows. Rejecting the "vulgar" definition of the notion, Trip advances a philosophical one: we speak of a vacuum "where there is no substance". This definition brings up at once the difference between "space" and "body". "Extension" is an essence of both space and body; for body, "mobility" is also an essence. The essence of a body does not lie in its form, nor – and here Mach should have read him – in its sensual qualities, since there can be – indeed, there are – bodies so small that there is no access to them through our senses. Thus there is a difference between space and body, but it is only a difference in kind (*modo*) and not according to the matter (*in re*). Nor is there any necessity for a vacuum at all, since all bodies move on circles, or as we would say today, on closed curves. (This, by the way, is the only statement in this thesis that approaches physics.) Trip refutes Epicurus and Democritus, that is, atomism. Finally, he eventually proves that there is no vacuum, nor can there be any: it is a contradiction in terms.

Trip's thesis is clearly and well written, and I wish I could show it to every philosopher of science as a warning of how a confusion or even a blurring of the frontier between philosophy and science can do harm to both. Here, in spite of his sound grasp of Newton's new mechanics, philosophical prejudices blinded Johann to some of its fundamental ideas. He has summed up his arguments again in his "Essai" of 1734, writing: "He who says body says nothing but extension, mobility and impenetrability; this is all that the idea of body contains".[14] Thus, we may note in passing, he has added "impenetrability" to Trip's list. Today we can see the cause of Johann's misunderstanding, indeed, of his error.

About twenty-five years later Euler, in his *Anleitung zur Naturlehre* [1862], mentions as the essential qualities of bodies: "extension, mobility, inertia and impenetrability".[15] By adding inertia to the list, Euler shows himself as the better reader of the *Principia*. The inertial force, the *vis inertiae* (and also *vis insita*, as Newton says), is fundamental for

---

[14] *Qui dit corps, ne dit autre chose que ce qui est étendu, mobile & impénétrable ; voilà tout ce que l'idée du corps doit renfermer* [1734 : XIII].

[15] See "Euler's *Anleitung zur Naturlehre*" in this volume, pp. 185-196.

Newtonian mechanics; indeed, it belongs to body and space. Thus while a space empty of bodies may be called geometrically empty, physically it is not: it still is the carrier of the inertial force. Newton had seen this clearly, and so did Euler after him. But how many saw it during the period following Euler and before Einstein discovered the inertial field from which the gravitational field is now inseparable?

Browsing through the literature suggests to me that at best only very few did so, and therefore here too we must beware of looking down too much on Descartes, the Cartesians and on Johann Bernoulli, who, if not an unconditional Newtonian, had, contrary to most of us, at least carefully and with no help read the *Principia* and grasped a good deal of Newton's principles.

To paraphrase the theologian Karl Barth, one should never use the word tragedy lightly, but only where there is a connection between greatness, guilt and downfall. Translating this into the sober world of science we can perhaps say: "where there is the connection between greatness, error and bad luck on a great scale". Greatness? Certainly, for today it is easy to see in the vortex theories the beginning, or at least a prehistory, of the question: "whence the solar system?", a question of which few would deny its great importance and to which we still do not have the answer. Error? Certainly, in so far as a better grasp of the fundamentals of Newtonian mechanics might have protected these scientists from a fatal prejudice. Failure because of bad luck? Certainly, since they asked a good question but they asked it too early, indeed, much too early, at a time when the observations had not yet been made and the necessary theoretical tools were not ready.

Conclusion

Here are two lessons to be learned. First, theories often remain unsuccessful because the time for them has not arrived; some of the indispensable ingredients are not yet ready. Second – and this is especially important for the historian –, good and bad luck also reign in the field of science, and already for this reason not everything that the historian observes admits an explanation, let alone a philosophical explanation. I am grateful to have learned this lesson from a friend in Italy, a shrewd observer of scientists. Sometimes in his endeavour to explain a development the historian must stop the explaining.

But it is easy to see that Daniel's appeal to the sun's atmosphere is a decisive step in the prehistory of the theory of the formation of the solar system. He recognized that while today gravitational attraction governs this system uniquely, a long time ago the viscosity of a gas was equally important, even on a cosmic scale. Furthermore, and this was radically new, circumstances determine which laws are of primary importance. Johann had not seen this last point clearly, but he too recognized that laws, albeit different ones from those that today govern the solar system uniquely, had determined the actual shape of the solar system and the motion of the planets. He too had grasped that the agent responsible for the shape must be tied to the sun, and he too used ordinary mechanics for explaining a special feature of the formation of the solar system.

In this perspective, I think, we must look also at Descartes's cosmology. Descartes has often received bad press among physicists. But he was not only Newton's most important teacher in physics, as John Herivel [1988] and others have argued (or was this Huygens?), even if Newton sometimes spoke angrily about him. True, his theory of gravitation as such is a failure, but it is unhistorical to throw it away simply for that reason, and then to speak of a "revolution". Sticking to what observations had told him, namely, that all planets move in the same plane, Descartes had asked for a mechanical cause, for which he had proposed his vortex. As we know today, a vortical motion is, in addition to gravitation, somehow responsible for the formation of the solar system as well. Descartes is not among the founders of the theory of gravitation, as are Kepler and Newton, but he does have an important place indeed, together with the Bernoullis, in the prehistory of the formation of the solar system. It was Descartes, and not Newton, who sensed the importance of the alignment of the planetary orbits and took the first step in what proved to be, until the present, the right direction, even if he was not aware of the true significance of this step.

But then, we must not forget to be grateful also to the *Membres de l'Académie des sciences*, who asked such a good question!

Acknowledgment

It is a pleasure to thank Drs. W. Klever and H. Breger for the invitation to this interesting *Arbeitsgespräch*, Dr. F. Niewöhner and Ms. B. Braun for the organisation and for their help, and all participants for the interesting exchanges. M. M. Mattmüller helped me with the sources.

Johannes Kepler

The Kepler Problem from Newton to Johann Bernoulli[1]

To Clifford Truesdell,
Notre Maître à tous!

Introduction

Newton solved what would be called for a short time afterwards "the direct Kepler problem" (*le problème direct*): given a curve (for example, an ellipse) and the center of attraction (for example, the focus), what is the law of this attraction if Kepler's second law holds?

The *problème inverse* (which we today call the direct problem) was attacked systematically only later, first by Jacob Hermann, and was then solved completely by Johann Bernoulli in 1710 and, following Bernoulli, by Pierre Varignon. This paper examines how Bernoulli solved the problem, what method he used for this purpose, and which of his accomplishments are still followed today.

The second part of the paper deals with various questions connected to the first part from the point of view of "Conflict and Cooperation," suggested by Jan van Maanen to the participants of the Groningen conference at which the original paper was presented.

The precursors: Galileo's *Discorsi* and Huygens's *Horologium oscillatorium*

On the third day of the *Discorsi e dimostrazioni matematiche intorno a due nuove scienze* [1638], Galileo first discusses the motion at a constant speed – the *motus aequabilis*, as he calls it – starting much in the classical, Euclidean way from four axioms.[2] He presents and discusses systematically kinematics, as we would say today.

But for treating the fall of a body he must proceed differently. He knows that it is a *motus naturaliter acceleratus*, a naturally accelerated motion. This means that the increase

---

[1] Originally published in the *Archive for History of Exact Sciences* **50** (1996): 103-116, from a paper presented at the international meeting *Conflict and Cooperation in Building the Scientific Estate: Johann Bernoulli, Other Cases and a General Perspective* (Groningen, July 5-6, 1995), organized by Jan van Maanen. The meeting was held to mark the tercentenary of the appointment of Johann Bernoulli as Professor of Mathematics at Groningen University.

[2] See "The Second New Science in Galileo's *Discorsi*" in this volume, pp. 17-38.

of the speed is proportional to the time of fall,

$$\Delta v \propto \Delta t\,.$$

This definition is proposed by Salviati, Galileo's spokesman in the *Discorsi*, but only after a lengthy discussion that brings them close to some of the very questions that would be solved half a century later by infinitesimal analysis.

The three participants in the *Discorsi* eventually accept this definition, which is repeated by Sagredo:

> We say that equally, that is, uniformly accelerated motion is that which, starting from rest, in equal intervals of time equal moments of speed are added.[3]

For our present purpose, it is really not important to decide whether Galileo arrived at this "definition" – he is really selecting the acceleration that describes the fall of a body – empirically or, as the text seems to indicate, by reflection (an alternative that seems more natural to Simplicio is rejected by Salviati as it leads to contradictions). What is important here is that Galileo lets Salviati reply immediately: "Once this definition is agreed, the author demands only *one principle* which he supposes to be true", namely, "I assume that the degree of speed acquired by the same body on variously inclined planes is always the same provided that the height of the planes is the same."[4]

In other words, on whichever plane the body descends – CA, CD, or CB, etc. –, when it arrives on the plane AB, its speed will in all cases be the same (fig. 1). Of course, this holds for any horizontal plane parallel to AB. The important, indeed fundamental, point to note here is that Galileo realizes that he needs a new – we would say, a dynamical – principle beyond Euclid's geometrical ones, beyond his own which are kinematical, and different as well from Archimedes' dynamical principle of the lever. All propositions that will be derived later on in the *Discorsi* will be derived from this principle alone.

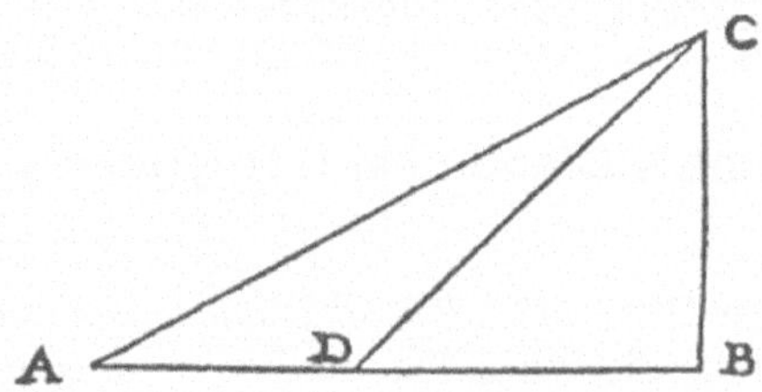

Fig. 1. From the *Discorsi* [1638: 166]

On the other hand, Galileo is not quite satisfied with the principle. It seems that he wished to derive the principle itself from another, deeper and more basic, one. In any case, as he says after a brief discussion during which he indicates an elegant experimental

[3] *Motum aequabiliter, seu uniformiter, acceleratum dicimus eum, qui, a quiete recedens, temporibus aequalibus aequalia celeritatis momenta sibi superaddit* [Galileo 1638 in *Opere* VIII: 208].

[4] *Accipio, gradus velocitatis eiusdem mobilis super diversas planorum inclinationes acquisitos tunc esse aequales, cum eorundem planorum elevationes aequales sint* [Galileo 1638 in *Opere* VIII: 205].

confirmation with the pendulum, "Let us therefore take this as a postulate, the absolute truth of which will be established for us from seeing other results, built on such hypotheses, as correspond and are comparable to experiment."[5]

What does this principle say from today's point of view? It certainly does not speak of a force, as any scientist trained today would expect. Rather, we may call it a "rudimentary energy principle", foreshadowing the notion of potential energy. Indeed, we know that for the kind of force which Galileo tacitly presupposes, the kinetic energy of a body depends only on the height of its fall and is therefore the same wherever the body lies on a horizontal plane. The same holds for the speed, since it is proportional to the square root of the kinetic energy.

Later scientists laid other, stronger, more powerful foundations for mechanics, but here we can already see the ingredients necessary for such an undertaking: new notions, a fundamental hypothesis formulated in a way adequate for deriving other results, and, of course, confirmation of these results directly and indirectly by experiments, that is, by observations and measurements.

We can observe the progress which science made in the time between Galileo and Huygens by comparing the *Discorsi* of 1638 with Huygens's *Horologium oscillatorium* [1673]. There in Part II the results are derived from a system of three hypotheses, in which we can easily recognize the precursors of Newton's three axioms.[6] Galileo's hypothesis is now proved as proposition VI, just as Galileo had expected such a possibility. It is likely, but we cannot be sure, that Johann Bernoulli had learned the theorem here rather than from the *Discorsi*.

Having noted this progress made by Huygens beyond Galileo, I find it historically interesting to observe that in Part IV when Huygens attacks the more difficult problem of the physical pendulum – that is, the oscillation of an extended, rigid body – his system of hypotheses used in Part II is inadequate, and he too must now have recourse to a rudimentary energy principle.

Newton

Newton has derived Kepler's three famous laws from his system of three *Axiomata sive Leges Motus* (Axioms, or Laws of Motion); how did he do this, and what exactly did he achieve?

In the second section of the first book of the *Principia*,[7] from his first two axioms and the additional assumptions that a force attracts a body towards a center, Newton proves his Theorema I: the motion of this body obeys Kepler's second law of planetary motion. Namely, the area covered by the line connecting the body to the center is proportional to the time of the motion; it is equal during equal times. Immediately after Theorema I Newton states and proves its inverse in his Theorema II: if a body moves in a plane on any

[5] *Prendiamo dunque ora questo come postulato, la verità assoluta del quale ci verrà poi stabilita dal vedere altre conclusioni, fabbricate sopra tale ipotesi, rispondere e puntualmente confrontarsi con l'esperienza* [Galileo 1638 in *Opere* VIII: 208].

[6] See "Huygens's *Horologium Oscillatorium* and Newton's *Principia*" in this volume, pp. 41-49.

[7] Cf. [Newton 1687], especially *Liber de Motu I*, Sects. II, III, IV.

curve such that a line from the body to a certain point covers areas that are proportional to the time of the motion, then this body is attracted to this point by a central force. Furthermore, since we know from our daily experience that there are forces between all bodies, we may conclude from Kepler's second law of planetary motion that all planets are attracted to the sun by a central force.

It is *this*, the inverse theorem, which Newton uses for deriving the results stated in the remainder of Section II and in Section III. Besides a few general theorems, these are the following:

1. Proposition VII: if a body moves on a circle and if the center of force lies on the same circle, then this force of attraction is inversely proportional to the *fifth* power of the distance from the center to the body;
2. Proposition IX: if a body moves on a logarithmic spiral and the center of force is at the center of the spiral, then the force of attraction is inversely proportional to the *third* power of the distance from the center to the body;
3. Proposition X: if a body moves on an ellipse and the center of the force is at the center of the ellipse, then the force of attraction is *directly* proportional to the distance from the center to the body.

Finally, the most important:

4. Proposition XI, section III: if a body moves on an ellipse and the center of force is at one of the foci, then the force is inversely proportional to the *square* of the distance from the center to the body.

Applying Proposition XI to the elliptical orbit of a planet, Newton later infers, in the third book of the *Principia*, the form of his law of universal gravitation. Because of "the dignity of the Problem" (*ob dignitatem Problematis*) he will not simply treat the cases of the hyperbola and the parabola by a reduction to the same case but prove each case separately, in Proposition XII and Proposition XIII respectively.

The reader trained by modern textbooks may ask if this is really all that Newton does. Does he not derive the whole mechanics of central forces? There is, of course, an enormous number of results in the *Principia* after Sections II and III, especially in Sections VII and VIII. If we mean by this question the explanation of various given laws of central forces $F=F(r)$, it is indeed essentially all, except for the general theorems which I have mentioned and the various corollaries. But even so, this is a great achievement which should not be underrated; indeed, it can hardly be *overrated.* Today we know that only three problems involving central forces are elementary: precisely those which Newton dealt with in Propositions IX to XIII. Given the question he asked, at that time he could hardly solve any other, except that he also found the answer to the problem stated in Proposition VII! But the greatest accomplishment in all this is how he set up and ordered his whole chain of reasoning, from the axioms and the additional special assumptions to the theorems and their proofs.

Before asking why he proceeded from the orbit to the law of force and not the other way around as we do today, a word may be said about the order in which the four cases are stated and solved. Does this order merely reflect the order in which Newton solved them? This is conceivable, but it is equally possible, or even probable, that Newton presented them artfully in increasing order of importance and in increasing order of the insight which he had gained.

Be that as it may, the other question of why Newton proceeded starting from the orbit is the more important one, and there are essentially two reasons for that. The first is simply that history, as it presented itself to him, had proceeded this way. Kepler's second law, proposed in 1609, was well established in Newton's day and had become common knowledge, while the idea of a central force of attraction was a new one, which few before Newton had even considered.[i] Thus, starting from a system of "Laws of Motion" – that is, his axioms – and from the hypothesis of a central force, Newton could now prove the form of this law:

$$F \propto 1/r^2 .$$

To the second reason I shall return later.

But first we must ask if Newton also solved the "inverse problem" (*le problème inverse*), that is, did he find all orbits in each of the four cases? Concerning the first two cases, propositions VII and IX, the answer is obviously negative, since there are clearly other solutions as well. For instance, in both cases there can be a circular orbit around the center. For Proposition X, on the other hand, the answer is clearly affirmative. Newton states it explicitly in the first of the two subsequent corollaries, and in the second one, he proves that the period of the motion is the same for all orbits.

How is it with the all-important Proposition XI? Jacob Hermann, Johann Bernoulli, and Pierre Varignon thought that in this case Newton had not answered the question. When I lectured for the first time publicly on the *Principia*, I expressed the opinion that he had not, but then I received a letter from a former teacher writing that in fact Newton had done so. I followed his opinion in the printed version of the lecture, only to receive a sharp objection from an American historian, who insisted emphatically that Newton had not given such a proof. The reader can find both opinions defended in the *Studia Leibnitiana* of the Leibniz-Bernoulli Symposium in Basel 1987. The affirmative one is defended by E. J. Aiton [1989], the opposite one by P. Costabel [1989]. In both articles the reader can find much valuable information.[ii]

I feel inclined today to believe that Newton gave a proof, even if it is indirect and rather well hidden. But I do not think that the proof itself is a matter of great historical importance. The importance of the achievement of Hermann and Bernoulli lies somewhere else, as will be seen presently.

Jacob Hermann

Jacob Hermann, the pupil of Jacob Bernoulli and author of the *Phoronomia* [1716], was the first to attack the inverse problem systematically [1710].

Although an important accomplishment, his paper is less transparently written than that of Johann Bernoulli [1712] for which, however, it provided the stimulus. And with respect to at least one essential point only Bernoulli's solution is complete. For this reason, I shall concentrate here entirely on Bernoulli's work.

Johann Bernoulli

In 1710 Johann Bernoulli, you will recall, not only had the mechanics of Descartes and of Huygens at his command, he had also penetrated deeply into the *Principia*, perhaps more deeply than anyone on the Continent at that time. Thus, we may ask, how did he attack this problem, and where do we still follow him today, for instance, in our teaching?

To begin to answer these questions it seems best to look at the procedure most often used today. Doing so, however, we must recall that Newton's equation

$$\mathbf{F} = m\mathbf{a}, \tag{1}$$

was written in this form, – that is, in a Cartesian, space-fixed frame, in which we write it today, – only 40 years later.

Starting from eq. (1), most authors today derive Kepler's first law in essentially four steps:[8]

1. One multiplies vectorially eq. (1) by v thus reducing the force equation to the power equation,

$$\frac{dT}{dt} = -\frac{dV}{dt}, \tag{2}$$

and then, by integrating it, into the energy equation

$$T + V = E, \tag{3}$$

or, using polar coordinates in the plane of the orbit and observing that the coordinate φ is cyclic,

$$\frac{m}{2}\left(\dot{r}^2 + r^2\dot{\vartheta}^2\right) + V(r) = E. \tag{4}$$

2. Using now the areal theorem, that is, conservation of angular momentum,

$$mr^2 d\vartheta/dt = L, \tag{5}$$

one replaces the second kinetic term with r by the kinetic potential L/r.

3. Using once more the areal theorem in the form

$$\frac{d}{dt} = \frac{L}{mr^2}\frac{d}{d\vartheta},$$

one replaces the time as the independent variable by the angle. Thus one obtains the equation of the orbit

$$\frac{L}{2m}\left(\frac{r'^2}{r^4} + \frac{1}{r^2} + V(r)\right) = E \tag{6}$$

where the prime denotes the differentiation with respect to ϑ.

4. Thus far every step was valid for any central force $F = F(r)$. For determining the form of the orbit, and indeed of all orbits, one must now assume one special law of force, or one special potential, namely $1/r$. In this case

$$r = 1/u, \; r' = -u'/u$$

transforms the integral after quadratic completion into the standard form of an elementary integral. Its solutions are *all* conic sections: ellipse, parabola or hyperbola.

Now, how does Bernoulli proceed? He does not know Newton's axiom in modern form

[8] See, for example [Sommerfeld 1947: § 6].

and so instead of the first step, he uses Newton's theorem XIII, section VIII of Book I,[iii] as a lemma (fig. 2), but simplifies its proof greatly by returning to Galileo and Huygens. Considering two orbits passing through the same point and comparing the fall of the body along each of them, he applies Galileo's principle to the infinitesimal triangle formed by the differentials on each orbit and the infinitesimal arc.

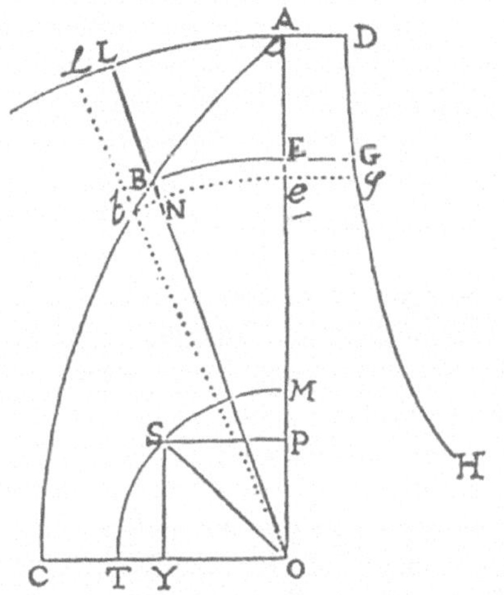

Fig. 2. From "Extrait de la Réponse de M. Bernoulli à M. Herman"
[Johann I Bernoulli 1712: 524]

Now, the principle says that the speed at the end of both infinitesimal orbits is the same, and continuing – or, as we would say, integrating – he finds that the speed of the body on each orbit is the same on all circles around the center of attraction. This holds especially for the vertical fall. For this case it is easy to see that if $\phi(r)$ is the force, the quantity $\int \phi(r)\, dr$ now appears in the equation. The vertical fall had been investigated by Newton in Section VII of the first book of the *Principia*, and Bernoulli quotes him.

He then takes the other three steps in essentially the same way as we do today. The small differences merely concern the writing. For instance, he does not have a special symbol for the angle and for angular velocity, both introduced only later by Euler,[iv] but indicates an angle by the length of the arc of an arbitrarily chosen circle divided by its radius, and he chooses the constants of integration somewhat differently from our modern procedure.

The formula obtained after the third step is:[v]

$$dz = aac\, dx : \left(abx^4 - x^4 \int \phi\, dx - aaccxx\right)^{1/2},$$

while we would write for the same

$$d\vartheta = \frac{L\, dr}{\sqrt{2Er^4 - 2r^4 V(r) - L^2 r^2}}.$$ [vi]

The main point to note here is that this expression is valid for *all* central forces.[vii] For *certain forms* of the law of force the equation can be integrated with the help of a known function, be it an elementary, an elliptic or a transcendental one; for others the integral can at least be qualitatively discussed.

Bernoulli does not pursue these possibilities. It cannot even be said that he had clearly grasped the potential,[viii] but the great progress due to Hermann and Bernoulli is now clear. Newton, like Huygens earlier and Jacob Bernoulli later, had always started from the geometry of the problem – a curve, a surface or a streamline, etc. In contrast, here the starting point is provided by the dynamic concepts – the energy, the force, the angular momentum. And this starting point leads to much more powerful and general methods. In short, *mechanics had now come into its own*. Geometry was still needed, but the required knowledge of it diminished with time, while mechanics became more and more an independent discipline. For this reason, the papers by Hermann, Bernoulli and Varignon[ix] now mark a turning point.

Bernoulli had reduced the problem of computing the orbit of a body which moves in the field of a central force to an integration. An equally big step was taken only later by the following generation, by his son Daniel and by Clairaut, Euler, and d'Alembert, namely, the formulation of many-body problems. This, as we know today, is literally impossible if we try to start from the orbits. But starting from the forces it costs nothing to write down at least the dynamical equations, and often enough we can easily find some integrals, such as the energy integral, and apply approximation methods.

We can sum up the achievement of Hermann and Bernoulli by saying that they had opened the door, if only a little bit, to analytical celestial mechanics.

Euler

I may be permitted to continue from Johann Bernoulli's achievement to the work of the greatest of his disciples, Leonhard Euler. If Johann was the first who penetrated deeply into the *Principia*, Euler was the first for whom Newtonian mechanics was the motor which incited and drove him to his great discoveries.

Earlier, in the *Mechanica* [1736], after Hermann, Bernoulli and Varignon, he had derived again, in his propositions 80 and 81, Kepler's ellipse from Newton's law of force. However, it is surprising to see that his presentation is, for once, less transparent than that of his teacher.

The great turning point, the second one in this account, came in 1750. In that year Euler wrote his paper, "Découverte d'un nouveau principe de Mecanique" [1752]. Here we find, for the first time, Newton's second law formulated in the form we know and use today, namely, in a space-fixed Cartesian coordinate system. The importance of this step was explained and underlined by Truesdell [1960a] and more recently by Maltese [1992]. At that time Euler believed, albeit erroneously, that he had found the common basis for the whole of mechanics. One immediate fruit was the definite foundation of hydrodynamics through the equations which bear his name.[x] Of hydrodynamics we shall hear more in the second part of this paper. The great role played by celestial mechanics in Euler's thinking and in his work is well known, but the series of three papers (E 301 [1766], E 328 [1767a], and E 337 [1767b]) which he devoted to the problem of a point moving according to Newton's law around two fixed centers of attraction may be mentioned here especially.

Bernoulli is not mentioned by name, yet the papers testify to his influence on his former disciple even more than a decade after his death. It is noteworthy that these three papers have exerted a scientific influence on mechanics even today.

Conflict and Cooperation

I should like to add to this presentation of a historic development a few remarks relating to "Conflict and Cooperation", the official subject of the symposium at which this paper was presented. This gives me an opportunity to say a few words about Bernoulli's *Hydraulica* [1744], as Dr. van Maanen had wished me to do.

Johann Bernoulli's attitude towards Jacob Hermann was not one of conflict, though it was certainly one of competition; his attitude towards his friend Pierre Varignon was one of collaboration. Bernoulli was steeped in many conflicts indeed, and always very much emotionally so. Thus it might be useful to consider first two scientific topics where he reacted emotionally to scientific ideas and to ask, whence these strong emotions?

We just saw one important achievement in Bernoulli's receptivity to, and his contribution to the development of, Newtonian mechanics. But there were ideas in Newton's mechanics with which he was in deep, passionate conflict. The most important ones were Newton's "action at a distance" and the vacuum. Concerning both ideas we can see his motives in his mémoire "Essai d'une nouvelle Physique Céleste" [1735] which won a prize in the year 1734. The Académie des Sciences had asked the question: What is the physical reason for the inclination of the planes of the orbits of the planets in relationship to the plane of the equator of the rotation of the sun on its axis, and how does it happen that the inclinations of these orbits are different?[9]

First we can see clearly in Johann's "Essai" that he was no "Cartesian" with respect to gravitation. He says:

> The system of the vortexes... are not impervious to great objections: it is known that the gravitation of the planets toward the sun, attributed to the effect of the centrifugal force of the matter of the vortex, should not be exactly directed towards the center of the sun, but perpendicularly toward the axis of the vortex, in the same way that heavy bodies on the earth tend perpendicularly to the axis, and not to the center of the earth.[10]

I have explained elsewhere why Bernoulli still uses Descartes's vortex, and why, doing so, he went in the right direction.[xi] He turned the question into the more appropriate one: what determines the *present* shape of the solar system? Thus his paper, together with the paper of his son, belongs to the prehistory of astrophysics.

---

[9] *Quelle est la cause physique de l'inclinaison des Plans des Orbites des Planetes par raport au plan de l'Equateur de la revolution du Soleil autour de son axe, & d'où vient que les Inclinaisons de ces Orbites sont différentes entr'elles?* (cited by Johann I Bernoulli in [1742b: § I].

[10] *Le systéme des Tourbillons ..., ne laisse pas d'etre exposé aussi à de grandes objections: on sçait que la gravitation des Planetes vers le Soleil, attribuée à l'effet de la force centrifuge de la matière du Tourbillon, ne devroit pas se faire directement au centre du Soleil, mais perpendiculairement vers l'axe de Tourbillon; de même que les corps graves sur la Terre devroient avoir une tendance perpendiculaire à l'axe, & ne point tendre au centre de la Terre* [Johann I Bernoulli 1742b: § VI].

But why does he fight Newton's "action at a distance"? After having insisted that Newton's attitude is moderate compared to the one of his followers, he writes:

> Now, impact is made via pressure; this is therefore an action that is the result of an effect. If one were to conceive of an action without an effect, he would be conceiving a chimera. … For this reason I don't see how separate bodies at rest can effect the will of God … it would mean eliminating the second causes of Nature; this would mean that … all that happens in the universe takes place directly … because of divine will.[11]

Thus Bernoulli, the *savant du siècle des lumières*, speaks. Descartes and Huygens, through introducing mechanisms for which they had discovered laws, had freed science from the *qualités obscures*, and one should avoid anything that might induce us to slip down again into obscurantism.

It is a different story with the vacuum. First I must say that at that time the vacuum was often an emotional subject. Earlier, in Groningen, Johann Bernoulli had directed a philosophical thesis by his student Scato Trip [1705].[12] The main point of the thesis is that "body" and "space" are different only *modo* but not *in re* and that therefore the notion of a vacuum is a mistaken one. Bernoulli had apparently not understood that Newton's vacuum was more than just an empty Euclidean space. In the "Essai" he says, "What is meant by *body* is nothing other than what has extension, mobility, and impenetrability; this is all that the idea of body must encompass".[13] But according to Newton, space-time carries the *vis insita*, or *vis inertiae,* as he called it. Later Euler understood this point clearly.[xiii] He added to Bernoulli's three properties a fourth one, inertia, and he showed that this force singles out from all transformations those which we today call "Galileo transformations".[xiv] That is why Bernoulli, for performing what I called the first step, had to fall back on the energy principle of Galileo and Huygens.

You will now see how this failure to understand Newton on this point indeed impeded him in another question as well, in spite of the indubitable progress which he had achieved.

Of the many conflicts and quarrels in which Johann Bernoulli was a party, I will restrict myself to the ugliest: the rivalry with his own son. Daniel Bernoulli had written in St. Petersburg a first version of the *Hydrodynamica* [1738], which he then revised in Basel. It should have appeared in 1734, but because of the usual difficulties it did so only in 1738. At about the same time, the father published his *Hydraulica* [1742] indicating 1732 as the year of writing. At that time the work certainly did not exist in the form in which it was printed, and therefore Johann has often been accused of fraud. His general boasting seemed

---

[11] *Or le choc se fait par pression; c'est donc une action dont il résulte un effet. Qui veut concevoir une action sans effet, il veut concevoir une chimère. ... ainsi je ne vois pas comment deux corps éloignés & en repos peuvent recourir à la volonté immédiate de Dieu, ... seroit bannir les causes secondes de la Nature ; il vaudroit autant dire que ... tout ce qui arrive dans l'univers, s'execute immediatement ... par la volonté divine* [Johann I Bernoulli 1742b: § IV].

[12] See "Johann Bernoulli's Work on the Theory of Gravitation and the Weight of the Atmosphere" in this volume, pp. 105-121.

[13] *Qui dit corps, ne dit autre chose que ce qui est étendu, mobile & impénétrable; voilà tout ce que l'idée du corps doit renfermer;* …[Johann I Bernoulli 1742b: § XIII].[xii]

to make the accusation plausible. Some went even so far as to accuse him of having plagiarized his son, and naively I myself repeated this accusation, an error for which I still feel sorry. Leaving aside for a moment the predating of the *Hydraulica,* I will deal first with the accusation of having plagiarized.

In fact, Euler's statement, printed in Bernoulli's *Opera* [Euler 1742], should have warned the historians. Euler praises Johann for his "... Theory of flowing water because of the true and genuine Method which You alone, Most Excellent Sir, first revealed for fully treating Problems of this type".[xv] What does Euler mean by "the true and genuine Method"? To answer the question, we must look at both books.

In the *Hydrodynamica*, twelve of the thirteen chapters are based on the energy principle ("balance of energy"); the last one is based on the momentum principle ("balance of momentum"). But we need only open the introduction of the *Hydraulica* and we see that Johann uses Newton's equation![xvi]

Daniel's book is by far the more carefully written and also the more understandable of the two; I doubt that there are many more carefully written books in the whole history of physics. But Johann's, while containing at least one serious mistake, was at the time the more advanced: *it introduced Newton's concept of force into hydrodynamics*. This is what Euler meant by "the true and genuine method", which he himself would explore later with such success. To quote Truesdell: "... he had calculated the force acting on an infinitesimal element..." [1954: XXXIII]. True, as Gleb Mikhailov observed [1983: 238], Johann, here too, has not fully grasped the notion of force, but Truesdell is right when he says that the roots of Euler's hydrodynamics lie in Johann's rather than in Daniel's work. Thus to speak of plagiarizing is absurd. Istvan Szabó has defended Johann in even stronger terms than Truesdell, and he showed that this accusation is, in fact, only of quite recent origin [1979: 171].[xvii] Ironically it originated in Basel from one whose very hero, and the favourite among all members of the family to boot, was Johann Bernoulli.

The predating is a different story. To compete with a son shows always questionable wisdom and taste, and Bernoulli chose a heavy-handed and miserable way to do so. But, however we judge his action, it is unlikely that he invented the date 1732 out of the blue; at any rate the burden to disprove him rests with us.

In fact, we know that during the period prior to 1734, when he wrote his great "Essai", he again studied hydrodynamic theories for explaining why the planets do not move in the same plane. His frequent references to his book, *Essai d'une nouvelle theorie de la manoeuvre des vaisseaux* [1714], testifies to this. I am glad that Mme. Jeanne Peiffer showed how important the ideas developed in the book were for him,[xviii] and that the prize offered by the Académie incited him to work on them again. Thus I suggest that the date 1732, although unacceptable according to our norms, refers to results obtained by these studies and then tucked away somewhere.

I wish to conclude by stating an impression, the subjective character of which I am quite aware. While preparing the plans for the edition of the various Bernoulli correspondences, together with Fritz Nagel, who knows the letters better, I looked through all those of Johann I.

With respect to his character and personality, two impressions stand out. One is, I need not insist, his love for disputes and fights. But one must note that almost all of his fights were directed against peers or even persons of superior standing. After all, in his fight for

Leibniz against Newton and the English he was the one of the three who was in the least exalted position, and therefore needed the greatest courage; the quatrain of Voltaire,[xix] who was a great admirer of Newton, testifies to this.

The other strong impression I receive is his great dedication to his students. He spent much time on them and showed them freely and generously indeed what they needed to learn. He always recognized unhesitatingly and without reservation the superiority of his greatest student, Euler. Indeed, recognizing that he was the teacher of one of the greatest scientists and teachers ever is the most fitting tribute which history possibly could pay to him.

Acknowledgments

It is a pleasure to thank Dr. Jan van Maanen for the invitation to the Symposium in Groningen, Mr. Martin Mattmüller for his help with the literature and the corrections, and my wife for linguistic advice. I am especially grateful to Diarmuid Ó. Mathùna for his extremely careful reading of the manuscripts as well as for his many suggestions. Clifford Truesdell honoured me with the invitation to submit this paper to *his* and Charlotte's *Archive*; his seminal article [1960b] which opened this more than successful journal was a sign post for many of us!

Daniel Bernoulli

Daniel Bernoulli [1]

Introduction

In the year 1684, the then forty-year old Gottfried Wilhelm Leibniz published "Nova Methodus pro Maximis et Minimis" [1684] in the *Acta Eruditorum* in Leipzig, which he founded, in which he discussed a new calculus that he had developed. Just a short time later two practically unknown brothers would throw themselves into this work with such aggressiveness and energy that the secrets of the new mathematical language were for the most part revealed.

The older brother, Jacob, had studied theology at the university, as his father had wished, but had taken time on his own to learn the new Cartesian mathematics. He had also dealt with astronomy and had, for example, worked out a theory that the big comet of 1680 was the moon of a yet unknown outer planet, just one year before the German priest Georg Samuel Doerfel from Plauen theorized that comets moved on parabolic orbits around the sun. He had stayed for a time in France, his second trip to that country, and before that in England, and at that time was giving lectures on physics that were open to the public and quite successful, so that he was soon nominated to the position of professor for mathematics at the university of Basel.[2]

Many of Leibniz's ideas had remained unclear to Jacob, and to his younger brother Johann as well, and so he wrote to him a letter. Unfortunately, Leibniz was travelling and didn't receive the letter until three years later. The absence of an answer forced the two brothers to derive the whole method of Leibniz anew and at the same time apply it to new problems. These were generally problems of differential geometry, but there were also problems of mechanics as well. Thus Jacob succeeded in calculating the form of a bent beam [1694].[3] He had submitted the problem to all "geometers", and when after three years he had not had not received any answers, he published his own solution. This had an influence on the work of his nephew, Daniel Bernoulli, as we shall see shortly. His most important work, *Ars Conjectandi* [1713], published after his death by his oldest nephew

---

[1] Lecture in honor of the 200th anniversary of the death of Daniel Bernoulli, given on 7 October 1982 at the Bernoullianum Basel on the occation of the convention of the Schweizerischen Physikalischen Gesellschaft. Originally published in *Helvetica Physica Acta* **55** (1982): 504-523.

[2] For the lives of Jacob and Johann Bernoulli, see [Wolf 1859], [A. Speiser 1939] and [Fleckenstein 1949] .

[3] See "The Beginnings of the Theory of Elasticity from Galileo to Jacob Bernoulli" in this volume, pp. 1-14.

and student, Nicolaus I (the son of his brother Nicolaus), was also to prove fruitful for Daniel's own research, though indirectly.

Jacob's most important student was his brother Johann, twelve years younger, together with whom he had studied Leibniz's work. Johann was a less deep and inventive mathematician than Jacob, but he was swifter and more brilliant, and he had a much greater gift for explanation.

While Jacob's first steps in science were slow, laborious and difficult, Johann made swift and seemingly effortless progress. After Newton and Leibniz, Jacob was the third mathematician to grasp the new infinitesimal calculus. When Jacob introduced it to Johann, he became the fourth. With ease, he grasped the new ideas, and with great elegance he applied it to new problems. The long-sought discovery of the form assumed by the hanging chain justified his fame; thanks to having discovered that the curve along which a weight slides down from one point to another in the shortest time is a cycloid, his name was associated for the first time with that of natural philosopher. He gave private lessons to Marquis de l'Hôpital, who then published the first textbook on infinitesimal calculus [1969]. However – as Johann Bernoulli always claimed and Otto Spiess has shown [1955] – Bernoulli's lessons were the basis for the book.

In 1695, the great Huygens, at that time the most eminent physicist on the Continent, had arranged Johann's appointment in Groningen, Holland, where Johann's second son Daniel was born in 1700. In that year, when, following its reorganization in 1699, the French Académie Royale des Sciences began nominating foreign members, the first four were Newton, Leibniz, Jacob Bernoulli, and Johann Bernoulli! In 1705 Johann returned to Basel at the insistence of his father-in-law, Mayor Falckner. Soon after he arrived he was offered the professorship left vacant only a few weeks earlier by the death of his brother, which he would hold until his own death in 1748. During the decades to come, he would make it the most prestigious chair in Europe. Among his students in Basel were Clairaut, Maupertuis, Gabriel Cramer, Albrecht von Haller, Samuel Koenig and, of course, Leonhard Euler. The autodidact d'Alembert, about whom we shall hear more, once said, "what I learned of mathematics, I owe to Johann Bernoulli".[i]

I have entered into this background so extensively, because without knowing them Daniel Bernoulli's life and works are difficult to understand.[4] He was literally born in the center of the world of mathematical science. In a certain sense, for some decades infinitesimal calculus was almost a kind of Bernoulli family secret; certainly, thanks to Johann Bernoulli, Basel was a Mecca for mathematicians of the age. Understanding the world in terms of geometry (*more geometrico*) was a distinct objective of the period. Even medicine, as we will see, acknowledged this aim. But more specifically – as Huygens and Newton but above all the Bernoullis showed – knowledge was gained not only in terms of geometry, but in terms of mechanics (we might say, *more mechanico*).

Overview of Daniel's scientific activities

Daniel Bernoulli spent his first five years in Groningen. Then, back in Basel, he attended school but also learned mathematics at home, in particular infinitesimal calculus,

[4] For the life of Daniel Bernoulli, see [Wolf 1859], [A. Speiser 1939], and [Straub 1970].

tutored not by his father but by his older brother, Nicolaus II. His father was demanding and thought rather derogatorily of him. At first he argued against Daniel's continuing his studies, and urged him to enter commerce, but he finally relented and agreed that he could study, not mathematics, as Daniel probably wished, but rather medicine, which he finally completed in Basel and Heidelberg. Thus nothing was handed to Daniel on a silver platter, and at a young age he had to assert himself at home.

Daniel's first works were in the area of medicine – more exactly, in physics applied to medicine, or iatrophysics, as it was called at the time. Galileo's student, Borelli, had tried to apply to the human body the new science being taught, and both Johann and Daniel Bernoulli followed his example. For this reason Daniel travelled to Padua, where he worked with Michelotti and Poleni. In Padua Daniel turned his attention to mathematics as well; his first interests were differential equations and questions concerning calculus of probability. Parts of this work were collected in a small volume printed in Venice, the *Exercitationes quaedam mathematicae* [1724]. This brought him quite a bit of fame, so much so that the Republic of Genoa offered him the presidency of its newly re-established Academy. This, however, he turned down, because in 1725 together with his older brother Nicolaus he was called to the Imperial Academy of St. Petersburg, where his fellow countryman Jacob Hermann worked, and where his friend Leonhard Euler would follow him some years later. The years in St. Petersburg are the first high point of his work. His first important works from that period are in the area of mechanics and here too he made the first discoveries in the theory of vibration. But the greatest number of his works in those years are in hydrodynamics, including the first draft of the *Hydrodynamica*.

Of Daniel's works in mechanics from this period, I have written extensively elsewhere [Radelet-De Grave and Speiser 1987a], so I won't describe them here.

After a few happy years in St. Petersburg, Daniel's brother Nicolaus, to whom he was very attached, died unexpectedly of an intestinal infection, and Daniel became more and more homesick. In 1733 he was offered a position in Basel as a professor of anatomy. (Only later would he be given a chair in physics, which he accepted, on condition that he could remain affiliated with the Petersburg academy.) His younger brother, Johann II, joined him in St. Petersburg. The trip back took them through Northern Germany, Holland, where they were received with honor in Groningen, and Paris, where they attended a session of the Académie royale des sciences.

From then on, Daniel spent the rest of his life in Basel. First, he was professor of anatomy, then botany, and only from 1750 on, a professor of physics. In what follows, I would like to speak primarily of four areas of Daniel's activity: hydrodynamics, where Daniel continued Newton's program with the help of the ideas of Huygens and of his father; the theory of universal gravity, where Daniel helped lead Newton's theory to victorious acceptance on the Continent; the theory of elasticity, where his work was almost completely original and independent; magnetism and electricity, where he is a forerunner of the physics of the nineteenth century.

Hydrodynamics: the relationship between hydrostatics and hydraulics

Newton had drafted a brilliant program for hydrodynamics in the seventh chapter of the second book of the *Principia*, which, as Truesdell says, became the stimulus for all his successors for the next fifty years [Truesdell 1976]. It should be noted that the first continuation of Newton's work didn't occur in point mechanics, the area for which Newton created a lasting foundation, but rather in an area where physical experiments were used to provide further support.

For most of his lifetime, and almost exclusively after his death, Daniel Bernoulli's fame was based on his *Hydrodynamica* [1738]. This book contains results of the work he had done in this area in St. Petersburg. He had published sections of it previously, such as, for example, in 1727, the part concerning the correct calculation of the force produced by a flowing stream [1729a]. In St. Petersburg, he had already written a first draft but it appears, however, not to have satisfied him, because after his return to Basel he rewrote it completely.

To start with, the title word *hydrodynamica*, which means roughly "the theory of the behavior of water under the influence of forces", was coined by Daniel. Before this, *hydrostatics* – the theory of liquids at rest (their equilibrium, pressure, etc.) – was separate from *hydraulics* – the theory of liquids in motion. Daniel wanted a word which encompassed both. Today we usually relate motion in physics with the word dynamics (from *dynamis*, Greek for force), but this is wrong, because statics also requires laws of force. A reading of the *Hydrodynamica* makes evident the interest in concrete problems that guided Daniel's research. It also immediately shows that the book is only partially a systematic exposition. A completely systematic exposition would progress from simple basic equations to ever more complicated theorems and applications. However, this was not possible at that time, because the actual basic equations would be discovered by Euler only fifteen years later (building, however, on Daniel's results). Some chapters give numerous, concrete, and calculated numerical examples, and we can assume that he found the way to his results through the solving of such problems.

Daniel's theory is based on a particular energy principle, which Huygens had formulated thus: the center of gravity of a body cannot rise higher than it can fall, and if friction plays no part, it climbs exactly that high. In a letter written to Daniel, his father, Johann I, extended Huygens's principle significantly [1729]; the son, who published it in the Petersburg Academy *Commentarii*, also properly emphasized his father's share in those passages [1729b].

The two high points of the book, which are still today connected with the name of Daniel Bernoulli, are located in the tenth and twelfth chapters. At the beginning of the chapter ten, we find his famous derivation of the laws of Townley, Boyle and Mariotte and of Amontons:

$$\text{pressure} \times \text{volume} = \text{const.} \times \text{temperature}$$

from the laws of mechanics and the assumption that air consists of small particles. Daniel's formulation of the law is equivalent to:

$$\begin{aligned}\text{Pressure} &= 1/3 \times \text{density} \times \text{the square of the average speed of the particles}\\ &= 1/3 \times \text{average energy density of the particles.}\end{aligned}$$

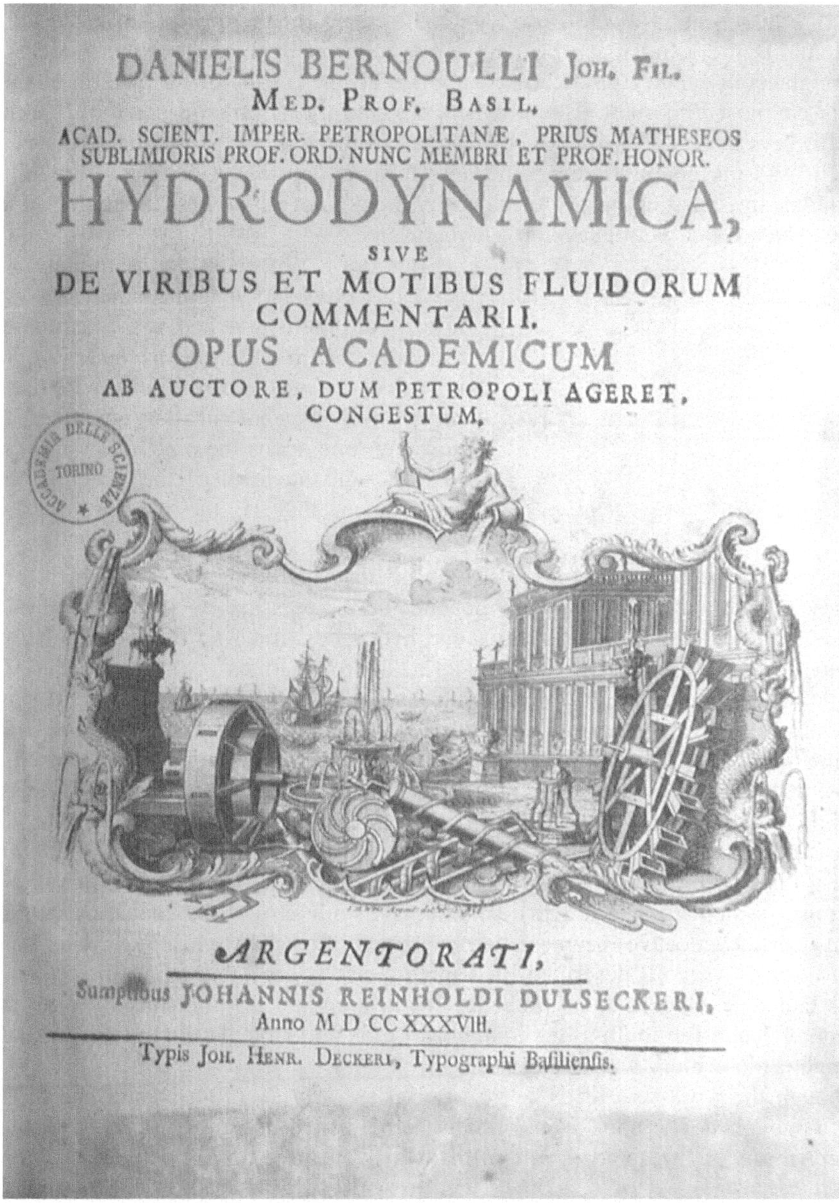
DANIELIS BERNOULLI Joh. Fil.
Med. Prof. Basil.
ACAD. SCIENT. IMPER. PETROPOLITANÆ, PRIUS MATHESEOS SUBLIMIORIS PROF. ORD. NUNC MEMBRI ET PROF. HONOR.
HYDRODYNAMICA,
SIVE
DE VIRIBUS ET MOTIBUS FLUIDORUM COMMENTARII.
OPUS ACADEMICUM
AB AUCTORE, DUM PETROPOLI AGERET, CONGESTUM.

ARGENTORATI,
Sumptibus JOHANNIS REINHOLDI DULSECKERI,
Anno M D CC XXXVIII.
Typis Joh. Henr. Deckeri, Typographi Basiliensis.

Frontispiece of the first edition of the *Hydrodynamica*

Even today, Daniel's derivation can be found in modernized form in all physics books. According to Truesdell, this was the only successful application of atomistic ideas to mechanics in the eighteenth century. Otherwise, all the successes in mechanics seem to be based on the idea of matter as a continuum [Truesdell 1954: XXIV]. Daniel also noticed that his formula is not valid for low temperatures, i.e., at the dew point. Even today this derivation of Daniel is rated as a brilliant achievement of the highest caliber.

For eighteenth-century physics it was above all the results of the twelfth chapter that would prove most important. This is where we find the "Bernoulli equation". In modern terms, this says that in an irrotational, steady flow the sum of the internal pressure and of the kinetic and the potential energy per unit mass is constant. However, Bernoulli could not formulate the equation in this form because he did not yet possess the notion of internal pressure, which Euler would develop ten years later.

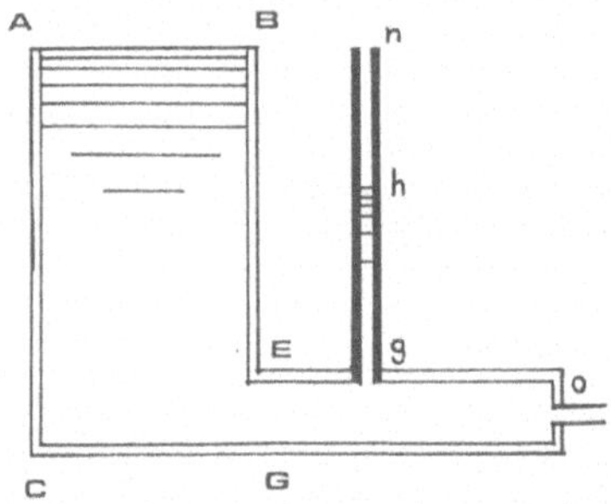

Bernoulli proceeds approximately in this way: one considers a full container, to which a horizontal pipe is attached; this is equipped with a small, open tube. As long as the horizontal pipe is sealed, the water in the tube will rise to the same level as that in the container. When, however, one opens the pipe, allowing the water to flow out, the level of the water in the tube goes down, that is, the pressure in the tube decreases.

Bernoulli's great achievement consists in having defined the level to which the water would rise, which today we would say depends on the pressure, as a function of its speed. With this, he finally united hydrostatics and hydraulics in a way that no one before him had done. This synthesis justifies the new term "hydrodynamics".

While here I limit myself to describing the high points of the book, numerous other new results are also reported, and everything is explained and calculated in the examples. Given the richness of the work, it is easy to understand the sensation that it caused among his contemporaries. However, the modern reader also senses a certain awkwardness and a lack of order. For example, the titles in the text don't agree with those of the table of contents, and in the title to the twelfth chapter we still find the term *hydraulico-statica* in the place of *hydrodynamica.*[ii] Daniel never willingly revised or proofread a draft! Furthermore, systematizing the representation and symmetry of the choice of symbols and notation had never have been his strong point. Nevertheless, fifty years later the *Hydrodynamica* was still described by Lagrange as "a work that shines moreover for an analysis that is as elegant in its procedure as it is simple in its results".[5]. With the sole exception of Jacob Bernoulli's *Ars conjectandi*, it is undoubtedly the most important book of the eighteenth century written in Basel.

Although the work was ready to go to print as early as 1734, Daniel had difficulties finding a publisher. The book would finally be published in Strassburg in 1738, thanks to the intervention of the famous Schöpflin, who we know from Goethe's *Dichtung und Wahrheit*.

---

[5] ... *Ouvrage qui brille d'ailleurs par une Analyse aussi élégante dans sa marche, que simple dans ses résultats* [Lagrange 1788, part. II, sect. VII].

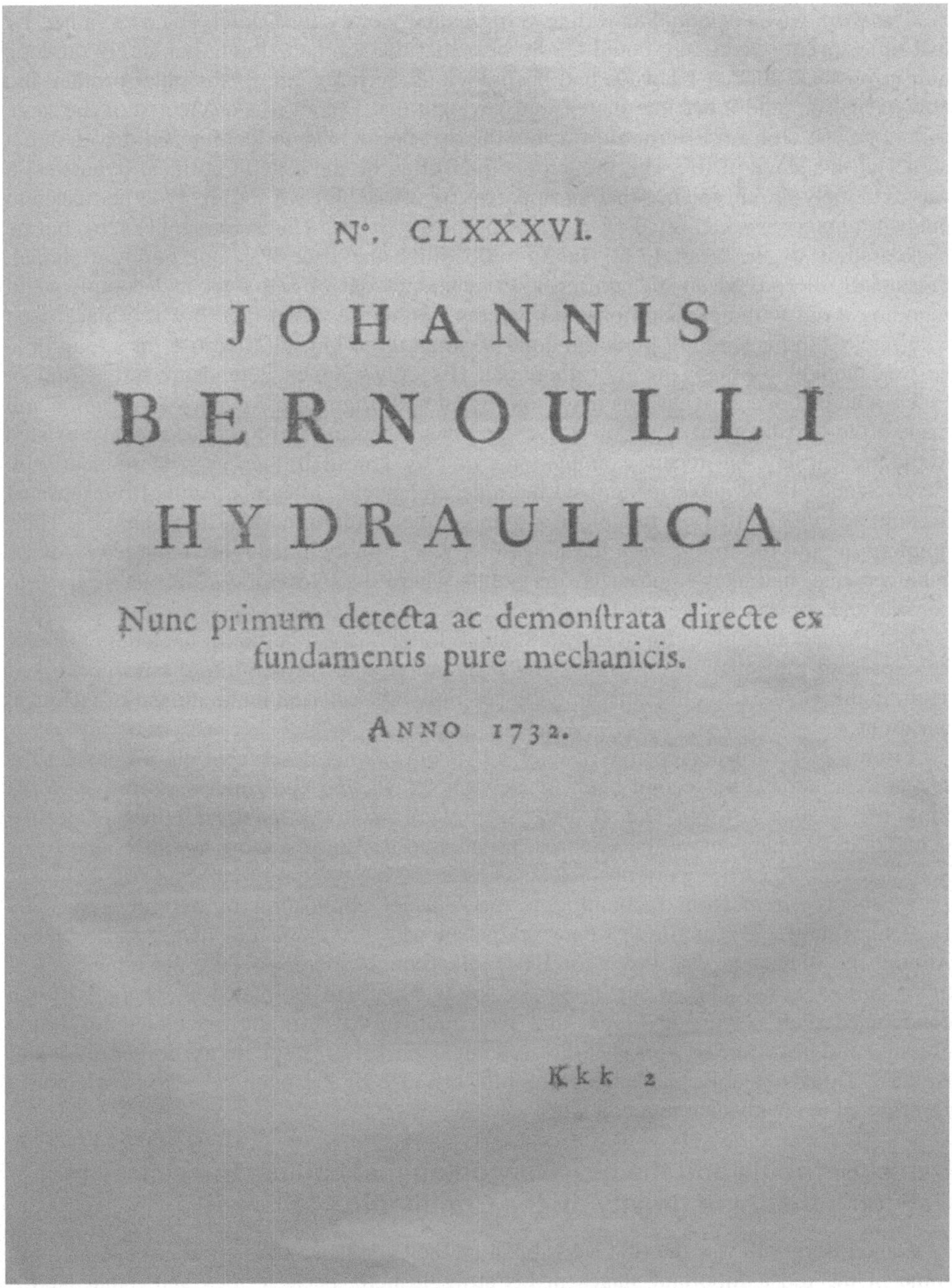

N°. CLXXXVI.

JOHANNIS

BERNOULLI

HYDRAULICA

Nunc primum detecta ac demonstrata directe ex fundamentis pure mechanicis.

ANNO 1732.

Kkk 2

Frontispiece of the *Hydraulica*

The work was a personal as well as professional victory for Daniel. His own father, by that time seventy years old, could hardly bear that the son had slowly but surely outdone him in physics. Johann I always had been envious, in his youth of his older brother and now of his son, and it became increasingly evident that Daniel was a scientist of the same calibre he was. Johann I Bernoulli was not the first nor the last to feel himself driven into a corner at this stage of life. But though it might often be the case in political dynasties, it was extremely rare in science that his apparent rival was his own son. In 1743 he rushed to publish a longer work entitled *Hydraulica* in the Petersburg *Commentarii* [1742b], but he backdated it to the year 1732! In all probability the date was not purely fictional. Presumably he pulled an old, unfinished manuscript out of a drawer and completed it, rounding it out with new additions, but we don't know this for a fact. It was a black deed with regards to his son, but the harm done was mainly to himself, because for a long time he was thought by many to be a plagiarist. His only staunch defender was his student Euler, who discovered what was new, significant and original in Johann's *Hydraulica* and made it the starting point of his own investigations. Accusations of plagiarism have existed to the present day, but recent examinations, first by Truesdell, then even more clearly by Istvan Szabó [1979], have recognized the independence of Johann's results from those of his son, and indeed that his treatment of some points is more modern than Daniel's. Their opinions are not dissimilar from those of Euler who, although he highly regarded Daniel's achievements, had also emphasized the points where the two works differ as he tried to console and appease his teacher.[iii]

But Daniel, struggling for acknowledgment, fought a hard battle. In his letters to Euler he expressed himself with bitterness and resignation, and furthermore, he never worked again in this field. He would enter into it again only indirectly, in the treatment of technical questions.

From a historical perspective, Daniel's *Hydrodynamica* deserves a place between the seventh chapter of the second book of Newton's *Principia* and Euler's definitive works from the 1750s. Newton, Daniel, and Euler were each the master of their respective periods, and as happens in the sciences, progressed beyond the work of those who came before, and made earlier work obsolete.

I cannot refrain from recounting a very special application of hydrodynamics by Daniel. Around 1737 in Basel a poor craftsman who had stabbed a man was beheaded. Among the spectators was Professor Bernoulli, brought there, not like the others out of mere curiosity, but out of the thirst for knowledge! He estimated the height of the condemned man's spurting blood, and with help of his formula calculated his blood pressure and heart rate. If you visit the zoological garden in Basel, do not forget to look at Auberg. Think not only of Daniel Bernoulli, but also of the poor lad who involuntarily contributed to an advance in medicine.[iv]

Daniel Bernoulli and the new Newtonian mechanics: the victory of Newton's theory of gravity on the Continent

Daniel Bernoulli was the first Newtonian on the Continent, and I would like to mention some of the most important stages in his struggle here, because they show that is was indeed a struggle. One of his earliest works was inspired by Newton's ideas [1728]. As

Szabó makes clear in his beautiful book about the history of the mechanics, this is the work in which he decided to demonstrate that the dispute between the Cartesians and Leibnizians about the true measurement of forces was a mere battle of words [1979: 71].

Some years later, he made a crucial contribution to point mechanics by introducing potential energy in two works for general systems of heavy mass points [1747; 1750]. This helped open the way to the analytic mechanics of the many body problems. At the end of the first work, he gave a first, albeit somewhat crude, application of his new results to the calculation of the orbit of the moon. The French Alexis Claude Clairaut would continue on the new way that led to the first real breakthrough in higher analytic celestial mechanics.

However, Newton's theory of gravity was met with the sharpest resistance, most of all from the Cartesians. Descartes's theory of gravity probably still had a very large number of supporters at that time. According to Descartes, space is filled with a fine substance, the "ether". This ether is in constant motion, not just any irregular motion, but motion that creates a vortex. Such a vortex of the ether surrounds the sun and drives the planets that are surrounded by it. The effect of the ether on the planets therefore occurred through immediate mechanical contact; effects through contact were considered as given in Cartesian physics, and didn't require any further explanation.

On the bases of Kepler's laws and of Huygens's calculation of centrifugal force, Newton had recognized that the attraction of the sun on the planets must decrease with the square of the distance (i.e., as $1/r^2$), so that the orbits are ellipses, as Kepler had shown. On the other hand, Newton had claimed that Descartes's vortex theory was in contradiction to Kepler's laws. But the Cartesians didn't abate their opposition to Newton's theory. On the question of how gravity is transferred to the planets by the sun, Newton remained guiltily conscious that he had not come up with any hypotheses. It seems that Newton too had searched for a medium that conveyed gravity; however, in his opinion, that he was unable to define the nature of such a medium did not invalidate his results. As we know, Einstein succeeded in filling this gap only in the twentieth century, with the general theory of relativity. However at that time, only few dared to believe mathematical formulas without the statement of a mechanical explanation, and as late as the 1730s – that is, fifty years after the appearance of his book on the Continent – Newton's theories were still not generally accepted.

In 1734, the Académie royale des sciences in Paris asked as a prize question why all planes of the orbits of the planets coincide almost exactly with the plane of the sun's equator. Apparently, what was hoped for was a decision between the theories of Descartes and Newton. Newton's attractions don't distinguish any one plane from all others, but the Cartesian vortex theory did: namely, the vortex plane, which is perpendicular to the axis of the vortex. The predominantly Cartesian-minded Académie probably purposely selected a topic on which Descartes appeared to have an advantage over Newton. Newton's theory offers no explanation for why the orbits of the planets coincide with that of the sun, more specifically, why they coincide with the plane of its equator.

Daniel's way out of this dilemma [1735] is to assume that all orbits were actually oriented completely arbitrarily in the beginning, and that over the course of time, through a physical cause, i.e., a mechanism, they were forced into the same plane. What is this mechanism?

Daniel assumes that originally the planets were not in empty space like they are today, but rather in a thin gas, a "sun atmosphere", that extended as far as the most outermost planets, or perhaps even further. As the sun rotated on its axis, they travelled around it; the planets, however, travel much more slowly than the sun atmosphere itself, so that they were exposed to a constant frictional force that slowly but surely dragged or pushed them into coincidence with the plane of the sun's equator. At the time that Daniel was formulating this hypothesis, he could not back this up with an adequate calculation, but his argument is ingenious. At first it might appear that he had fallen victim to a flaw in reasoning, but as the American J.L. Pietenpol has shown, his argument holds up, and computerized numerical calculations made by L. Bossy in Brussels confirm this [Radelet-De Grave and Speiser 1987a]. Daniel expressly emphasized that his sun atmosphere is not to be confused with the Cartesian vortex, and that all his arguments are based on Newtonian mechanics. On the other hand, however, he conceded that, even if there could be no doubt about the validity of Newton's gravity, it had not yet been completely certified and that perhaps a Cartesian vortex as well as other forces are necessary. Even Daniel himself, in 1734, was not yet completely convinced about Newton's theory of gravity!

Even today the formation of the planetary system is an unsolved problem. Of the many theories that have been devised, Daniel's has the honour of being, if not the absolute first, at least one of the first. (Kant's famous *Allgemeine Naturgeschichte und Theorie des Himmels* (*Universal Natural History and Theory of Heaven*) was first published in 1755.)

So how did the Académie decide? For Descartes or for Newton? No decision was made, that is, the prize was shared. One half was awarded to a Newtonian, Daniel Bernoulli; the other half to a Cartesian, Daniel's own father, Johann.[6] For the son, triumph and honour were enormous, all the more because the majority of the members of the Académie were Cartesians. For the ever jealous Johann, however, the decision was a bitter pill to swallow. He had to share the prize equally with his own son. Not only that, but that son was in the camp with the Englishmen who Johann had hated ever since the dispute between Leibniz and Newton. We have already mentioned the consequences for the son. Personal consequences aside, the work is still notable for a quite another reason.

Daniel himself was thoroughly aware that the pressure of friction that the sun atmosphere exerts on the planets is obviously extremely weak, and thus an enormously long time is required to bring all orbits into the same plane. We should keep in mind that biblical chronology, according to which the world was approximately 6000 years old, was commonly accepted in this age by scientists such as Newton, Euler, and Johann Jacob Scheuchzer, the universal scholar from Zurich. Daniel leaves no doubt that he thinks that a much longer period of time was necessary for the formation of the planetary system. He says:

> I myself am persuaded therefore that a very long time ago, the bodies that move around the Sun described the orbits, making angles with the solar equator that were much bigger than they now make, & that these angles between the different orbits varied much more than in our times: but that

[6] For a detailed comparison of the two memoirs of Daniel and Johann, see "Johann Bernoulli's Work on the Theory of Gravitation and the Weight of the Atmosphere" in this volume. pp. 105-121.

> these Orbits have been reduced little has little to within the narrow boundaries where they are now, & that after an infinite amount of time, they will meet entirely in a same plane, which will be that of the solar equator.[7]

This may be the first time that a scientific theory demanded a temporal expansion of the biblical concept of the world, on the basis on a concrete physical theory. The only trouble, Daniel said, was that for this reason he didn't dare to write down a number! The issue was much too controversial for him, and perhaps he was afraid of ruining his chances of winning the prize. The idea that that the world is much older than previously believed on the basis of biblical study would only become generally accepted in the second half of the eighteenth century, approximately about the time of Buffon.

Since the prize question of 1734 had not resulted in any decision being made between the systems of Descartes and Newton, in 1740 the Académie des sciences posed a question regarding "ebbing and flowing", or as we would say today, the tides.

According to Newton, the tides are a quadrupole effect: by their means a liquid ball is distorted into an oblate ellipsoid. Newton's theory thus immediately explains why the tides occur twice each day. He had included the basic ideas of this explanation in the *Principia*.[v]

Daniel apparently studied and noted the *Principia* thoroughly, perhaps using the copy that Newton himself had sent his father, and then, as was his habit, he tried applying Newton's ideas to complicated cases, such as the multifarious form of the ocean and the inland seas, to see how they would pan out. Daniel's work on this question was his most extensive and most scientifically valid after the *Hydrodynamica*.

The titles of the eleven chapters not only provide a good overview of the work itself, but they also give a good picture of Daniel as a scientist, what interested him and how he grasped problems.

- I. Contenant une Introduction a la Question proposée
- II. Contenant quelques Lemmes sur l'Attraction des Corps
- III. Contenant quelques Considérations Astronomiques & Physiques, préliminaires pour la détermination du Flux et Reflux de la Mer
- IV. Qui expose en gros la Cause des Marées
- V. Contenant quelques Propositions de Géométrie préliminaires pour l'explication & le calcul des Marées
- VI. Sur l'heure moyenne des Marées pour toutes les Lunaisons
- VII. Qui contient a l'égard de plusieurs Circonstances variables, les Corrections nécessaires pour les Théorèmes & pour la table du Chapitre précèdent, & une explication de plusieurs observations faites sur les marées
- VIII. Sur les différentes hauteurs des Marées pour chaque jour de la lune

---

[7] *Je me persuade donc qu'aux temps fort reculés, les corps qui se meuvent autour du Soleil, ont décrit des Orbites, faisant avec l'Équateur solaire, des angles beaucoup plus grand qu'ils ne font à présent, & que ces angles ont varié beaucoup plus entre les différentes Orbites, que dans nos temps ; mais que ces Orbites ont été réduites peu à peu dans les bornes étroites où elles sont à présent, & qu'après un temps infini, elles se réuniront entierement dans un même plan, qui sera celui de l'Équateur solaire* [1735: § XVII].

IX. Sur les Hauteurs des Marées corrigées, suivant différentes circonstances variables
X. Dans lequel on examine toutes les propriétés des Marées, qui dépendent des différentes Déclinaisons des Luminaires & des différentes latitudes des Lieux
XI. Qui contient l'Explication & Solution de quelques Phénomènes & Questions, dont on n'a pas eu occasion de parler dans le corps de ce Traité, surtout a l'égard des Mers détachées, soit en partie, soit pour le tout de l'ocean.

We can see here that Daniel proceeds from simple assumptions (such as, for example, that only the moon has an effect on the ocean) to more complicated ones. He calculates the date and time of events, the different effects corresponding to the various phases of sun and moon in different seasons, etc. The numerical calculation is always carried out to its final result, and these are condensed in tables for practical use.

This work strikes a happy medium between the terse, choppy style of his early works and the often long-winded style of his later years. Few of his works are so pleasant to read, and one senses the enjoyment he took in preparing it. This was particularly beneficial for the presentation and the notation, both of which are considerably more systematic here than in other works of his. Only the last chapter about the inland seas is somewhat meagre, presumably because the due date had almost arrived. This time, in 1740, Daniel now confesses to being an unconditional supporter of Newton. In the third section of the first chapter he has this to say about the Cartesians:

> As for the vortices, I admit that it is very difficult to demonstrate their fallacy to those who want to persist in defending them: but on the other hand, it is not the same in Physics as it is in Geometry. In [Geometry] one neither admits nor rejects anything which one can absolutely demonstrate the truth or falsity of, while in Physics it is often necessary to base oneself on a certain natural instinct for sensing falsity & truth, after having carefully weighed all reasons on both sides. As for me, I don't at all find this character of truth in either the hypothesis of the vortices, or in the consequences that are drawn from it. Enough said and done about this matter; since it will always fall to the followers of the sect of Descartes to prove the effect of the vortices on the ocean with the same clarity that one can prove it by means of Kepler's principle, a principle which is no longer contested; that is, that the Earth & all celestial Bodies have a mutual tendency to approach each other.[8]

[8] *Quant aux Tourbillons, j'avouë qu'il est bien difficile d'en demontrer le faux à ceux qui veulent s'obstiner à les défendre ; mais aussi il n'en est pas de la Physique, comme de la Géometrie. Dans celle-ci on n'admet, ni ne rejette rien, que ce dont on peut absolument démontrer la vérité, ou la fausseté, pendant que dans la Physique il faut se rapporter souvent à un certain instinct naturel de sentir le faux & le vrai, après avoir bien pésé toutes les raisons de part & d'autre. Quant à moi, je ne trouve point ce caractere de vérité, ni dans l'hypothese des Tourbillons, ni dans les conséquences que l'on en tire. ... En voilà assez & trop sur cette matier ; car ce sera toujours aux Sectateurs de Descartes de montrer l'effet des Tourbillons sur l'Océan, avec la même clarté qu'on peut le faire,*

However, as we know, his father was one of the Descartes's "sect".

This time the Académie prize was shared among four authors: one Cartesian, Antoine Cavalleri, and three Newtonians, Leonhard Euler, Colin Maclaurin, and Daniel Bernoulli. Cavalleri, a French Jesuit priest, is almost completely forgotten today. The work of Euler is not ranked among his most important. That of the Scot Maclaurin [1741] is justly famous still today, because it contains the proof of an important theorem.[vi] Of all these, Daniel's contains the largest number of individual results, particularly of the application of the theory to practice, which was always very important to the Académie. However, there is no doubt that this work made the most substantial contribution to the confirmation of Newton's theory.

This decision was a partial but not a complete victory for Newton's theory of gravity. Nevertheless, this paper of Daniel's enhanced his fame in a completely unexpected way. Despite the Catholic Church's condemnation of the Copernican theories in 1616, the learned Pope Benedict XIV instructed French priests Jacquier and Le Seur to publish a new, annotated edition of Newton's *Principia* [1739-1742]. The two fathers, who complied by providing a commentary that expressed due reservations regarding the Copernican theories, included the three works by Euler, Maclaurin and Daniel Bernoulli in an appendix, but omitting Cavalleri's! The book was then printed in, of all places, Calvinistic Geneva and was republished in several editions, so that this work became Daniel's most widely distributed.

The theory of elasticity: The basis for the theory of vibration

While Daniel Bernoulli's work in the field of hydrodynamics never lapsed into oblivion, it is a different story with his works regarding the theory of elasticity, which deserve to be ranked equally highly.[9] Today only one of all his achievements in this field is still associated with his name. We now want to turn to these works.

Daniel mainly investigated three forms of oscillation, namely, in the order in which he dealt with them: the hanging chain, the vibrating rod, and stretched string. The study of each of these three problems led him to at least one fundamental realization.

Before Daniel, the only definitive result known was Brook Taylor's formula for the frequency of the fundamental vibration of a string [Taylor 1714], and Taylor's work inspired Daniel's investigations. In two works from the years 1732-1734 [1738; 1740] he finds the following: The double, triple, and quadruple pendulums exhibit exactly two, three, and four stationary vibrations respectively: if and only if the amplitudes satisfy an nth degree equation with n real roots, the form of the vibration will remain constant. The amplitudes themselves depend on the size of the masses and their distances. Finally, he passes to the limit of a continuous mass distribution, that is, to the hanging chain. He finds that the solution is an infinite series, whose zeros determine the stationary vibrations of the rope. Thus, in terms of physics, Daniel had discovered the simple modes. In terms of

---

moyennant le principe de Kepler, principe d'ailleurs qui n'est plus contesté ; sçavoir, que la Terre & tous les Corps célestes ont une tendance mutuelle à s'approcher les uns des autres [1741: § III].

[9] For a detailed discussion of Daniel Bernoulli's works on the theory of elasticity, the reader is referred to [Truesdell 1960a].

mathematics, he had solved the first eigenvalue problems; in particular, he had solved the first *infinite* eigenvalue problem and discovered the proper frequencies. As a by-product, among others, he found the first Bessel function.

The search for the conditions under which the vibrations of a mechanical system are stationary pushed and pulled him to various examinations, for example, to the analysis of the vibrations of an elastic rod. For the elastic rod, he developed the so-called "linear theory" and found a fourth-degree differential equation, which he solved generally. His main result can be summarized in the following statement: for each mechanical system, there are exactly as many stationary vibrations as the number of degrees of freedom possessed by the system. In particular, for an elastic body, there are infinitely many.

In order to discuss his third great discovery, I must go back a bit. Daniel's work concerning the hanging chain found two important readers: Euler and d'Alembert. D'Alembert was led to the partial differential equation of the hanging chain through Daniel's work, the first partial differential equation in physics (which, however, he left unsolved), published in 1743 in d'Alembert's *Traité de Dynamique*. Later, he published the equation of the vibrating string with the solution by means of progressive waves, which was taken up and elucidated by Euler. As for the polemic between d'Alembert, Euler, and Bernoulli that ensued, we need to go into it only as far as Daniel's ideas on the discussion stood; we can leave aside the dispute between d'Alembert and Euler, as important for the development of mechanics as it was. While these two looked for the solution by means of progressive waves and attributed total generality to them, Daniel looked for a solution by means of trigonometric functions. Much earlier, he had recognized that trigonometric functions in fact describe the stationary vibrations of the vibrating string, as well as, incidentally, that of an organ pipe. But from them, and this is his third great discovery, all remaining solutions can be built. We therefore owe to him the discovery of the *superposition principle.*

I cannot go into the question of whether or not Daniel recognized that not only the amplitudes but also the phases can be arbitrary. However, we know today that his solution is in principle as general as that of his opponents. I will content myself with the following remarks: Daniel allowed himself to be guided, to a much greater extent than his opponents by phenomena. He owes his successes to this, but his blind mistrust towards the world of formulae also set limits for him. Therefore we can blame neither d'Alembert nor Euler for their criticism, and it is somewhat understandable that Daniel Bernoulli's results received so little attention at first; he himself is not completely without blame for that.

One must also say, however, that mathematics missed something here: the theory of the eigenvalues and eigenfunctions, the theory of orthogonal transformations of quadratic forms in infinite-dimensional spaces and several others were within reach, yet remained unnoticed, and had to wait over one hundred years to be discovered. No doubt that the immediate progress of physics as well as mathematics was decided by d'Alembert's partial differential equation and its solution, and especially by Euler's extension of his theory, and not by Daniel's solution by trigonometric series. And not only Daniel paid a high price for his failure to participate in this crucial development; it is probable that this completely "unBernoulli-like" aversion to the progress of mathematics was a main reason for the decline of science in Basel at the end of the eighteenth century.

But in the history of science, not only preferences but requirements change constantly, and so we can nevertheless finish this section with a positive observation: to the physicists of the twentieth century, who were often formed through the study of quantum theory, Bernoulli's ideas for eigenfunctions, eigenvalues, the superposition principle, and so on, are often much more familiar than those of d'Alembert and Euler. We cannot therefore deny Daniel Bernoulli the fame of having founded the theory of vibration.

Other fields of work

In later years, Daniel turned above all to questions of probability and statistics, as well as to questions of technology. The papers in probability are now available in volume 2 of *Die Werke von Daniel Bernoulli* edited by B. L. van der Waerden (1982). Daniel Bernoulli took up what his uncle Jacob had begun and carried it on in own way; he was above all interested in applications. Among these, the statistical proof that the smallpox vaccination promised considerable protection against an epidemic, which was rightly feared at the time, is noteworthy. He and his brother Johann II earned great fame for themselves with their educational activities in Basel and throughout Switzerland.

His research into technical problems was driven again and again by prize questions posed by the Académie des sciences. Ten times he was awarded the prize. Only Euler, who won twelve prizes, surpassed him. Some of these works, which have been edited by Patricia Radelet-de Grave and Lucien Bossy in Brussels, are:

> *Discours sur la maniere la plus parfaite de conserver sur Mer l'égalité du mouvement des Clepsidres ou Sabliers* (Discourse on the best way of conserving at sea the uniformity of motion of hourglasses, Prize for 1725) [1725];
>
> *Réflexions sur la meilleure figure à donner aux Ancres* ... (Reflections on the best shapes for anchors, Prize for 1737) [1737]);
>
> *Mémoire sur la manière de construire les Boussoles d'inclinaison...* (Memoir on the manner of constructing inclination compasses Prize for 1743) [1748a];
>
> *Nouveaux Principes de Méchanique et de Physique, tendans à expliquer la Nature & les Propriétés de l'Aiman* (New principles of mechanics and physics, intended to explain the nature and properties of magnets, together with his brother Johann II, Prize for 1744-46) [1748b];
>
> ... *La meilleure maniere de trouver l'heure en mer* ... (The best way to tell time at sea, Prize for 1745-47) [1750];
>
> *Recherche sur la maniere la plus avantageuse de suppléer à l'action du Vent sur les grands Vaisseaux,* (Research on the most advantageous way to replace the action of wind on large ships, Prize for 1754) [1769];

> *Mémoire sur la maniere de diminuer le Roulis & le Tangage d'un navire...* (Memoir on how to diminish the rolling and pitching of a ship, Prize for 1757) [1772].

One of the most interesting of these works concerns the construction of a precision device to measure the inclination of a magnetic needle. The inclination had been discovered in 1544, and it was hoped that this effect could be used to regulate the location of a ship at sea. Daniel carefully examined the mechanical conditions for such a measuring device, and with Diterich, a craftsman from Basel, he constructed some of these devices. He sent one to Euler, who, in *Lettres à une Princesse d'Allemagne*, wrote:

> One artist alone, Mr. Diterich, of Bâsle, has succeeded, having actually constructed a machine proper for the purpose, under the direction of the celebrated Mr. Daniel Bernoulli. He sent me two of the machines, by means of which I have observed, at Berlin, this inclination of 72 degrees; and however curious in other respects the English and French may be in prosecuting such inquiries, they have put no great value on Mr. Diterich's machine, although it is the only one adapted for this purpose. This instance demonstrates how the progress of science may be obstructed by prejudice; hence Berlin and Bâsle are the only two places on the globe where the magnetic inclination is known.[10]

The work on the needle of the magnet, as Patricia Radelet-de Grave has remarked [1994], is the first in which Daniel Bernoulli appraises the errors in measuring systematically with the aid of his theories of probability and statistics. He tries to indicate margins of error and a standard of the deviation of a measurement from the true value. One might say that in doing so he has "mathematized" observation, because this has turned it into what we would today call an "experiment. In this, together with his last work in the theory of probability in which he presumes the existence of a universal law of errors, we see his most personal contribution to science. The significance of this contribution can be hardly overrated.

This now brings us to electricity. In 1751 a small book of letters from Benjamin Franklin to Peter Collinson had been published in London, in which for the first time the conservation of the electrical charge was clearly stated and numerous experiments were described. Bernoulli studied the book carefully and with his students, including his nephew Johann III and Abel Socin among others, repeated the experiments. In *A History of the Theories of Aether and of Electricity*, Whittaker writes: "That electrical attraction follows

[10] *Il n'y eut qu'un artiste de Bâle, nomme Diterich, qui y a réussi, ayant construit une machine propre a ce dessein, suivant les vues du célèbre M. Daniel Bernoulli. Il m'avait envoyé deux de ces machines, par le moyen desquelles j'ai observe ici cette inclinaison de 72 degrés; et quelque curieux que soient d'ailleurs les Anglais et les Français sur ces sortes de découvertes, ils ne firent pas grand cas de la machine de M. Diterich, quoiqu'elle soit la seule propre a ce dessein. C'est un grand exemple qui nous fait voir combien les préjuges sont capables d'arrêter les progrès des sciences. Par cette raison on peut soutenir que Bâle et Berlin sont encore les seuls endroits sur la terre où l'on connaît l'inclinaison magnétique* [vol. 2, letter CLXXV (Eng. Trans. Vol. 2, letter LX)].

the law of the inverse square had been suspected by Daniel Bernoulli in 1760; cf. Socin's Experiments, *Acta Helvetica* iv. p. 214" [Whittaker 1951, vol. 1, p. 53].[vii]

That is a great discovery! One must consider that Coulomb's experimental proof of this law – published in 1785 – is still not excessively exact.

However, it must be confessed that we know all too little about Bernoulli's experiments. It would be worthwhile for an experimental physicist to reconstruct them in details. We have no biographical information about the unexplainable and conspicuous gap in his investigations: none of is works deals with optics, in spite of the fact that we know that he knew and admired Newton's *Opticks* because he cites it. Why didn't he ever work in this field? Did he want to leave optics completely to his younger brother Johann II, perhaps in order to avoid new family arguments? We don't know, at least not at present.

I mentioned earlier Daniel's refusal to learn the new mathematics of partial differential calculus. This refusal had unfortunate, perhaps even dire, consequences. Physics grew solid here after his death. But this refusal must now be considered contrary to his pioneer-like attitude as an experimentalist. Of the great scientists who mastered the mechanics of the eighteenth century, Daniel was the only one who was as interested in the art of experimentation as he was in theoretical works, and this interest led him to works in fields that were quite new. In short, he is the only one in this generation in whom we find the beginnings of a new physics, both empirically and experimentally oriented, which would come to dominate the field in the nineteenth century.

Conclusion

To conclude, I would like to make a few remarks about Daniel the man. In the diaries of the three Teleki counts [Spiess 1936], who probably came from Hungary to study with him, he comes across as a kind, accessible and affable scholar, always ready to discuss every scientific question, but also not ill-disposed to the usual social customs. He was certainly a sensitive human being. He didn't have his father's bellicosity, and he entered unwillingly into polemics. Nor did he have the dark aloofness of his uncle, with whom he otherwise had much in common; he was a livelier, more sociable, and usually good-tempered man. He enjoyed great prestige in Basel: children were instructed to take off their hats in the presence of the high authorities, parish elders, and Professor Daniel Bernoulli. He never married, because, as he said, he didn't want to "lose all his freedom in a single day".[viii] But in his old age the confirmed bachelor set up a fund to cover travel expenses for poor students.

For the most part he passed his life in and around the "Bernoullianum." At the end of Bernoullistrasse is Petersplatz where, as the Counts Teleki describe, he strolled with his brother and his students while discussing lessons. On the right is the Stachelschützenhaus (House of the Crossbowmen), today a biological institute, where he conducted his experiments. On the other side of Petersgrabens, next to Peterskirche, is the "Kleinen Engelhof", a rather narrow building with Gothic windows, where he lived as a bachelor. Next to it, in the "Grossen Engelhof", lived his younger brother Johann II with his family. This is also where poor Maupertuis died, who Johann brought with him on his return trip from Berlin. Just a little further, on the left side of Nadelberg, is the house named "Zur alten Treu", which belonged to his father Johann I, where Daniel spent his youth. Finally, in the Peterskirche, of which the state had nominated him as Canon, Daniel is buried alongside three other famous mathematicians of this unique family, Johann I, Nicolaus I, and Johann II.

Acknowledgments

I would like to thank Mrs. I. Zschokke-Gränacher and Mr. P. Dinichert for the invitation to speak at the convention of the Schweizerischen Physikalischen Gesellschaft. Mrs. P. Radelet-de Grave assisted me in producing bibliographical material and with the corrections. Mme. N. Van Olmen, Fr. H. Speiser and my wife produced the manuscript and took care of the corrections; I thank them all very cordially

Translated from the German by Kim Williams

Leonhard Euler

Leonhard Euler: Mathematician, Physicist, Engineer[1]

In Basel Leonhard Euler, born there on 15 April 1707, first learned the rudiments of mathematics from his father, the priest Paul Euler, who had been a student of Jacob Bernoulli. Later, as a student of Johann Bernoulli, he acquired all the knowledge of mathematics and physics available at the time, before leaving the city never to return again. The cities of his where he wrote, taught, and worked were St. Petersburg and Berlin, far north of Basel.

The external course of events of Euler's life was relatively unremarkable and simple. At the age of twenty, the student of Johann Bernoulli, who had recommended him to the Imperial Academy founded by Peter the Great in St. Petersburg, left Basel. His friend Daniel, Johann Bernoulli's son, and his compatriot Jacob Hermann, a student of Jacob Bernoulli's, were already there (Johann's eldest son Nicolaus II had died a short time before). It was at this time that Euler's cooperation with Daniel Bernoulli, which would continue for many years by letter, began. Euler worked as mathematician, physicist and astronomer at the Petersburg Academy. He also, as we recently learned thanks to Russian historians [Nevskaya 1983a, 1983b], conducted experiments and executed twice-daily astronomic observations. As an astronomer, he was also responsible for the preparation of Russian maps, and was involved in two expeditions into Russia's interior.

The political circumstances that affected his life in St. Petersburg were less favourable and became increasingly difficult, so that in 1740 he gladly accepted Frederick the Great's call to the Royal Academy in Berlin, where he worked for the next twenty-seven years. Then, it appears that mainly because of differences in opinion with the King, he answered a call from Catherine the Great to move back to St. Petersburg. This last phase of his life, as far as outward recognition was concerned, which in a singular manner certainly did not depend on him, was to have been his happiest. After sixteen years of work in St. Petersburg, on 18 September 1783, after a full day's work, he had his dinner and discussed the motion of Uranus, the planet discovered two years previously by Herschel, he was relaxing when a stroke carried him away.

---

[1] Address held at the academic celebration of the bicentennial of Euler's death the Eidgenossische Technische Hochschule, Zurich, 19 November 1984. Originally published as "L. Euler (1707-1783): Mathematiker-Physiker-Ingenieur" in *Vierteljahresschrift der Naturforschenden Gesellschaft* 129 (March 1984): 325-338 (in German).

To speak of Euler's life means speaking of his work. A brief list of his most important books best expresses this, and gives an idea of the scope of his work.

From the first Petersburg phase are the two-volume *Mechanica* [1736], which after Newton's *Principia* and Jacob Hermann's *Phoronomia* was the first systematic representation of mechanics, the first easily comprehensible representational exposition of this area in the new analytic language, and which would become the textbook for dynamics for a long time to come; the treatise on the new theory of music, *Tentamen novae theoriae musicae* [1739], which in recent years has become once again the focus of attention; his two-volume *Rechenkunst* [1738-40], an elementary school book that demonstrates his genius for teaching, about which we will speak below.

Among his publications during his years in Berlin are the *Methodus inveniendi lineas curvas* [1744a], his textbook of so-called calculus of variations; *Theoria motuum planetarum et cometarum* [1744b], his celestial mechanics; the annotated translation from English of Robins' *Gunnery* [1745], which contains one of his most important discoveries; the two-volume *Introductio in analysin infinitorum* [1748], the first of three great textbooks on analysis; the two-volume *Scientia navalis* [1749], a theory of shipbuilding and navigation; *Theoria motus lunae* [1753], his first moon theory, written not only to address purely astronomical problems but also the application of the determination of longitude on the ocean; the four Latin works on hydrodynamics, which were planned as a book, but appeared only as individual treatises in the *Novi Commentarii Academiae Scientarum Petropolitanae* [1761; 1770a; 1771b; 1772]; the two-volume *Institutiones calculi differentialis* [1755], the second mathematics textbook which, like the *Introductio*, was filled with new discoveries as well as parts of entirely new fields; the *Theoria motus corporum solidorum seu rigidorum* [1765], the general exposition of his theory of rigid bodies, that he conceived almost entirely on his own, inspired by the work of Jacob Bernoulli.

While in Berlin he also wrote the three-volume *Institutionum calculi integralis* [1768-70], the third analysis textbook, printed when he was in St. Petersburg; the *Lettres à une Princesse d'Allemagne* [1768-72], his most-read work, translated into eight languages, published in over one hundred editions, and still available in print today; and the *Anleitung zur Naturlehre*, printed posthumously in 1862, which, as Hermann Weyl said, wonderfully summarizes the whole scope of scientific knowledge at the time it was written.

From the second Petersburg period, finally, we have the three-volume *Dioptrica* [1769; 1770b; 1771a], a technical handbook of the optics of lenses ; the extraordinary, often praised *Vollständige Anleitung zur Algebra* [1770a]; *Theoria motuum lunae* [1772], his second moon theory; *Theorie complette de la construction et de la manoeuvre des vaisseaux* [1773], his last great work, the title of which is remarkably similar to that of a work by his teacher, Johann Bernoulli.

One can see that Euler deals with mathematics, physics, and technology, and even art and philosophy are marginally represented. The first three of these are closely related, as I will show.

The works listed above are merely representative of all the material that Euler produced, and for the most part they are filled with original discoveries. Grouped around these major works are an immense number of smaller articles. The catalogue compiled by the Swede Gustav Eneström comprises some 850 numbers.

Whether the major books or the shorter works had the greatest effect on and aroused the greatest admiration – so unanimously handed down to us – of his successors, is still for history to decide. We find mathematics, physics and technology side-by-side, each penetrating and stimulating the others to a degree that it is almost impossible to find anywhere else. Showing how this mutual stimulation is present in the work of just one man will be perhaps the most important and interesting task of future Euler scholars. The great complete edition, which is approaching completion, has created the crucial prerequisite, but this task still lies almost completely ahead of us. Even the most knowledgeable and well-versed Euler scholar has up to now penetrated into only part of this far-reaching, enormous project. No researcher so far has successfully attempted an overview of Euler's entire oeuvre.

I would like to mention some particularly remarkable characteristics of Euler's work now. In his article in Basel's Euler *Festband* on number theory, André Weil says:

> Perhaps its most salient feature is the extraordinary promptness with which he always reacted even to casual suggestions or stimuli: a question from Goldbach …; Fermat's statement which at first he did not even read through…; a problem … by Naudé; a theorem by Lagrange; a few … theorems by [Fagnano]. … Every occasion was promptly grasped; each one supplied grist to his mill, often giving rise to a long series of impressive investigations [Weil 1984: 283-4].

What Weil says here with reference to one single field is true as well with respect to all others. The favorite student of Johann Bernoulli, the former protégé of Huygens and the correspondent and defender of Leibniz, who was the opponent of everyone and everything English, absorbed all that he could without discriminating from Fermat, Descartes, Huygens, Newton, Jacob Bernoulli, Leibniz, the remaining Bernoullis and many other, less well-known mathematicians and scientists. One might say that the actual achievements of his youth demonstrate the best characteristics of all his predecessors: with Fermat's originality, Descartes's algebraic vision, Huygens's clarity, and Newton's acute understanding of the bases and his power to synthesise, he welded together the mathematical languages of Leibniz and the Bernoullis into a new unity, formulated mechanics in new language that is still valid today, and extended physics into new fields.

Later, he turns the works of Bradley, Daniel Bernoulli, Clairaut, d'Alembert, Segner, and Lagrange, among others, into the starting points for new research. As he does so, he always explicitly mentions others, expressly underlining their merits. Hardly ever can there be felt a trace of animosity, that animosity that makes so many other researchers blind and dooms them to sterility.

We find a similar case in the field of technology, where it was often external events, such as imperial or royal desires and commands, fulfilled with the greatest willingness, which led him to a discovery. For example, the translation of Robins' *Gunnery* led him to the discovery of the internal pressure, which, according to Todhunter [1873: § 301], at once rendered clear and intelligible formulae and demonstrations in fluid mechanics.

The activity of problems of hydraulics and an invention of Segner's led him to the invention of the first modern turbine. However, at the time it remained in the project phase.

However, almost two hundred years later, during World War II, Jakob Ackeret built it and demonstrated that its efficiency amounted to over 70 percent [1944], which, I have been led to understand, had been achieved only about the turn of the nineteenth century. This readiness and willingness that usually characterises youth, lasted in Euler literally until the last day of his life. When news arrived from Paris of Montgolfier's ascent, Euler immediately sat down and calculated this. The not uninteresting results that he obtained on that occasion became the object of his last work [1787].

Another thread that the reader senses immediately is that Euler is an enthusiastic explainer. Among others, we owe to this enthusiasm the large number of textbooks which, as Gauss, Jacobi, and others have noted, formed for a very long time the absolutely essential basis for all students. Euler's enthusiasm finds expression in his textbooks and, similar to Kepler, he candidly allows us to participate in his search as well as in his errors. But, what is the secret of his great art of exposition? One can merely try to find the answers.

First, once certain of the theorem on the one hand, and interesting examples and tasks on the other, the general theory and the concrete calculations are always balanced.

Next, Euler is never afraid to begin by explaining the simplest, indeed even the first, most trivial case that the beginner has to have complete command of. He then proceeds with a series of new difficulties one after the other, always explaining each with an example; each one leads to a generalization of the initial claim, until the general theory and the principles on which it is based, finally appear to us in all their beauty.

Further, Euler brings a deep sense of structure to each theory and its symmetries. Regarding the mechanics, we can always at least partially follow as Euler as he penetrates into its exact mathematical and physical structure. The dispute between Newton and Leibniz, as well as their successors, as to the nature of space gave Euler no peace. One fruit of this reflection was that he was able to formulate more exactly and with greater acuity the so-called Galilean principle of relativity than anyone had done before, or would do for a long time after him. Another fruit was that in Euler we already find clearly modelled in coordinate form the fields which today we call tensor algebra and tensor analysis. This was in part merely intuitive, but was also already in part a conceptual understanding of symmetry groups of space and time and their meaning for the formulation of the laws of nature, the framework within which all his other works on physics are naturally arranged. This has probably contributed more than anything else to making his writings on physics still readable today. A comparison of the *Theoria motus corporum solidorum* with the *Principia*, equally wonderful though in different way, makes this clear.

Still another aspect is Euler's innate sense of a systematic – that is, a symmetrical – notation that makes it so easy for the reader to find his way through his work. Eulerian pedagogy is therefore always first based on a keen understanding of the deeper structure of the material. It is no miracle then that his works have had such a lasting effect, as Laplace, Cauchy and others have attested. Euler wrote the first textbooks and technical books that are still readable by a modern reader.

I would like to comment on the juxtaposition of the mathematician, the physicist and the engineer in the personality of a single researcher. In Euler we find these aspects closely related from the onset of his work through the very end of his life.

In the period of his youth, the works in physics predominated. The physicist and historian Clifford Truesdell, to whom we owe the essentially new discovery of Euler the physicist, writing about Euler's physical dissertation on sound (*Dissertatio physica de sono* [1727]), commented that rarely has a twenty-year old physicist created such a masterpiece, and compared its significance with Newton's seventh chapter of the second book of the *Principia*, his hydrodynamics, which was the program for decades to come. Euler penetrated more and more deeply into Newton's *Principia*; he wrote numerous mechanical works that finally led him to his own systematic representation of point mechanics, the first modern representation, written in the mathematical language that we still use to formulate it today.

Also during his youth, we find his first discoveries in the fields of analysis and algebra. As early as 1727, Euler submitted a technical work on the masting of ships to the Académie royale des sciences in Paris [1728].

Earlier I mentioned his work on Russian maps. Later, in Berlin, he assumed the management and supervision of the preparation of maps for the German schools and post office. These tasks, as he himself admitted, sorely strained his eyes.

Out of these activities arose his discoveries concerning stereometric representation, and in particular the discovery of the so-called conformal representation.

How very closely interrelated all three aspects were during his middle and later years of work is clearly seen in the chronological sequence of his major works listed above: he alternates between works in mechanics, works in pure mathematics, and works in technology. However, a reading of these books makes this even clearer. For example, the *Scientia Navalis* contains passages that are crucial for his theory on rigid bodies. Results that are significant for his hydrodynamics are located in the comments about Robins' *Gunnery*. In contrast, his turbine is based on his basic equations for hydrodynamics, which are still called "Eulerian" today. Finally, both directions of research contributed to the formation of analysis. The theory of differential equations is an unwieldy field and that of non-linear partial differential equations is particularly thorny. Problems that appear in the textbooks are sometimes only mentioned because as an exception, seemingly by chance, a solution exists. This branch of knowledge would hardly have developed if certain partial differential equations had not been known, just as the Eulerian equations of hydrodynamics had such fundamental significance for our understanding of natural occurrences. But their role was also to advance analysis. A similar case is found in Euler's examinations of acoustics.

In his introduction to vol. II.15 of the *Opera omnia*, Jakob Ackeret described the mutual stimulation of the individual branches of research, writing:

> Looking at the treatises contained in this volume, one must reach the conclusion that the most important bases of present-day principles of hydraulic engineering were created here. It is likely that before Euler's time there was already a technology that had achieved empirically astonishing results despite the lack of theory, but the wonderful development of the

> turbine (as well as steam and gas turbines), which began much later, was only possible under the guidance of Euler's theory [Ackeret 1957: VIII].

Ackeret also says:

> What, however, underlies these new theories? It is the Eulerian hydrodynamic field equations, their formulation, as perhaps a later analysis may still show, not without stimulation from the technical problems, that we see being treated here [Ackeret 1957: IX].

Of Euler's discovery of negative pressure, Ackeret writes:

> One has to consider Euler's anticipation of such phenomena on the sole basis of theory as an extraordinary achievement, all the more so since he even stated the constructive measures that one might encounter in this phenomenon [Ackeret 1957: LI].

He adds that it would be one hundred and fifty years before science noticed Euler's negative pressures!

In addition to the *Introductio* [1748], the *Institutiones* [1755] also dates from the middle years. Euler also attempted a textbook on number theory published posthumously, the *Tractatus de numerorum doctrina* [1849b], which Rudolf Fueter dated and discussed in detail [1941: XIX]. These years saw the great discoveries of the analytic functions, formal series, the *partitio numerorum*, the discovery of the law of reciprocity, the research on prime numbers and the discovery of their asymptotic distribution, the Zeta function, its functional equation, and thus, as Weil said, the birth of analytic number theory [1983: 127]. On the other hand, it is precisely his work in the area of so-called pure mathematics that benefited from his work in the other two areas. One can probably claim that the dazzling clarity, the distinguishing characteristic of all of Euler's writings, is one of the fruits of his never-ending struggle to grasp the laws of mathematics more deeply and to formulate them more clearly.

The last decades of Euler's life were immensely fruitful. Over one third of his works, almost three hundred of them, originated in the last years of his life, and their publication was carried out for more than four decades after his death. Euler had produced so many works that for twenty years after his death the academy was able to fill its volumes with them. But even forty years after his death fourteen finished, unpublished papers still "awaited" publication, as the German historian Biermann put it [1983: 489]. Because of this, as Truesdell said, Euler, who began as a student and colleague of Johann Bernoulli, became an intellectual contemporary of Gauss, Cauchy and Fresnel.

Anyone who writes about Euler's work runs the risk of ending up like poor Marco Polo, whose descriptions of China upon his return from there only earned him the scorn of the Venetians, who mockingly called him "Marco Millioni", Marco of the million stories. As usual, in his last years Euler had completely mastered his means and shifted easily from one of the three activities to the other.

It is not only the quantity of what Euler accomplished that so impresses us here. As André Weil said,

> Hardly less striking is the fact that Euler never abandoned a problem after it had once aroused his insatiable curiosity. ... Once he had taken up a question, not only did he come back to it again and again ... but also he loved to cast his net wider and wider with never failing enthusiasm... Even when a problem seemed to have been solved to his own satisfaction ... he never rested in his search for better proofs, "more natural" ..., "easy" ..., "direct" ... [Weil 1984: 284].

In other words, even when he had achieved the greatest results Euler never said, "Stop were you are! You've arrived!"

For example, it was in his last years that he came to his deepest comprehension of the law of reciprocity, the discovery of which dated from some fifty years before.

What Weil said with reference to the works on arithmetic is just as true for all other areas. This can be seen, to cite one example, in the *Anleitung zur Algebra*, which is the result of Euler's supreme vision of all that was essential to the *Schatzkästlein,* treasure chest, as Andreas Speiser, the long time general editor of the *Opera omnia*, once called it. His last period of work would open with the three-volume *Institutionum calculi integralis*, which was in large part inspired by mechanics, but which on the other hand rejected much of mechanics and technology. And from his later years also comes the complete exposition of the theory of rigid bodies, which contains his deepest investigations about the physical properties of space and time, as well as the final explanation of the so-called Galilean principle of relativity, which I mentioned earlier, and the use of its consequences for the clear representation of the theory.

There were a few other great scientists, although *not* many, in whom one also finds the mathematician, the physicist and the engineer; Newton, Kepler, and Huygens come to mind, and to a lesser extent Poncelet, Maxwell and Poincaré. However, where all of these carried on with interests or activities in contemporary, it is only in Euler that we see a true interpenetration of these activities.

If we take an overview of these activities, we can perhaps hazard the following conclusions. The discovery of the ground rules for the sequence of movements, and thus for mechanics, which Newton had achieved for simple systems of mass points, was achieved by Euler, based on the work of Newton and Jacob Bernoulli, for continua: liquids, gas and rigid bodies were carried forward. Somewhat later Cauchy did the same thing for elastic bodies, and so set forth the definitive basis for the general formulation of classical physics, which is still valid today. In recent decades, Truesdell was the first to show the historic significance of this development.

Euler's work owes its determination and purposefulness to this objective, great in its very nature. In addition, his ideas on optics, electricity and magnetism, based on those of d'Alembert and the concept of field that he developed, would influence development of these areas, and with it, modern thinking in physics. Here it must suffice to mention Faraday's writings on the *Lettres a une Princesse d'Allemagne* and the comments in his diaries.[2]

---

[2] See "Euler and Physical Optics" in this volume, pp. 197-209.

But Euler was not satisfied with the search for basic principles, which, as we have seen many a time, can lock the researcher in a prison of sterility. It is probably the engineer's inventiveness, some of the fruits of which are, for example, the turbine, the gear, the lens systems, that lent Euler his eternal youthfulness.

But most of all, without a doubt mathematics was the engine that drove him. Mathematics was the main source of his inexhaustible imagination, which was literally brought to a halt only by his death.

The philosophers of the generation after Euler spoke of "productive imagination" as the source of our progressive understanding. A scan of the lecture schedules of universities and technical colleges makes it clear that our understanding of nature (and today, for that matter, of the economy as well) involves mathematics in an essential way. This is not least thanks to Euler!

With the *Opera omnia* now available, we are in a position to study his works. Many of his writings have been dealt with in excellent technical and historic essays. In the *Opera omnia*, such essays dealing with his scientific works serve as *introductions,* and this is a great contribution, especially for the history of science. But the most important and probably most interesting historical question – namely, how have Euler's results in any given individual field, for instance, in physics or in mathematics, influenced his results in others – still remains to be answered. Today, the *Opera omnia* allows us to dare to tackle it. Once this question has been answered, even if only partially, then we will have learned about, not only about the production of the human spirit in the area of science and technology, but about all the decisions of an epoch.

I will permit myself, in conclusion, to speak about Euler the human being; not so very much about the public man, whose traits have often been portrayed: his happy family life, his accessibility as a teacher, his affability in company; his quickness to become heated in discussions and his equal quickness to cool down afterwards and bear no grudges. His generosity and serenity regarding priority questions in the middle of the belligerent eighteenth century was proverbial. Ralph Raimi once remarked, "There is ample precedent for naming laws and theorems for persons other than their discoverers, else half of analysis would be named for Euler" [1976: 522]. Along with Kepler, Euler belongs among those of whom science can be proud as a human being as well as a scientist.

It is less well known that besides his research he was an able administrator. First in St. Petersburg and then above all in Berlin, where he was de facto president of the Académie, he managed a large share of administrative work efficiently. In his last years, presumably because he wanted to get them over with quickly, he took care of such matters in an authoritarian manner. It is also little known that he often worked in volunteer capacities in his private life as well, for instance, as an extremely active member of the French-reformed community (it is well known that Euler was an ardent defender of Christianity).

The essential question we should now ask is: What does Euler mean to us as a human being today?

Every scientist requires constant intellectual, moral, and material support if he is not to lose the courage necessary to invest time and energy in pursuit of a goal that he can envision only tentatively. "Young people will once again have the courage to think of the new", said Andreas Speiser as he welcomed state funding for the *Opera omnia*, and even if no scientist will ever say that there is sufficient funding, I think that everyone will at least admit that the state didn't treat him badly. Above all else, the scientist must always start from the decisive impulse; what researcher doesn't reproach himself, out of inhibition, or false respect for authority, or even out of fear, for having contravened against the words of Goethe:

> Don't let yourself be robbed of an original idea!
> What the masses believe, is easily believed.[i]

There is hardly a more impressive example of unshakable trust in the value and the success of one's own research than Euler's work in number theory. Before Euler, only a single significant number theoretician, Pierre de Fermat, had worked in this area a hundred years earlier. After Fermat, almost no one felt that such questions were worth the effort. Jacob Bernoulli was one, but the scope of his work constitutes only a small exception.

In 1730 the diplomat and amateur mathematician Christian Goldbach submitted such problems to Euler; if for no other reason, his name should be remembered for that. Euler immediately pounced on these questions and created completely new branches of mathematics from them.

I have mentioned some of his works. But Euler worked virtually alone for almost forty years, without any collaborator who was on his same level; with the exception of Goldbach, there was not another single mathematician seriously interested in number theory. In all likelihood, over and over again he met with a lack of understanding and mockery on the part of his colleagues as well his employer, Frederick II of Prussia. Only in 1769 came the first creative echo, from the mathematician from Turin, Lagrange, who was quite young at the time.

Johannes Oeschger, the sensitive and discreet editor of the beautiful volume *Briefe von und nach Basel aus fünf Jahrhunderten*, who was himself, as we say, "unmathematical", wrote aptly of Euler's handwriting: "With Euler's letters, before it begins to read at all, the eye is aware of the uncanny sternness and consequence with which line follows line ... one thinks to himself that behind the writer of such things is an uncannily secure man" [Oeschger 1960: Einleitung].

This "uncanny security" allowed him to recognize the value of "a matter" and to go far on his solitary ways. None of his contemporaries had a presentiment of this "matter"; later, however, countless others saw it thanks to him, and made it the object of their own work, from which they derived satisfaction and fulfilment.

This security alone justifies counting Euler among the greatest. Spontaneously we think of the unwavering vision of a Columbus, as well as of the young Einstein, who at the beginning of the twentieth century put a new face on physics and conquered new areas for mathematics.

May Euler's courage be ever contagious!

Translated from the German by Kim Williams

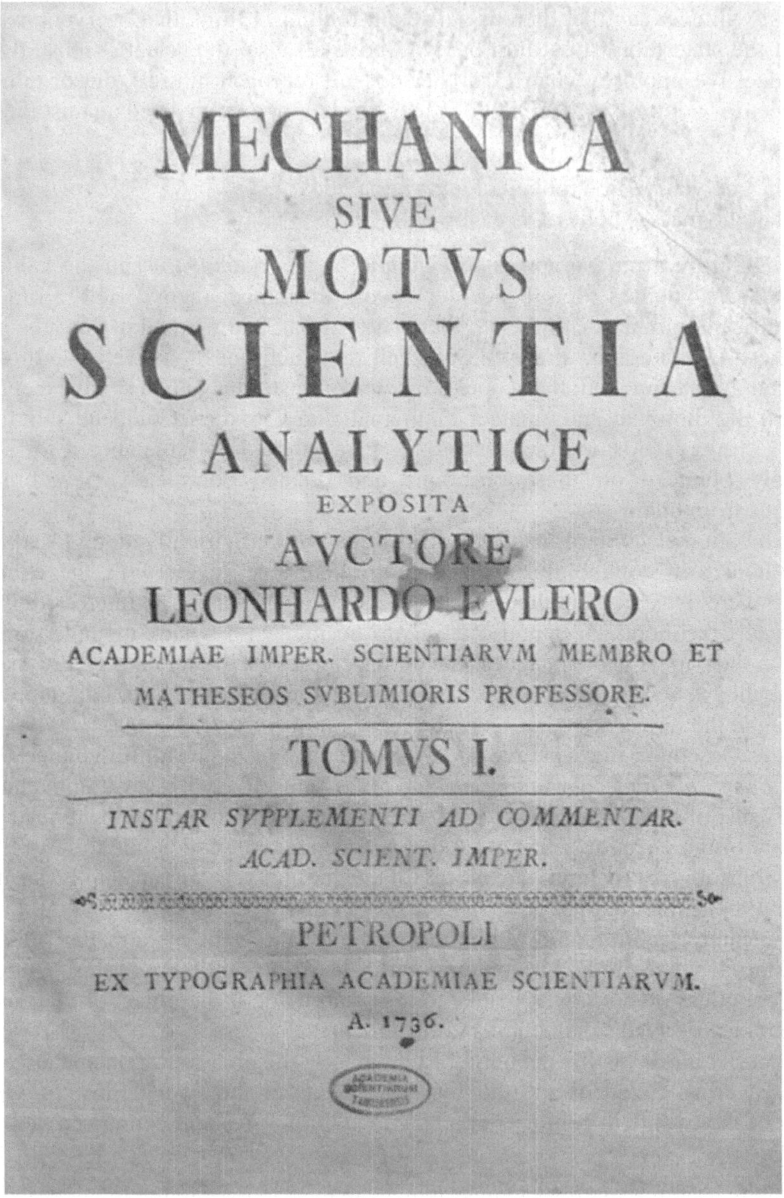

MECHANICA
SIVE
MOTVS
SCIENTIA
ANALYTICE
EXPOSITA
AVCTORE
LEONHARDO EVLERO
ACADEMIAE IMPER. SCIENTIARVM MEMBRO ET
MATHESEOS SVBLIMIORIS PROFESSORE.

TOMVS I.

INSTAR SVPPLEMENTI AD COMMENTAR.
ACAD. SCIENT. IMPER.

PETROPOLI
EX TYPOGRAPHIA ACADEMIAE SCIENTIARVM.
A. 1736.

Frontispiece of the first edition of the *Mechanica*

The Principle of Relativity in Euler's Work[1]

Introduction

It was a great pleasure for me to talk at the University of Barcelona about the principle of relativity. It was in Barcelona, exactly 60 years before I gave this lecture, that Hermann Weyl delivered his lectures on the "space problem"; these lectures had grown out of his work on the general theory of relativity [1923]. Thus, it was not out of place to talk about a chapter of the history of the principle of relativity in a year in which we also commemorated the two-hundredth anniversary of Euler's death; even more so, since this chapter – certainly studied at one time – has been long forgotten.

The principle of relativity: a work of many hands

If we are to talk about the genesis and development of a great idea in physics and the historical discovery of laws of nature, it seems best to state these ideas and laws first in the form in which modern science expresses them today. The great changes that physics underwent in this century have taught us that history of science never expresses more (but often less!) than a relationship between a former period and the present time. For instance, for a long period of time people looked at the principle of relativity quite differently than it is looked at today and they treated it, if they did so at all, with neglect. Perhaps in a short while, it will be looked at differently again, but at the moment we cannot know how.

For the physicist, the principle of relativity says that two observers are at rest or in a state of uniform linear motion with respect to each other, then they describe a process by the same equations. Stated in mathematical language, the principle of relativity of classical physics says that space and time are a homogeneous manifold on which a linear group acts. Up until the early twentieth century, everybody believed this group (today called the "Galileo group", after the man who took one decisive step in this direction) to be a rather complicated one.

The principle of relativity is fundamental. From it follows that the laws of nature are relations between vectors and tensors that transform according to the laws of this group; and in the whole domain of physics, from the simplest laws of motion to the most refined selection rules or representations of transition amplitudes in terms of so-called "form

---

[1] Originally published in *Symmetries in physics (1600-1980) : Proceedings of the 1st International Meeting on the history of scientific ideas held at Sant Feliu de Guíxols, Catalonia, Spain September 20-26 1983*, edited by M. G. Doncel, A. Hermann and L. Miguel, Barcelona: Universitat Autònoma de Barcelona, 1987, pp. 33-47.

factors", we make powerful use of this basic law. It is certainly the most important symmetry principle that we know of.

But by and large this was not understood by the physicists before the twentieth century, because the Galileo group, as I said, is a complicated mathematical structure. Curiously enough, only after its replacement by the Lorentz and the Poincaré group – that is, precisely when the Galileo group ceased to be considered a fundamental entity – did physicists once again take up and investigate the problem of relativity systematically. Felix Klein, to whom Dr. Radicati drew my attention, says:

> The different role played by the variable *t* in the Galileo-Newton group has had a decisively inhibiting effect on the historical development of mechanics. Although Lagrange had already called mechanics a 4-dimensional geometry, one has only recently made real use of this interpretation. All the older authors always considered only the Euclidean group and did not connect to it neither time translation nor the proper Galileo transformations although they knew them. This also happened to me when I wrote the *Erlanger Programm*. I remember distinctly that I had not overlooked Lagrange's remark, but I believed that I could not find a place for it in my group principle. Only the appearance of the Lorentz group has lead the mathematicians to a correct appreciation of the Galileo-Newton group [Klein 1926-27: chap. 2, part A, § 2].

Of course the principle of relativity itself was present in physics long before it was adequately formulated. So it will be a surprise to no one, when I say, that it is not the discovery of one single man. This most fundamental idea cannot possibly be said to have caused or to be the effect of a revolution, if by this we mean a change in a comparatively short time. It is, to paraphrase a well-known title: "a work of many hands". Here I can mention only a few of the contributions; the following list is by no means complete.

Some of the contributors

The beginnings are told and discussed by A. Koyré in his *Etudes galiliénnes* [1939]. As I am not a specialist concerning these questions I cannot say how much and what exactly is due to Galileo and what exactly and how much to Descartes. At any rate, Huygens states the principle of relativity clearly in his *Tractatus de motu corporum ex percussione* [1704: Huygens 1704, appendix ii: 141-142]; for the details, see [Szabó 1979: 449].[i] What is more, for Huygens it was not only a property of nature, one among many other ones, though a surprising one. In his hands it became a tool for solving problems and for finding new laws: he used it for proving his famous laws of collision. However important and impressive Huygens's various accomplishments were, and indeed I do think they are usually underestimated, these accomplishments do not form a comprehensive scheme of mechanics.

Shortly afterwards, Newton found a coherent scheme for point mechanics that could not only be applied to some simple ordinary everyday mechanical processes, but also and with overwhelming success to astronomy. His scheme was based on only a few axioms. Thus the principle of relativity again became a matter not only of investigation, but also of intense dispute. This dispute remains associated mainly with the names of Newton and Leibniz. I cannot go into their quarrel here; a few remarks must suffice. For Leibniz space and time are nothing but the relations between the objects which they contain; hence all geometrical and kinematical notions are relative. Newton's position is more complicated and, you might say, contradictory. There is of course a relativity principle implicit in Newtonian mechanics, yet Newton himself held firmly to an "absolute space" and an "absolute time". He went even so far as to state that "the center of mass of the solar system stays at rest in the center of the universe".[2] For this reason Hermann Weyl flatly denied Newton the principle of relativity [1949: 2]. Indeed, to state "it rests" contradicts Galilean relativity; to say "in the center of the universe" is meaningless because of Euclidean relativity. Whatever we think about Newton's mechanics, the question remained unresolved, and together with other related questions it remained a matter for dispute between physicists as well as between philosophers for the whole of the eighteenth century.[3]

It is at this moment, at the beginning of the second quarter of the eighteenth century that Euler enters the scene. He was the disciple of Johann Bernoulli who had been the *protégé* of Huygens and the correspondent and friend of Leibniz, whose principal defendant he later became in the famous – or better, infamous – priority dispute with Newton, concerning the discovery of calculus. Moreover, Johann Bernoulli's accomplishment in physics lay mostly in the tradition of Huygens. All this should have lead the young Euler into the tradition of Huygens and Leibniz. Yet, already as a student, Euler had penetrated deeply into the *Principia* and he had at once become its ardent advocate. After his friend, Daniel Bernoulli, the son of Johann, who was a few years his elder, Euler may be called the second Newtonian on the Continent. But while Daniel Bernoulli's interest in Newton was focused mainly on the theory of universal gravitation and its applications, Euler was in interested in all branches of Newtonian mechanics as an all-embracing structure. Thus we shall not be surprised to find that the problem of the relativity of motion intrigued him throughout his whole life. Again and again he takes it up, trying to understand it better and to make the issue even more transparent. Besides, science and philosophy are linked in a peculiar way through this problem. In what follows I shall discuss several works in which Euler deals with this problem. Therefore it is unavoidable that I repeat myself frequently, and I apologize for this in advance.

Euler's project

In 1736, Euler had published his great work on point mechanics, *Mechanica, sive motus scientia analytice exposita.* At the end of the first section, he announces that this book, which covers the motions of mass points, is only the first of a series of six. The second will be devoted to rigid bodies, the third to flexible bodies, the fourth to gases, the

---

[2] In the first edition we find: *Hypoth. IV. Centrum Systematis Mundani quiescere* [Newton 1687: 402]

[3] For much of what I say in this conference, I am indebted to the basic articles by Truesdell [1954, 1956 and 1960a]. It was Truesdell who discovered again Euler's true position in the history of physics.

fifth to bodies linked mutually by constraints, and the sixth, finally, to liquids![4] This is a proud announcement for a young man twenty-seven years old, especially when we recall that the laws of most subjects which he intended to treat – and Euler knew this – had not yet been found! Yet Euler was able to execute two-thirds of his program, and it was in pursuing this goal that his ideas about principle of relativity emerged. But I must add at once that we find statements that refer to this question throughout his work; since this comprises more than seventy volumes, not all of them yet published, it is impossible for me to present to you a complete survey of all relevant sources. Moreover, to the passages contained in his works on physics, we must also add some in his mathematical papers, for instance some on geometry or even on number theory, which refer to problems of group theory, as we would say today. I underline this because, like all historians of science, the historian of physics always runs the risk of remaining enclosed in his own branch, neglecting and overlooking ideas that flow into it from other branches, even though these influxes are often as important and decisive as the ideas that were formed in his own field. Moreover, continuum mechanics are more and more forgotten by physicists, yet for the development of the whole of physics their problems were just as, or even more important than, point mechanics.

Relativity in Euler's point mechanics and in optics

In the *Mechanica* we find Euler's first systematic investigation of the problems of relativity. While it may be said that somehow everything is already there in a nutshell, it must also be said that the accent is always on aspects of kinematics rather than on aspects of dynamics of the question, which seems not to be really discussed. However this may be, these early investigations do not yet show the penetration of the later ones and therefore I will pass over them here.

In 1739 Euler stumbled over the principle of relativity in a most unexpected context. He was dealing with a concrete problem, namely the aberration of the light of stars, discovered by Bradley in 1725 [1728],[5] and he was eager to compute this effect from first principles. Most people at that time accepted Newton's corpuscular optics, which held sway in the scientific world for over a century. Euler, it seems, was almost the only one convinced that light was a wave phenomenon and, a few years later, he was to present his views systematically.

In his paper "Explicatio phaenomenorum quae a motu lucis successivo oriuntur" ("Explanation of the phenomena that arise from the motion of light") [1750a], he wants to derive the aberration of light from the kinematics of wave optics. The aberration is a systematic error in observing the stars, of the magnitude of 20″, which occurs through the earth's movement in relation to the stars. But it is characteristic of Euler that he is not only interested in the special problem to which this examination gives rise, namely the

[4] [Euler 1736, I: § 98]. See also [Truesdell 1954: IX].

[5] Cf. [Grant 1852: 338].

calculation of the required correction, but rather he attacks the question in its fullest generality as a physical problem. I have discussed this in detail elsewhere.[6]

Here we can see the instant where – probably for the first time – a completely new point of view emerged from the confrontation of mechanics with optics. To be precise, Euler had discovered that the decision between a wave and a corpuscular theory of light reflects on the validity of the principle of relativity! From the following remark, it is also clear that Euler was well aware of the significance of what we call today "inertial frames" as well as of their peculiar problems: "According to the safe laws and rules of mechanics, it is therefore allowed to impress on an entire system of bodies a uniform straight motion, for nothing in the phenomena will be changed, but to give it a non-uniform or a curvilinear motion is strictly forbidden".[7]

I must say that I doubt that Euler's paper produced any echo whatever. Now we must go next to the branch of mechanics with which Euler's investigation on absolute and relative motion is mainly intertwined. This is the theory of rigid bodies. I may mention that this theory, apart from a contribution from d'Alembert and one from Segner, is entirely due to Euler.

An application of the principle of relativity

The results of Euler's investigations are contained in a long series of papers where he attempted to come to grips with the problem of three-dimensional rotation. The sheer complication of the kinematics alone seemed to defy any decisive progress. Of his many papers I will quote here only the one which he called "Découverte d'un nouveau principe de mecanique" [1752b], since it was crucial for his own understanding. The "new principle" is the claim, that all equations of mechanics can be derived from the equation:

$$\text{mass} \times \text{acceleration} = \text{external force} \tag{1}$$

provided they are written in a "space fixed" Cartesian coordinate and in an adequate inertial frame.

Of course, what Euler is driving at here is again the distinction of the Cartesian coordinates from all other coordinate systems. The essence of Newton's great discovery of the geometrical composition of forces is that forces are vectors.[ii] This is Newton's most important contribution to symmetries. Cartesian coordinates are the closest surrogate for an abstract vector calculus, and indeed henceforth Euler will almost always write his formulae in space-fixed Cartesian coordinates; translating his major writings into the abstract vector language would be a routine exercise. But behind this also lies the realization, never before stated so clearly, that for formulating mechanical equations we need an inertial frame. Only with respect to an inertial frame are the laws of mechanics valid in the simple form in

---

6 See"Euler, the Principles of Relativity and the Fundamentals of Classical Mechanics" in this volume, pp. 179-183.

7 *Licet enim salvis legibus et regulis mechanicis universo cuidam systemati corporum, motum aequabilem in directum imprimere, ita, ut nil in phaenomenis mutetur, at vel motum inaequabilem vel curvilineum tribuere minime licet* [1750a: § 35].

which we quote them. But then, there are an infinity of these frames and this freedom is now used efficiently by Euler.

The great variety of possible motions of even a free rigid body, their intricacy and their complication had defied everyone before. Euler now uses the fact that the law (1) is valid, also in any other frame which is in uniform straight motion relatively to – or, as he says "with respect to" – the first one. Take a freely moving rigid body, its motion as complicated as one wishes: the center of mass moves always uniformly on a straight line. Therefore, it is permissible to attach the origin of the system of the space fixed coordinates at the bodies' center of mass: we are still then in an inertial frame, for the system remains "space fixed".[iii] Thus the motion of the rigid body has been reduced to a purely rotational one, and this restricted problem he is now able to solve. In this way, Euler found the equations of the spinning top.

Here I will not describe them, for they lie outside the scope of this paper; but it is worthwhile to stress the fact that again a concrete mechanical problem had stimulated a scientist to think through the whole cluster of questions around the inertial frames and that it was the principle of relativity that had lead him to the solution of this problem just as it had led Huygens, 75 years before, to the solution of another one.

Relativity in Euler's hydrodynamics

Before continuing to follow Euler's work on rigid bodies further, I would like to draw, in accordance with chronology, attention to a remark he makes in his great paper on hydrodynamics, "De principiis motus fluidorum" [1770a]. Euler wrote two systematic expositions of the theory of hydrodynamics. The first is a series of three memoirs written in French [1755a; 1755b; 1755c]. They are among the most beautiful in his work and could still be used today as an introductory text. The second is a series of four memoirs written in Latin [1761; 1770a; 1771b; 1772]. Together, they form in fact volume six of the planned *magnum opus*. They could now indeed become a single book only, since Euler had discovered in the meantime that the motions of gases and liquids are described by the same equations, the equations which are still today named after him. That they were published as a series of four long papers rather than as a book seems due to the fact that he could not find a publisher, a misfortune that happened to him several times in his life.

In sections 58 and 59 of the "Sectio secunda" ("Problema 28"), Euler wishes to determine the pressure field if the fluid moves under arbitrary external forces, but with constant speed and he computes the answer.[8] But then he adds in a scholium:

> This case could have been derived from the previous solution where the whole fluid remained at rest, by means of the principle, accepted in mechanics, that in the motion of bodies everything remains the same, if the whole system is held to move uniformly on a straight line".[9]

---

[8] The significance of this section was unearthed and first discussed by Truesdell [1956: XV].

[9] *Hic casus ex praecedente, quo tota fluidi massa in quiete perseverabat, deduci potuisset secundum principium in Mechanica receptum, quod in motu corporum omnia manent eadem, si totum eorum systema insuper uniformiter in directum proferri concipiatur* [1770a: § 59].

Thus Euler recognizes the validity of the principle for the entire field of mechanics, indeed for the whole of physics. But in this case, it seems that he did not use the principle for finding the solution but checked it afterwards.

The final clarification of the problem of relativity

But Euler's most thorough investigations into the question of absolute versus relative motion are found in his *Theoria motus corporum solidorum* [1765a] and in his *Anleitung sur Naturlehre* [1862].

The first, the *Theoria*, written in Latin, is part two of Euler's great project which I have mentioned, and it contains his final exposition of the mechanics of rigid bodies.[10]

The first part of Euler's *Theoria motus corporum solidorum* is a new, systematic and, we might say, more penetrating discussion of the fundamentals of mechanics. In particular he now wants to decide once and for all the old dispute between Newton and Leibniz, opened up again and again by their followers: does motion have an absolute or only relative meaning? For, even at that late date, this question was still intensely disputed among scientists as well as among philosophers. Now, he again examines systematically the whole cluster of questions as they present themselves from all sides, first in geometry, then in kinematics and eventually in dynamics. The importance of this examination lies in its being systematic, especially also from the point of view of dynamics, and in this latter lays its novelty. This is his answer:

First: clearly, no absolute location can be fixed in Euclidean space. We can attribute such a location only "relatively" or, as he says, "with respect to" a given system, which must consist of at least four points that do not lie in the same plane. Second, the same holds if we add time, that is, if we consider motions; not only location but also direction, velocity, acceleration, etc., have no meaning unless defined "with respect to" or, as we would say, "relative to", a given space-time frame. So far, *Leibniz* and his followers are right.

But if we now turn to mechanics, that is if we study motion under the influence of forces, the whole situation changes radically. According to mechanics, the accelerating force exerted by one body on another has an absolute meaning: an acceleration is always the effect of a force. Here then *Newton* and his followers are right.

However, that much must now be considered too. Still according to Newtonian mechanics, a body that is not subject to any force is either at rest or in a state of uniform rectilinear motion, but the two – whatever Newton may have said about the center of mass of the solar system, which supposedly stays at rest in the center of the universe – cannot be distinguished. If an observer wants to determine the forces that several bodies exert on each other, he must be aware of the restriction due to the equivalence: two observers, which are relative to each other in uniform rectilinear motion, must express their observations with exactly the same formulae.

---

[10] The work is extensively quoted and discussed by the philosopher Ernst Cassirer in his *Das Erkenntnisproblem in der Philosophie und Wissenschaft der neueren Zeit* [1906-07]. Cassirer, by the way, is also the author of one of the few interesting books written by a philosopher on Einstein's theory of relativity [1911].

All this is demonstrated either geometrically or through explicit analytic computation. The same is explained in somewhat less technical language in the *Anleitung*. It is uncertain when this book was written,[11] but it cannot be very far from the date of the former. The book was published only posthumously in 1862, and then only incompletely, since a few pages had been lost in the meantime.

The relevant sections in the *Anleitung* are in chapter 10 (*Von der scheibaren Bewegung*). There he says:

> If the observer advances uniformly on a straight line and if he determines the distances and directions correctly, then all bodies that rest or move on a straight line appear to him to remain unchanged in their state; thus this apparent motion could indeed exist without the effect of some force.[12]

He adds further,

> If the observer moves (advances) with constant velocity (speed) along a straight line and estimates his surroundings correctly, i.e., according to parallel straight lines, then exactly the same forces are required for maintaining the apparent motion, however much it may differ from the true one, as are needed to maintain the true motion.[13]

Both assertions are proved geometrically and analytically by comparing three inertial systems just as we do it today, and the computation makes clear that this equivalence is restricted to relative linear uniform motion and tied to the addition theorem of velocities (that is of course to "absolute" time).

But then he adds:

> If the observer does not move uniformly on a straight line... then besides those forces that act on the bodies, other forces are needed, which produce the changes on the location of the observer, but in the *opposite* direction (my emphasis).[14]

---

[11] As the date of writing N. Fuss proposed the year 1731, Eneström and Rudio thought it could have been written at the earliest in 1745, Truesdell proposed replacing 1745 with 1755; see [Truesdell 1954: C, footnote].

[12] *Wenn der Zuschauer gleichgeschwind in einer graden Linie fortrücket und die Gegenden richtig schätzet, so scheinen ihm auch alle Körper, welche entweder ruhen oder gleichförmig nach grader Linie fortlaufen, unverrückt in ihrem Zustande zu verharren, und könnte also diese scheinbare Bewegung in der That ohne Wirkung einiger Kräfte bestehen* [1862: chap. 10, § 80].

[13] *Wenn der Zuschauer gleichgeschwind in einer graden Linie fortrückt und die Gegenden richtig, das ist, nach gleichlaufenden Linie schätzet, so werden zur Unterhaltung der scheinbaren Bewegung, wie sehr dieselbe auch von der wahren unterschieden sein mag, eben diejnigen Kräfte erfordert, als zur Unterhaltung der wahren Bewegung* [1862: chap. 10, § 81].

[14] *Wenn aber der Zuschauer sich nicht gleichförmig in einer graden Linie bewegt, dennoch aber die Gegenden richtig schätzet, so werden, um die scheinbare Bewegung aller Körper zu bewerkstelligen, noch ausser der Kräften, welche wirklich auf dieselben wirken, solche Kräfte erfordert, welche in einem jeden Körper alle Augenblicke eben die Veränderung hervobringen, welche in dem Orte des Zuschauers vorgeht, aber nach einer ungekehrten Rictung* [1862: chap. 10, § 82].

One sees that the determination of the "true force" is in the focus of Euler's interest. He shows the applications to astronomy and how these theorems are the foundation for the mechanical explanation of all observations.

The text also makes it clear that it was observational astronomy and the question of how observations must be linked to Newton's formulae that pushed Euler to these investigations. Only recently have Russian historians of science unearthed experiments and observations that he had made during his first stay in St. Petersburg [Nevskaya 1983a; 1983b].

Incidentally, we may also have found the moment here when the "observer", so important in today's discussions on quantum mechanics, but also essential beyond it, enters into the discussion of the principles of mechanics. At this point, I should also remark perhaps that it was Euler's capacity for intuitive reasoning that appealed so strongly to Faraday, who read and quoted him.[15]

I think it is apparent from what I have said that Euler had clearly perceived the idea of the transformation group, although he had not, of course, conceived the notion of abstract group. At any rate, through his incisive reasoning, the old dispute between Newton and Leibniz was definitely settled. However, when one puts the accents on the historic development that led to the understanding of space-time symmetry and to the principle of relativity, one thing is clear: they had never before been stated and explained so clearly, nor would they be again for a long period to come, as I shall briefly indicate.

The origin of group theory

Richard Feynman, in an appendix to his book *The Character of Physical Law* [1965],[16] expresses his surprise that the notion of group was not found earlier and not through physics. He is quite right, for this is exactly what happened. Groups *were* discovered in physics, but once again the physicists have forgotten their own results and their own history, and they have even forgotten that they have forgotten.[iv]

The omission noted by Feynman may also have to do with the erroneous apprehension, often held in the literature, that the abstract axiomatic formulation of a branch of science, since we present it at the beginning of a theory, also stands chronologically at its beginning. For this reason, we usually let the history of group theory start only towards the late 1830s. But of course, an axiomatic formulation becomes possible only when a large part of the field is already covered, and when the experience gained by long and persistent work in a field, allows the scientists to determine its structure safely. So group theory does indeed have its origin also in physics!

Perhaps I ought to mention too – although it is a bit outside the scope of this investigation – that we find in a paper on optics by Euler, "Examen d'une controverse sur la loi de réfraction des rayons de différentes couleurs par rapport à la diversité des milieux transparens par lesquels ils sont transmis" [1755c], stimulated by his (alas, erroneously directed) search for a universal law of dispersion, the explicit formulation of the two

---

[15] The reader will find Faraday's quotations of Euler in paper "Euler's Work in Physical Optics" in this volume, pp. 197-209.

[16] Dr. Pestieau directed my attention to Feynman's statement.

axioms that state the unique composition of two elements and the possible inversion of each one. From these he then proves that the logarithmic law that follows from these postulates is unambiguously determined. Furthermore, we should not forget what we find in his purely mathematical writings; that the idea of a group played a big role in his thinking was pointed out frequently by Andreas Speiser and again by André Weil in his beautiful survey on Euler's number theory [1984].[v]

These examples show that the most interesting task of the historian lies largely unanswered before us, namely to find out how Euler's mathematical ideas shaped his work in physics and how physics in turn stimulated his mathematical investigations.[17]

The fall of an important problem into oblivion

We may ask now how it happened that what was once so clearly understood was forgotten again. Precisely at the moment, when all the mathematical tools – that is, group theory, transformation theory, etc. – were now ready for systematic use, the scientists abandoned the pursuit of these questions. They did not avail themselves of what the mathematicians had found to formulate the fundamentals of physics rigorously, or to make all these problems transparent and the difficulties disappear. I can only indicate a few points.

Perhaps the beginning of the neglect of this problem may be traced to Lagrange's formulation of the foundations of dynamics, applied systematically in his *Méchanique analitique* [1788]. As is well known, Lagrange had succeeded in combining the old principle of virtual *déplacements* with d'Alembert's principle, thereby achieving a powerful starting point for the whole of classical mechanics. Not only its simple formulation, but the power it has whenever it is applied, made his method enormously popular. It is easy to grasp, and is an almost foolproof method for writing down the correct equations of motion. As everybody knows, this is often the most difficult task in physics. As a tool for formulating the equations it remained unsurpassed. Its power is due to the fact that, unlike Newton's equations, Lagrange's is scalar, for the virtual work is a scalar quantity and a great deal of the difficulties due to a vector formulation disappear. It puts the work and the energy into the center of mechanics and I need tell no physicist who has been formed through quantum mechanics how much easier his task thereby becomes. But indeed, the good is often the enemy of the better, and sometimes progress demands a price: Lagrange's equations permit an identical elimination of all holonomic constraints. They are valid in all frames, and thus also in curved spaces. Inertial frames are no longer distinguished and so the group property that relates them is lost. It seems to me that Euler's formulation of mechanics, based on Newton's concept of force and expressed in Cartesian coordinates (as a prelude to a modern formulation in terms of vectors, tensors, forms etc., if you wish), is the most adequate introduction to mechanics.

The door to mechanics opened by Lagrange was the one most often used later. This, for the reason which I just mentioned, is understandable enough, yet without a discussion of inertial systems and the principle of relativity, the foundations of classical physics are not really complete. As I am not a specialist in the history of physics of the nineteenth century,

[17] See "Leonhard Euler: Mathematician, Physicist, Engineer" in this volume, pp. 157-165.

in what follows I can make claims of only restricted validity, but I could not find any discussion of these problems in the line of Euler, not in Lagrange's *Méchanique analitique* itself, nor in Poisson's *Traité de mécanique* [1811], nor in Jacobi's *Vorlesungen über Dynamik* [1866], which, according to Felix Klein, became the model for most of those who followed, nor in the work of Kirchhoff [1876]. Only Thomson and Tait [1890] talk about what we today call in the restricted sense "Galileo transformations", thereby correcting Newton's mistake about the center of the universe. But in the years when mathematical progress would have allowed a much more concise formulation of what Euler had to say with many words, this was no special achievement. Hertz's most interesting, posthumously published *Mechanik* [1894] deserves to be studied closely, yet on this point we are left with few remarks.[vi] And still in 1928, Max Planck could write:

> An observer in uniform motion can never determine anything concerning his speed and one is never in a position to indicate a point in the universe, of which it may be stated that it be in absolute rest... This theorem is called the classical "principle of relativity" (to be well distinguished from the modern principle of relativity due to Einstein, which unities the invariance of the equations of motion under a rotation of the coordinate system, discussed before, and the invariance with respect to a uniform motion of the origin of the coordinate system, from a higher point of view) [Planck 1928: 327].[18]

Surely no one will accuse the illustrious discoverer of the quantum of action to have been overly infected by the "Gruppenpest".[vii] Sommerfeld, the close collaborator of Felix Klein, of course, presents these questions with admirable clarity [1947].

But the main reason for this strange omission was simply the change in the interest and the attitude of the physicists. The nineteenth century, especially its second part, was above all empirically minded, and mostly believed that the direct observations through the senses were the only valid bases for natural science. Investigations like those I have sketched here were often called "only formal", a characterization that seems a bit silly now. Today's physics with its renewed faith in mathematics, is in many ways much closer to the eighteenth than to the nineteenth century.

Symmetry of space-time

Did Euler realize the connection between the principle of relativity and symmetries? In a way, I think one may answer in the affirmative. Thanks to the introduction of Andreas Speiser, I came across the fourth chapter of the appendix to the second volume of Euler's famous *Introductio in analysin infinitorum* of 1748, entitled "De immutatione coordinatum". There he discusses the behaviour of algebraic curves and surfaces when one or several of the variables change signs, that is, he discusses their possible discrete symmetry properties. Indeed, as Speiser points out, there we find the first "group table", a device now usually associated with the name of Cayley. In the book on Euler's number

[18] These remarks are interesting precisely because they are made by the very first supporter of Einstein's ideas on relativity. Vague understanding of the principle of relativity and the Galilean group by an opponent of the theory of relativity was common enough but never interesting.

theory [1984] that I mentioned earlier, Weil underlines how strongly Euler's ideas on discrete and on continuous mathematics parallel each other and how deeply linked they are in his mind. I suggest that further inquiry into this question would be fruitful.

At any rate, this understanding of space and time had an important consequence for Euler's presentation of mechanics, that is, for his style. Its formulation in a space-fixed Cartesian frame is the most symmetrical possible; as I said, it is the closest substitute for a vector and tensor calculus. If one adds the sense and taste of a man who knows how to use his symbols and letters – be they Latin and Greek, Roman and Arab, big and small, etc. – in a systematic (that is, in a symmetric) way, one can see the major reason why his writings are so transparent and easy to understand and, I might add, so beautiful. Symmetries are not the least of the reasons why Euler can be so much more appreciated today than he could a hundred years ago.

Be that as it may, I hope to have shown one of history's unexpected twists where the development that actually took place was totally different from what one, looking back today, would expect.

There are no symmetries in history; these are reserved for mathematics, the arts and the sciences. It is rather the unexpected turns, these twists and breaks, that make history so exciting.

Acknowledgments

Thanks to an invitation by Dr. L.A. Radicati, part of this work was done at the Scuola Normale Superiore in Pisa, whose rich library I could use. Mrs. P. Radelet-de Grave helped me with the historical references, and Miss H. Speiser, Mrs. N. Van Olmen and my wife helped me with the manuscript and its numerous revisions. I inserted some corrections due to Drs. L. Michel and R. Jost. It is a pleasure indeed to thank all of them. But I should especially like to thank Dr. M. Doncel and the organizers of the conference for having invited me.

Euler, the Principle of Relativity and the Fundamentals of Classical Mechanics[1]

For Wolfgang Pauli

Introduction

The "principle of relativity" contains two independent statements. First, it says that for two mechanical systems that differ from each other by a linear uniform motion, such as in the case of a man walking in a train, the laws of nature are the same. This statement goes back to Galileo,[i] and therefore scientists speak sometimes of "Galilean relativity". The second statement concerns the exact mathematical formulation of the law. Important insights are due to Huygens and Newton, but they did not lead to a complete formulation of this problem. Later scientists worked for over 150 years until they found a transparent formulation. For each new field of physics the problem had to be solved again. Einstein discovered in 1905 that it was the symmetry group underlying Maxwell's theory of electrodynamics that yielded the solution of the problem, and accordingly he formulated, as Poincaré called it, the "new mechanics", which must be used for describing systems that propagate almost with the speed of light. Thus for most mechanical processes "Galilean relativity" is used in a calculation.

Euler's paper "Explicatio phaenomenorum quae a motu lucis successive oriuntur", written in 1739 but published only in 1750, marks the moment when mechanics was for the first time confronted with the question of the propagation of light. It came to my attention when I was editing vol. 10 of series III of Euler's *Opera omnia.*

When this work was complete, J. R. Oppenheimer, the director of the Institute for Advanced Study in Princeton where I was working at the time as a physicist, advised me to show it to Alexandre Koyré, the famous philosopher and historian, who was regularly in residence there. When I showed him my introduction, pointing out especially the "Explicatio", he was delighted and said at once, "You should make a special publication about this paper, it's worth it!" Of course, I did so, but Koyré was not easily satisfied: he made me rewrite the article three times!

---

[1] An earlier version of this paper was first published *Nature* 190, 4778 (May 27, 1961): 757-759. This is a later, corrected version, augmented by an *Introduction* and privately re-published by David Speiser.

Oppenheimer and A. Pais both congratulated me and the former sent the article to *Nature*. As it was my first historical paper, I was of course very proud.

It was only much later that I discovered Euler's greatest discovery in the field of wave optics: the wave equation in three dimensions, connected today with the name of d'Alembert.

Euler, the Principle of Relativity, and the Fundamentals of Classical Mechanics

Since they received their definite form through Newton, the significance of the fundamentals of classical point mechanics has always remained a controversial subject, at least up to 1905. Where Newton had spoken of absolute space and time, and referred to experiments, some of his successors pretended that they did not need such hypotheses,[ii] or even went so far as to reduce his famous second axiom to a mere definition.[iii] Thus the boundary between mathematics and physics as an experimental science was strongly displaced at the expense of the latter, from which the so-called pure mechanics was separated.

Others, on the contrary, insisted on the basically experimental character of this axiom.[iv] There have been many attempts to display the long series of arguments from all sides of this very involved discussion. It is the aim of this article to direct attention to a single point of this process, namely, the instant where probably for the first time there emerged a completely new point of view: the confrontation of mechanics with optics. It will be shown that the result of this confrontation was the recognition that the principle of relativity, closely connected with the fundamentals of mechanics, was called into question. It is therefore appropriate to dedicate this article to the memory of Wolfgang Pauli who, forty years ago, in December 1920, published his famous article on relativity theory in the *Mathematical Encyclopedia* [1921].

The discovery that the principles of classical mechanics are entangled with the principles of optics was made by Euler in 1739. Or, as we can put it more precisely: he discovered that the decision between a wave and a corpuscular theory of light reflects on the validity of the principle of relativity. (Already in his *Mechanica,* published three years earlier, he had devoted the first chapter to an ample explanation of the difference between relative and absolute motion [1736] In a paper written in that year, "Explicatio phaenomenorum quae a motu lucis successivo oriuntur [1752a], he examines the aberration, discovered by Bradley in 1725 [1728], mainly in order to calculate corrections for astronomical use. The aberration is a systematic error in observing the stars, of the magnitude of 20″, which occurs through the Earth's motion in relation to the stars. Presuming that the fixed stars stay at rest, then three motions remain, of which the influence on the observation may be calculated: the daily rotation of the Earth, the yearly rotation of the Earth, and the motion of the planets. But, it is characteristic of Euler that he is not only interested in the special problem to which this examination gives rise, namely, the calculation of the required correction, but rather he attacks the question in its fullest generality as a physical problem. The astronomical calculation is only the starting point for investigating basic questions in mechanics, where the astronomical language is a mere fashion of presentation. Euler distinguishes four cases, which he examines one after the other: (1) the observer and the object are at rest; (2) the object moves; (3) the observer moves; (4) both object and observer move.

For the second case he then derives the aberration formula. In the course of this, he notes that one may ask whether the velocity of light depends on the velocity of the radiant object (always measured in the rest system of the observer). But he himself gives the answer, that this is not the case, since the propagation of light is a wave process in the stationary ether, which is independent of the motion of the emitting body, and he finds for the correction (I use Euler's notation):

$$tg\alpha = \frac{ms}{c - \mu s}$$

where α is the angle between the observed and the true location of the object, c and s are the velocities of light and the object respectively, and m and μ are the sine and cosine respectively of the angle between the light ray and the velocity vector of the object. He applies his formula to the simplest forms of motion (motion on a straight line and on a circle). Thereby he reaches the conclusion: to an observer at rest, the speed of a body in uniform and rectilinear motion seems to diminish more and more with increasing distance.

Now consider the third case, which for practical application is the most important one: the object rests and the observer moves. Since Euler was one of the few authors who allowed the reader to participate in his investigations and to watch his paths closely, it might be worth while to report his reasoning in detail. For here it is possible for once to observe the single steps through which a fundamental discovery is reached and to pinpoint as closely as possible the exact moment at which a crucial insight is gained.

He starts from the remark that one may reduce this case to the previous one by giving the observer-object system a speed which is equal but opposite to the original velocity of the observer. Therefore one may take over the formulae derived above (that is, those derived in case 2) and apply them also to this case:

$$tg\alpha = \frac{mr}{c - \mu r},$$

where r is now the velocity of the observer, and m, μ are the sine and cosine respectively of the angle between the light ray and the direction of the observer's motion. This formula seems to settle the problem.

Euler, however, is not yet satisfied and wishes, as he says, "to investigate the effect which the light ray exerts on the eye in motion and the difference between the true and the apparent place according to the rules of motion".[2] This investigation is essentially independent of the former and therefore a new derivation of the correction formula.

The velocity of light in the ether and the velocity of the observer are now added according to the rules of kinematics. Euler explains hereby his short trigonometric calculations amply with the help of his vector diagrams. But now one is surprised to find the astonishing result: the correction computed with this method differs from that found before! One gets:

$$tg\alpha = \frac{mr}{\sqrt{c^2 - m^2 r^2}},$$

instead of:

---

[2] *Investigemus igitur effectum, quem radius lucis in oculum motum exerit, et in discrimen situs apparentis et veri secundum regulas motus inquiramus* [1750a: § 20].

$$tg\alpha = \frac{mr}{c - \mu r}.$$

Euler asks whence this difference arises, and which of the two answers, both the result of what appear to be rigorous derivations, is the correct one?

First he remarks: it is clear that the difference arises from the fact that in the first derivation because of the reduction to case 2 we had implicitly assumed that the light velocity remains c, independent of the speed of the observer, whereas in the second derivation the light velocity becomes

$$y = \mu r + \sqrt{c^2 - m^2 r^2}.$$

The decision between the two correction formulae therefore is reduced to the question of which is the correct velocity of light? Euler now observes that the system under consideration comprises not only object and observer, but also the ether, which hence ought also to have received the same speed! But this, of course, is excluded. In other words: the ether distinguishes an absolute system, and cases 2 and 3 are not equivalent but basically different! Equivalence between both cases, however, would hold in a corpuscular theory, as he points out, for in such a theory no ether distinguishes an absolute coordinate system and the principle of relativity (Euler says, *haec hypothesis* ... motus compositionem *recipit* [1752a: § 28][3]) prevails.[4] Therewith, Euler found what he had looked for: an *experimentum crucis* which permitted, at least in principle, a decision between the two theories, wave optics and corpuscular optics. However, as he notes explicitly, the difference is too small to be susceptible to experiment, in this case, the astronomical measurement. Euler therefore calculates in his further investigations with his simpler formulae which follow from the corpuscular hypotheses, since they coincide with those of the wave theory to within a difference of the second and higher order in v/c, that is, he makes full use of the principle of relativity (linear addition of the velocities).

This piece of work seems to me to be one of the most beautiful that Euler ever wrote; perhaps one may regard it as one of the classical papers in physics. This is not only because of its clarity and elegance, but also much more so for the particular manner in which a seemingly simple question gives rise to a discussion of the most basic principles of physics. At first one believes that Euler has been caught in an error, and then one is even more surprised by the infallibility with which he aims at the point that would become, more than a hundred years later, the center of discussion about the fundamentals of mechanics. On the other hand, how surprised he would have been to learn that the decision had been made in favour of the principle of relativity and against the ether, and that this decision was made just on the grounds of the wave optics confirmed meanwhile by the Michelson experiment, which is closely related to aberration! This might be another example of what Pauli used to call "the life of the ideas", "showing", as he formulates it in the preface to the second edition of his article, "how a fundamental scientific discovery, sometimes even against the resistance of its creator, gives birth to further fruitful developments following its own autonomous course" [1921]. Though

---

[3] *Quamvis enim haec hypothesis minus sit probabilis quam altera, qua lumen instar soni propagari statuitur, tamen magis est accomodata ad nostrum institutum atque motus compositionem recipit, cui us altera hypothesis minus est capax* [1750a in *Opera* III, 5: 56].

[4] This is the point of departure for W. Ritz's attempt at corpuscular optics, in order to explain the negative result of the Michelson experiment; cf. [Pauli 1921].

perhaps in this case one rather should say "even in a direction opposite to the intention of its creator".

From the following remark it is clear that Euler was well aware of the significance of inertial frames as well as their peculiar problems:

> According to the safe laws and rules of mechanics, it is therefore allowed to impress on an entire system of bodies a uniform straight motion, for nothing in the phenomena will be changed, but to give it a non-uniform or a curvilinear motion is strictly forbidden.[5]

This demand, namely, that the question not be regarded only as a kinematical but also as a physical one (*legibus et regulis* mechanicis ...), should have saved him from being decried as having only a rather formal attitude towards physics. In this particular fashion of looking at those problems he remains the "true son of Newton", to quote Clifford Truesdell [1954: CXVIII]. However, later on the mathematicians and sometimes even the physicists often overlooked this point, whereas Euler thought of the foundations of mechanics as subject to experimental examination. His explanations of how the speed of a moving object appears to an observer at rest are a first tentative prelude to Einstein's later criticism of time measurement.

There is an amusing question to be asked concerning the response to this paper. It was written by Euler in 1739, but its publication by the Academy of St. Petersburg was delayed until 1750. Meanwhile, he had published a continuation to it, the "Mémoire sur l'effet de la propagation successive de la lumiere dans l'apparition tant des planetes que des cometes" [1748b]. There, however, he did not enter into the question of wave versus corpuscular optics, but applied without further explanation the simpler formulae of corpuscular optics as an approximation in the above-mentioned sense. A careful reader could have objected that Euler was in contradiction to the principles of his own wave theory, proposed and discussed in detail in his "Nova theoria lucis et colorum" [1746b].

Was there such a careful reader?

---

[5] *Licet enim salvis legibus et regulis mechanicis universo cuidam systemati corporum, motum aequabilem in directum imprimere, ita, ut nil in phaenomenis mutetur, at vel motum inaequabilem vel curvilineum tribuere minime licet* [1750a: § 35].

Euler's *Anleitung zur Naturlehre*[1]

For Gianna and Luigi A. Radicati

Introduction

Euler attempted to formulate in one unified system, as we would say, the whole science of inorganic nature as it was known in his time. In his great *Anleitung zur Naturlehre* (*Guide to the Science of Nature*), he indicated how he thought such as ambitious goal could be realized. Unfortunately, the book was never published during his lifetime, but only in 1862, and by that time parts of two sections had been lost.

For an understanding of the ideas that motivated and guided Euler's theories of light, electricity, magnetism and gravitation the *Anleitung* is indispensable, but so far it has been examined only very little.[2] For this reason, here I will give a few indications concerning the notion of field as it had evolved from Euler's hydrodynamical investigations. This idea was to guide him in his views on the subjects mentioned above. As we shall see, some of Euler's ideas again have a certain pertinence today. There can be no question of giving a detailed account of the whole work here, let alone an analysis. I will concern myself with only some of its chapters, but a short overview over the whole work is necessary to clarify the significance of what Euler is driving at.

The work in its present form fills 163 pages of vol. I, series III, of the *Opera omnia*. There are nine missing sections that may have filled another seven to nine pages, so the whole work would have been about 170 pages. Even if we had the missing sections, there are reasons to think that the work as it stands is not complete. At the end of the paper we shall ask why Euler did not finish it.

The book is divided in twenty-one chapters, which may be grouped as follows:

- Chapters 1-6. Introduction and discussion of the four fundamental properties of all bodies: extension (*Ausdehnung*); mobility (*Beweglichkeit*); persistence (*Standhaftigkeit*); and impenetrability (*Undurchdringlichkeit*). The end of chap. 5 is missing, and so are the title and beginning of Chapter 6. Rudio's guess was

[1] Originally published in *Symmetry in Nature,* a volume in honour of Luigi A. Radicati di Brozolo, on the occasion of his 70th birthday, Scuola Normale Superiore, Pisa, 1989, pp. 721-733. This paper is a slightly extended version of a section of the Introduction for vol. 10, series III, of Euler's *Opera omnia* [Radelet – de Grave and Speiser 2004].

[2] One exception is H. Weyl who expressed his high regard for the *Anleitung* in his book *Philosophy of Mathematics and Natural Science* [1949].

that this section may have been called "*Von den Kräften im Allgemeinen*", but the title may have been also "*Von den Veränderungen der Körper und ihren Ursachen*". At any rate, it looks like the introduction to the next part;
- Chapters 7-11: General introduction to the principles of mechanics;
- Chapters 12-14: Statement of Euler's main hypothesis concerning the structure of matter. There are only two kinds of matter: *die grobe* (gross, coarse, heavy) and *die subtile* (subtle, fine). All materials that we observe are built up of only those two elementary kinds of matter;
- Chapters 15-18: A discussion of the different kinds of bodies found in nature – fluid, solid, elastic;
- Chapter 19: A speculative digression on gravity;
- Chapters 20-21: A thorough exposition of the principles of hydrostatics and hydrodynamics. This theory is at the basis of his speculation.

Symmetries play a double role in the *Anleitung*. In chapters 7-11, Euler discusses explicitly what we call today the principle of relativity. Thereby he settled the dispute between Newton and Leibniz concerning the role played by space and time in nature. I have written about this question elsewhere and I shall not return to it.[3] But, throughout the *Anleitung*, symmetries play a hidden but none the less decisive role. Since 1750 and his *Découverte d'un nouveau principe de mécanique* [1752] Euler had slowly begun to grasp the most important consequence of the principle of relativity and (in modern terms) of the action of a group on space and time, namely, the fact that the laws of nature are relations between tensors. Euler was the first to see this. For this reason after 1750 he used what we now consider a manifestly covariant form (with respect to the usual metric of classical space-time) of the fundamental equations in most of his books and papers in mechanics, that is, he used Cartesian coordinates, which are the closest substitute for the use of vectors and tensors. The first fruit of this new understanding was the equations of hydrodynamics based on his own discovery of the internal pressure (a scalar density), which one used to formulate the effect through contact of one part of a fluid on another without the explicit a priori knowledge of the nature of the forces between these parts.

Without this new understanding the view of nature proposed in the *Anleitung* would be meaningless; this is really at the bottom of his idea.

This short survey shows that the *Anleitung* is not a *cosmology* in the sense of Leibniz and Wolff and of some of their followers (cf. [Schramm 1985]), that is, a deduction of the fundamental laws of physics from metaphysical principles. It is also distinct from Descartes's speculations, which in some respect it resembles. While Sects. 6, 7 and 11 contain a clear exposition of the basic notions following a metaphysical tendency that once prevailed in the scientific world, and a thorough discussion of it, the *Anleitung* itself is a rather easy to understand presentation and discussion of the fundamental laws of mechanics, especially hydrodynamics, for specialists and non-specialists alike, followed by a grandiose speculation about the structure and composition of all matter. But this speculation is a strictly scientific one. Euler hoped to show that from these laws of

[3] See in this volume "Euler, the Principle of Relativity and the Foundations of Classical Mechanics", pp. 179-183 and "The Principle of Relativity in Euler's Work", pp. 167-178.

dynamics and from his simple hypothesis concerning matter, all other hypotheses of physics can be derived. More about this speculation will be said later.

The *Anleitung* and Euler's equations for hydrodynamics

The introduction in the first sections, of the four properties (*Eigenschaften*) common to all bodies – extension, mobility, persistence and impenetrability – may strike the reader who glances through the book for the first time as a purely philosophically-minded and even as a slightly scholastic procedure. But if we read these sections carefully and if we observe what Euler is driving at in the later ones, we see that he presents and discusses simply the various elements that enter into his equations for hydrodynamics

$$\rho\frac{d\mathbf{u}}{dt}+\nabla p=\mathbf{f},\qquad\qquad (1)$$
$$\partial_t\rho+\nabla\cdot(\rho\mathbf{u})=0.$$

Here extension corresponds to space: $\mathbf{x}, d\mathbf{x}$; mobility to time and velocity: $t, dt, \mathbf{u}, d\mathbf{u}, \frac{d\mathbf{u}}{dt}$; persistence (or inertia) corresponds to mass m and mass density ρ; impenetrability to Euler's own creation: the internal pressure p.

Indeed, the left hand side of equations (1) is constructed out of these elements. So we may see in the importance attributed to these "properties" a mere concession to the style of his time. What about the right hand side, that is the external forces (or to be precise, force-densities)? In most of the usual applications of these equations **f** represents the gravitational field or, for example, the constraint due to a tube, etc. This, as we shall see presently, is where Euler's speculation enters.

Matter and material

To understand Euler's idea the reader must recall two limitations of classical physics. First, classical mechanics speaks only of the propagation of bodies through space and time. Classical physics cannot tell what these bodies are or what their structure is without making some further ad hoc assumptions. Today we know that quantum mechanics is of the essence here. Second, classical mechanics – hydrodynamics, for instance – assumes the existence of forces. For example, with Newton it introduces into his computations gravitational forces, without, however, having an answer as to the mechanism that transmits them. Euler tries to fill both gaps.

Let us begin with the second point. According to Euler the force-densities **f** in eq. (1) result, in the final analysis, always from the *impenetrability* of some matter, that is, from the pressure that this matter exerts on some other bodies. In other words, according to the *Anleitung*, what we observe in nature is (as we would say today) *a system of dynamically coupled fluids*. The coupling is the result of the impenetrability of each matter and all observed phenomena can be resolved into such interactions.

The big question now is how many different materials there are in nature. Infinitely many, as it would seem to the ordinary observer, or only a few, basic ones, out of which all other ones are built up? To begin with, surely a fluid – or perhaps more than one – is responsible for the phenomena of optics, electricity, magnetism and gravity. This fluid is the ether. We shall come back to it, and to the question whether there is only one ether or several.

But what about the ordinary materials: the metals, the gases, the insulators, to say nothing of organic matter? Is there really an endless variety of them, or can we hope to bring some mathematical and conceptual order into what we find through the known laws of mechanics? This is the question that Euler tries to answer with his speculation. But before presenting it to the reader we must recall that chemistry as we know it today hardly existed at that time; almost no law of chemistry was known and but a few elements had been identified. Thus there was not much that could guide or stimulate Euler, at least nothing comparable in importance and general applicability with his own equations for fluids.

From Euler's paper "Dissertatio de magnete [1748c], and especially from the *Lettres* on electricity [Euler 1768-72: CXXXVIII-CLIV] and on magnetism [Euler 1768-72: CLXIX-CLXXXVII], we know that Euler conceived all ordinary matter as permeated by "pores" and "canals" through which the ether flows. Obviously the size and form of these canals and pores are characteristic of any given material, and may vary considerably from one to another. Euler now conjectures that the various materials as we know them really only differ from each other through the configuration of these canals and pores, their size and form etc., but that the *matter* itself, the *stuff* of which they are built, is the same in all of them!

This one fundamental matter fills only a very small part of a body. It consists of tiny but very heavy specks and droplets whose density is always the same. Between these heavy droplets run the canals through which the ether flows. Of course, Euler can't say how and according to what laws this structure that characterizes a material is formed; he must leave this to future research. However, this basic matter, like the ether, obeys the equations of hydrodynamics (1).

Lest we should give the reader who does not know the *Anleitung* the impression that Euler's ideas are pure verbiage, we recall here that his verbal explanations are underpinned throughout with a thorough presentation of the fundamentals of mechanics, especially of hydrodynamics. It is indeed his long and enduring occupation with this continuum theory that guided his speculation. These underpinnings also show that Euler's ether was not just another new matter, similar or the same as those which so many of his predecessors had proposed: it was a matter *that obeyed established laws of continuum mechanics*. Thus its properties were not ad hoc assumptions, but could be deduced from these laws. It is the knowledge and use of these laws that separates Euler's speculation from older ones it may bring to mind. Thus his speculation, at least, is not lacking consistency and coherence. We may sum up these ideas as follows.

There are only two kinds of matter in the universe: the gross (*grobe Materie*) and the fine (*feine Materie*). The structure of all bodies and their qualities are determined by them. The heavy matter is the same in all bodies; it is porous, made up of *drops*. All drops have the same density and if the *observed* weights of bodies are different, this is due to the varying number of drops per unit volume as well as the different number and size of drops of heavy fluid contained in these bodies. The constitution of bodies is fixed by their stereometric configuration, and the distribution of the drops in the body is also determined by the fine matter, the ether. That is, the same ether which we encountered in the theories of heat, optics, electricity and magnetism is partially responsible for the structure and for some of the macroscopic properties of all bodies (cf. [Whittaker 1951]).

Ether, according to Euler, is also responsible for gravitation. Thus only the gross matter – i.e., the heavy drops – contribute to the body's weight. He leaves undecided whether there is only one ether or several. Empirically we cannot say, but he states that it would be against the principles of sound research to suppose the existence of more than one, unless such an assumption is made unavoidable by experience:

> We must leave aside here the question of whether there are further different types of this subtle matter of which one is more dense than the other, and desire no more than that all these types, if several exist, are comprised in the generic name of subtle matter. Since for a very long time the explanation of the occurrences of the nature would not admit of several such types, it would be bold and against the rules of a healthy science of nature if we wanted to increase the number of subtle matters merely for our own conceit.[4]

Before discussing the most interesting feature of the *Anleitung* we should try to evaluate what Euler achieved by asking where he went in the right direction, and where he went wrong; what, with hindsight, must be judged as promising; and finally, which of his ideas were eventually followed up on by successors.

That all ordinary materials are made up of one and the same matter and that these materials differ only by their stereometric structure must have seemed a fantastic speculation at the time. Even the 92 elements in Mayer's and Mendeleev's periodic table of more than 100 years later, the crowning achievement of a long line of research, are a far cry from one single heavy matter! Yet Euler's heavy droplets bear a curious resemblance to today's nuclear matter; the tiny nuclei are, after all, responsible for over ninety-nine percent of the weight of all bodies. And even though it would be silly to claim a tangible connection between Euler's drops and nuclei, this resemblance shows, at least, that Euler's vision was not absurd, and that he was headed in the right direction.

But then, what determines the specific structure of ordinary bodies and what is the interplay between matter and the ether? Euler does not say, and of course he cannot. A huge gap is manifest here. The whole world of the atom, of chemistry and of crystallography and what determines it – namely, the electron – are, of course, entirely missing. Modem chemistry, as we, said, was hardly born; though crystals may have occupied Euler, they could not lead him very far. The electron, to say nothing of quantum mechanics, lay in the far distance beyond any dream. No wonder then that nothing of all this can be found here.

---

[4] *Ob es von dieser subtilen Materie weiter verschiedene Arten gibt, von welchen eine dichter sei als die andere, müssen wir hier an seinen Ort gestellt sein lassen und wollen zum wenigsten alle diese Arten, wenn je mehrere vorhanden waren, unter dem allgemeinen Namen der subtilen Materie begreifen. Den so lange die Erklärung der Begebenheiten der Natur nicht mehrere solche Arten erheischet, so würde es verwegen sein und gegen die Regeln einer gesunden Naturlehre laufen, wenn wir bloss aus unserer Einbildung die Anzahl der subtilen Materien vermehren wollten* [Euler 1862c: § 96].

The ether

It is a very different story with the ether, "the subtle air of heaven" [Lucretius, Bk I, line 1087]. Here, as Faraday's testimony shows,[5] Euler was definitely on the right track and his ideas had a positive influence. We have to ask how Euler had arrived at such a profound and powerful insight. Indeed, today optics, electricity and magnetism are united by the same electromagnetic field, as we call it, of which the ether was a forerunner. Perhaps Euler went wrong when he hoped to make this ether responsible for gravitation as well. After the innumerable attempts by Weyl, Einstein and many others which all failed too, we now doubt that there is a direct way to bring gravity and electromagnetism together. Maybe this goal can be reached if the theory also includes the *heavy matter*, as is the case, for instance, with the *supersymmetric* field theories that also comprise spinor fields.

But what does all this amount to? These single ideas, interesting and fruitful as they were, are not in fact the most interesting aspects of the *Anleitung*. What strikes us most today is the totality of his ambition and the way he tries to realize it. Here Euler attempts neither more nor less than to conceive a coherent system capable of describing the whole of inorganic nature starting from a few principles. Newton, in the famous last queries 28-31 of the *Opticks* [1704], had sketched a few ideas in this direction, yet they remained queries. Newton lacked the new powerful tool that is the basis for field theories: partial differential equations. Here, in the *Anleitung*, the ambition is greater and the realization goes further. It shows how much more powerful mathematics and physics had become when continuum mechanics had been mastered. Especially, this new branch of analysis had permitted d'Alembert and Euler to create the field concept (see [Truesdell 1954: LXXXI]).

Thus Euler was the first to conceive of and imagine a unified field theory, an idea that today stands in the center of the physicist's aim. Indeed, according to Euler we can reduce all phenomena to the interplay of four scalar and of two vector fields: the densities and pressures of the heavy matter and of the ether and of their two respective velocity fields (i.e., their streamlines) and to the partial differential equations that connect and govern them. As an attempt, Euler's speculation is in principle not all that different from present day speculations that aim at unified field theories, in spite of the enormous additional amount of accumulated empirical knowledge that we possess today as well as our much more penetrating insight into the structure of matter. A well established differential equation, basic in one domain of mechanics, is declared the fundamental principle for all other observed processes as well, and this principle is combined with a radical hypothesis that must complete it. Here, as we said, the basic principle – the field equations – are the Euler equations of hydrodynamics.

This is another one of the many instances that allow and even compel us to say that by and large the physics of the twentieth century, where mathematics again plays a dominant role, is much closer to the physics of Newton, of the Bernoullis, of Euler and of Lagrange than was the physics of the nineteenth century. Therefore we are often much more apt to

[5] Faraday read Euler's "*Lettres à une princesse d'Allemagne*" and refers to them repeatedly in his diary; see "Euler's Work in Physical Optics" in this volume, pp. 197-205.

appreciate the former period than the scientists of the nineteenth century were. For today we have learned again that such speculations are not idle but can be the motor for progress.

Theory of optics

We conclude this report on the *Anleitung* by pursuing the subsequent work of Euler a bit further. In December 1759, inspired by the work of Lagrange, Euler again took up his work on acoustics and wrote a sequence of three remarkable papers on mathematical acoustics [1766a; 1766b; 1766c]. In the second of these papers he discovered the linear differential equation, today named after d'Alembert:[6]

$$\mathbf{u} \equiv \left(\frac{1}{c^2} \partial_t^2 - \Delta \right) u = 0 . \tag{2}$$

by deriving it from the equations (1) of hydrodynamics. Euler had first despaired of solving the wave equation in three dimensions since in two it had been shown to be intractable. But then, as he says in the third paper:

> It is precisely here, as I have since remarked, that the difficulties are not invincible, and it is the that place has a similar case has those that the Count Riccati has proposed once, or a certain equation becomes integrable, while in general it is not.[7]

For Euler, the theory of sound had always been the model for the theory of light (cf. "Nova theoria lucis et colorum" [1746]). Now finally after 1759, he had in his hands all the tools for explaining and formulating almost all optical phenomena, especially interference and diffraction, but with one significant exception to which we shall return. But then after that date other subjects attracted him more strongly, and in his last years he returned only twice to the questions discussed in this volume. We may regret this but we cannot quarrel with Euler.

However that may be, a few months after this discovery, on 24 June 1760, Euler explained his ideas concerning the propagation of light in a letter sent to Turin. To "M. de la Grange" he wrote:

> Dear Sir,
> I am very flattered by the approval of which your illustrious Academy, and you in particular, have wanted to honor my essay on the vibrations in an elastic medium; the honor of these deep researches is solely due to your sagacity, and in this I have done nothing more than to take advantage of the light provided to me by your excellent mémoire (that is, [Lagrange 1759]). There you have opened a quite new course, where all geometers who will come after us will always find an abundance of things to occupy their

[6] As is well-known, in 1746 d'Alembert had discovered the 1-dimensional d'Alembert equation valid for the vibrating string.

[7] *Cependant c'est précisément ici, comme je l'ai remarqué depuis, que les difficultés ne sont pas invincibles ; et c'est là qu'a lieu un cas semblable à ceux que la Comte Riccati a proposés autrefois, ou une certaine équation devient intégrable, pendant qu'en général elle ne l'est pas* [1766c: § 3].

talents, and as to the degree that they are successful, analysis will make some very considerable progress. Now, this argument itself is probably the most important in Physics: not only all phenomena of the propagation of sound depend on it, but I am sure that the propagation of light follows the same laws. One only has to substitute the ether instead of air, and the vibrations that are diffused there will give us the propagation of light. Now we should hope that it is possible to determine the changes that the vibrations excited in a medium are subject to when they pass to another medium of which the density and elasticity are different. I don't know if one can hope for the solution of this problem, but I am very convinced that one there one would discover infallibly not only the true laws of refraction, but also the more complete explanation of reflection, with which refraction is always accompanied: one would see that it is impossible that the rays pass from one medium to another without a part turning back on themselves. Maybe this consideration can facilitate the development of analysis and furnish at least some particular solutions; but here one will encounter a new difficulty. How should we estimate the density as well as the elasticity of other transparent media, as for example glass, the density being so large in relation to that of the ether, without being able to presume that its elasticity is greater [, so] that the speed of the rays in the glass, will become extremely small; however, I believe that the refraction itself proves sufficiently that the speed of the rays in the glass with respect to the one in the ether must be in the ratio of 2 to 3.[8]

---

[8] *Monsieur, Je suis bien flatté de l'approbation dont votre illustre Academie, et vous en particulier, aves bien voulu honorer mon essai sur les ebranlemens dans un milieu elastique ; l'honneur de ces profondes recherches est uniquement dû à votre sagacité, et je n'y ai rien fait que profiter des lumieres que votre excellent Memoire m'a fournies. Vous y aves ouvert une carriere tout à fait nouvelle, où tous les geometres qui viendront après nous trouveront toujours abondamment de quoi occuper leur addresse, et, à mesure qu'ils y reussissent, l'Analyse en acquerra des accroissements très considerables. Or la matiere même est sans doute la plus importante dans la Physique : non - seulement tous les phenomenes de la propagation du son en dépendent, mais je suis assuré que la propagation de la lumière suit les mêmes loix. On n'a qu'a substituer l'ether au lieu de l'air, et les ébranlemens qui y sont répandus nous donneront la propagation de la lumiere. Maintenant il seroit à souhaiter qu'on pût déterminer les alterations que les ebranlemens excités dans un milieu souffrent lorsqu'ils passent dans un autre milieu dont la densité et [l'] élasticité sont differentes. Je ne sais pas si l'on peut esperer la solution de ce probleme, mais je suis très convaincu qu'on y decouvriroit infailliblement non seulement les veritables loix de la refraction, mais aussi la plus complette explication de la reflexion dont la refraction est toujours accompagnée : on verroit qu'il est impossible que les rayons passent d'un milieu dans un autre, sans qu'une partie rebrousse chemin. Peut-être que cette consideration pourroit faciliter le developpement de l'analyse et fournir au moins quelques solutions particulieres ; mais on rencontrera ici une nouvelle difficulté. Comme il faut estimer tant la densité que l'élasticité des autres milieux transparens comme par exemple du verre, la densité étant si grande par rapport à celle de l'ether sans qu'on puisse supposer plus grande son elasticité [, ainsi] que la vitesse des rayons dans le verre, deviendront extrêmement petite;*

But this is not all. In a previous letter (1 January 1760) to Lagrange on the same subject [1762], destined for publication, Euler summarizes the three memoirs on acoustics cited above that he had just presented to the Berlin Academy, writing:

> Since my last letter I have succeeded in reducing to a calculation the propagation of sound, attributing to air all three dimensions, and although I don't doubt than you have arrived there in a better way, I don't believe that I can better to testify to my attachment towards your illustrious society that by presenting it with my research on this same topic.[9]

Note that this letter is *not* entitled: "De la propagation du son", as the first memoir of the series had been, but "Recherches sur la propagation des ébranlemens dans un milieu elastique". In view of what Euler wrote to Lagrange in his letter of 24 June 1760 (and of what we find stated throughout Euler's work) it is clear that *"milieu élastique"* refers here not only to air but also to *the ether,* in other words *to light*. The derivation of the wave equation is now shorter, and again we find the solutions that describe spherical waves.

So we may justly consider the letter to Lagrange printed in the *Mélanges* of the Turin academy as *the first paper on mathematical wave optics.*[10]

What we said just now allows us to put our finger more precisely on one limitation of Euler's physics. Equation (2) is a *scalar* equation: thus while interference and diffraction phenomena can easily be derived from it, one optical phenomenon cannot, and therefore resists Euler's attempts: polarization. Indeed Euler never speaks of polarization, although it is possible that he knew this phenomenon. In any case there is no room for polarization in a theory based on a *fluid* (i.e., scalar) ether: what was needed here is an *elastic ether*. From the point of view of field theory: a scalar field, like pressure, cannot produce transverse effects, but such effects (shearing) are obviously needed for making a material cohesive!

The modern theory of elasticity (as Truesdell has shown in his history of elasticity [1960a]), goes back to Jacob Bernoulli. Daniel Bernoulli created the linearized theory and discussed oscillations; later his brother Johann II Bernoulli had discussed transverse vibrations [1736]. Euler had taken up the subject independently of Daniel Bernoulli, exchanging ideas with him, and his numerous papers must be considered as one of the greatest contributions to the subject in the eighteenth century. Yet they were almost all limited to one-dimensional problems. Only once did Euler overcome this restriction successfully when he found the equation of the oscillating membrane (i.e., the drum). To formulate a true three-dimensional theory of elasticity was impossible at the time due, as Truesdell showed, to the lack of sufficiently powerful mathematical, especially differential geometrical, tools. This gap in mechanics was closed only in the 1820s by Cauchy through

---

cependant, je crois que la refraction même prouve suffisamment que la vitesse des rayons dans le verre à celle dans l'ether doit être dans la raison de 2 à 3 [Euler 1862b : 561].

[9] *Depuis ma derniere lettre j'ai réussi à ramener au calcul la propagation du Son, en supposant à l'air toutes les trois dimensions, et quoique je ne doute pas que Vous n'y soyés parvenu plus heureusement, je en crois pouvoir mieux témoigner mon attachement envers Votre Illustre Société qu'en lui présentant mes Recherches sur ce même sujet* [1762: 1].

[10] Huygens's *Traité de la lumière* [1690a], a book that cannot be praised enough, deals with geometric, not with wave (i.e., mechanical) optics.

his creation of the stress tensor T_{ik} [1823; 1827]. Thereby Cauchy laid the general foundations of continuum mechanics. Euler's internal pressure p is a special case of Cauchy's stress tensor valid for a non-viscous fluid:

$$\sigma_{ik} = p\delta_{ik}\,. \qquad (3)$$

Through the creation of this concept the way was opened for further progress, and the fluid ether was now superseded by its successor, the elastic ether. From Cauchy to Kelvin many worked on these theories, yet polarization once more proved to be the stumbling block. While the fluid ether could not accommodate it, the elastic ether proved to be too generous: the longitudinal polarization, predicted by every elastic theory, failed to show up. Eventually Maxwell's theory gave the correct solution. But even Maxwell conceived his theory of electromagnetism in the context of the elastic ether [1873: §§ 782-4], and this view was certainly also held by H.A. Lorentz. This may seem surprising since in the vacuum, at least, Maxwell's theory is a linear one, while an ether theory necessarily must contain non linear terms. At any rate the modern field concept emerged only after the investigations of Poincaré and Einstein.

Furthermore this new electromagnetic field, although a classical concept, is already very close to the photon, which is a true quantum notion, far outside anything that classical mechanics could think of.

Conclusion

Leaving aside the missing chapters five and six, is the *Anleitung* a complete work? If so, why was it not published? If not, what stopped Euler from pursuing the subject?

We cannot exclude trivial reasons, such as Euler could not find a publisher, or he was interrupted by more urgent business, for example, an order by Frederick II, who in those years was involved in the Seven Years War. Yet there may have been a deeper reason. Somehow Euler must have become conscious that without a theory of elastic materials he could not achieve his goal. With a scalar pressure only, it is not possible to generate shear forces, etc., and those are badly needed for a solid world.

True, most of Euler's work on two- and three-dimensional elasticity,[11] especially the "Genuina principia doctrinae de statu aequilibrii et motu corporum tam perfecte flexibilium quam elasticorum" [1771c] and "De gemina methodo tam aequilibrium quam motum corporum flexibilium determinandi et utriusque egregio consensus" [1776b], were written long after 1760. But his one successful work, "De motu vibratorio tympanorum" [1766b] also dates from 1759 and the subject certainly must have occupied him earlier. It may well have been the lack of a three-dimensional theory of elasticity which stopped Euler from pursuing his speculation.

Did the *Anleitung* find an echo?

I have mentioned already that Faraday had read and quoted repeatedly Euler's *Lettres à une princesse d'Allemagne*. Once, speaking about magnetism, he says:

---

[11] See [Truesdell 1960a: 320] and also [Truesdell 1964].

> There are at present two, or rather three general hypotheses of the physical nature of magnetic action. First, that of aethers, carrying with it the idea of fluxes or currents, and this of Euler has set forth in a simple manner to the unmathematical philosopher in his Lettre; in that hypothesis the magnetic fluid or aether is supposed to move in streams through magnets, and also the space and substances around them"... "My physicohypothetical notion does not go so far in assumption as the second and third of these ideas, for it does not profess to say how the magnetic force is originated or sustained in a magnet; it falls in rather the first view [i.e., Euler's], yet does not assume so much [1855a: 81-82].

But Faraday could not know the *Anleitung*.

There is a puzzling sentence in the posthumous work of B. Riemann, Concerning his own philosophical investigations, a note in his papers says:

> My main work involves a new opinion of the known laws of nature, expression of the same means others fundamental ideas, through which the use of the experimental data became possible to the exploration of her/its/their connection over the interaction between heat, light, magnetism and electricity. I was led mainly to it through the study of the works Newton, Euler and – on the other hand – Herbart.[12]

This note is not dated but it is in a close relation to what he writes to his brother in a letter written 28 December 1853 [1892: 547].

What did Riemann allude to in Euler's work? One thinks first of the *Lettres à une Princesse d'Allemagne*. The second, and especially the third, of Riemann's essays on natural philosophy stand in a close relation to Euler's sequence of three papers on mathematical acoustics mentioned above. In the third one, Riemann discusses Euler's scalar wave equation and applies it to motions that cause phenomena of optics and gravity [1892: 536-538, § b, c].

These notes were probably written before Euler's *Anleitung* eventually appeared in 1862. The Riemann Archives in Göttingen seem not to show traces of this book.[13] But given Riemann's interest in Euler's work and ideas, it is not unlikely that he may have read the *Anleitung*. He may even have done so in Pisa, where he went twice in 1863 and in 1864, since, as is well known, he had many friends among the Pisan mathematicians and physicists. His biographer R. Dedekind calls these stays "a true point of light in his life" (*einen wahren Lichtpunkt in seinem Leben*) [Riemann 1892: 556]. No one who knows Pisa, Pisan scientists, and especially the Directors and their wives, will doubt this statement!

---

[12] *Meine Hauptarbeit betrifft eine neue Auffassung der bekannten Naturgesetze – Ausdruck derselben mittelst anderer Grundbegriffe – wodurch die Benutzung der experimentellen Data über die Wechselwirkung zwischen Wärme, Licht, Magnetismus und Electricität zur Erforschung ihres Zusammenhangs möglich wurde. Ich wurde dazu hauptsächlich durch das Studium der Werke Newton's, Euler's und – anderseits – Herbart's geführt* [1892: 507].

[13] I am indebted to Dr. U. Neuenschwander for this information.

Acknowledgments

It was a great honor and even more so a special pleasure to be invited to contribute to a volume dedicated to Luigi Radicati. Our friendship (or should I say my discipleship?) dates from fall 1960 when we shared an office at the Institute for Advanced Study in Princeton. There we discussed many problems connected with symmetries, internal and other ones, and we even struggled bravely for several days and evenings with SU(3)-Clebsch-Gordan coefficients.

Since 1962, I have had the privilege of spending many weeks each year in Pisa, first in Piazza Torricelli, then in the even more illustrious Piazza dei Cavalieri. At both places, I spent only delightful hours and in addition, I spent many more in Via del Risorgimento and in Via del Capannone!

Mrs P. Radelet helped me with the manuscript and the references. My wife Ruth herself a sometimes guest in the Timpano, and my friend David Finkelstein provided good and much needed linguistic advice. To all of them go my best thanks.

Euler's Work in Physical Optics[1]

Introduction

The discovery of Ole Rømer in 1676, according to which light propagates at a finite speed, was decisive for all of Euler's work in optics. However, the name of Rømer is not mentioned in either Euler's works on physical optics, nor in the *Letters to a German Princess*. As we will see, Euler's starting point was the aberration, discovered by Bradley in 1725. Still, Rømer's discovery forced physicists to consider light as a kinematic process, directed from the object towards the eye. The question of what is responsible for its propagation was then posed; in other words, what mechanism lies hidden behind optic phenomena?

For Euler, who was born more than thirty years after Rømer's discovery and who lived entirely in the eighteenth century, the finite speed of light was an acquired notion that no longer required discussion. Rather, what was necessary was to try to comprehend this fact and understand its consequences. The same thing applied to the mechanics proposed by Newton for explaining the planetary system: as an explanation it was certainly not sufficient for Euler, but it could be accepted unhesitatingly, it could be defended if necessary, and it could serve to achieve new goals. As for the foundations of hydrodynamics laid by Newton, Daniel and Johann I Bernoulli, as well as for Jacob Bernoulli's foundations of elasticity, which were more important for Euler's optical works, he might have found them insufficient, but they constitute an attempt that he accepts as such unreservedly.

What characterizes Euler's work in all domains of physics is the search for the principles of dynamics from which, according to him, the explanation of all phenomena must derive. The point of departure for his investigations was always mechanics, for which, while still very young, he had projected a program of systematic works for the treatment of point mechanics, rigid bodies, hydrodynamics, elasticity, etc., a project that was carried out only in part. But in all these domains, notably in hydrodynamics, Euler arrived at decisive results published in various *mémoires*. His research in hydrodynamics and elasticity was admirably presented by C.A Truesdell in his introductions to volumes 11, 12, and 13 of the second series of Euler's *Opera omnia* [1954; 1960a].

---

[1] Paper given at the conference "Roemer et la vitesse de la lumière", 16-17 June 1976, Centre National de la Recherche Scientifique, Paris. Originally published as "L'oeuvre d'Euler en optique physique", in *Roemer et la vitesse de la lumière*, R. Taton, ed. Paris: Vrin, 1978, pp. 207-224.

Beginning with the study of these basic principles, Euler then turned to other domains of physics and wrote several works that outline a program for research. One such work is, for example, "Dissertatio physica de sono", his dissertation on sound [1727]. We also have some "programs" for optics, thermodynamics and magnetism. Each time, the program is based on the same assumption: all phenomena must be derived from principles of dynamics. Truesdell [1964] has made clear that Euler succeeded in completing the foundations of Newtonian mechanics, adding to the principles of linear moment that of moment of momentum, while stressing the independence of the second in the case of continuous media.

Euler studied optics throughout his whole life. Not taking into account his works on geometrical optics, which by themselves fill six volumes of the *Opera Omnia*, there are still fifteen works dedicated to "physical optics", whose realization extends over a period of about forty years.[2] To give an exact idea of the interest and the extent of his activities, I mention his "Expériences pour déterminer la refraction de toutes sortes de liqueurs transparentes" [Euler 1758], where we discover Euler as an experimental scientist, proposing experiences achieved by means of an experimental device of his invention. His son Johann Albrecht carried out some of these and thus discovered that the refractive index of a liquid depends on the concentration and the temperature.[3] This fact clearly shows that, contrary to a current affirmation, for Euler the study of the physics could not be reduced to a mere application of analysis. However, in what follows I will limit myself to the works dedicated to the propagation of light. As it is not possible here to give an overview of all aspects of it, I will focus on the points that permit me to show how the finite speed of light forces optics to assimilate principles of mechanics that, because of this, can be extended to all domains of physics.

What follows will be dedicated to a comparison of optics and mechanics, to the principles of wave optics, to the applications of optics to astrophysics, and to the influence of Euler's work of on his successors.

The comparison of optics and mechanics

In his first work in this domain, "Explicatio phaenomenorum quae a motu lucis successive oriuntur" [1750a], optics is compared to mechanics. It is interesting to note that Euler's work contains a discovery very close to Rømer's, who provided the stimulus for it.

In effect, Euler had learned about Bradley's discovery of aberration. As was his habit, he immediately tried to calculate this effect in complete generality in the setting of a wave theory. He first calculated the effect in the case where the observer is at rest and the emitter moves; he then reduced the second, more interesting case, where the emitter is at rest and the observer moves, to the first one, by making a translation of the entire system with constant velocity.

---

[2] These are found in *Opere omnia* III, 5.

[3] See also "Euler's Writings on Optics, Electricity and Magnetism" in this volume, pp. 211-223.

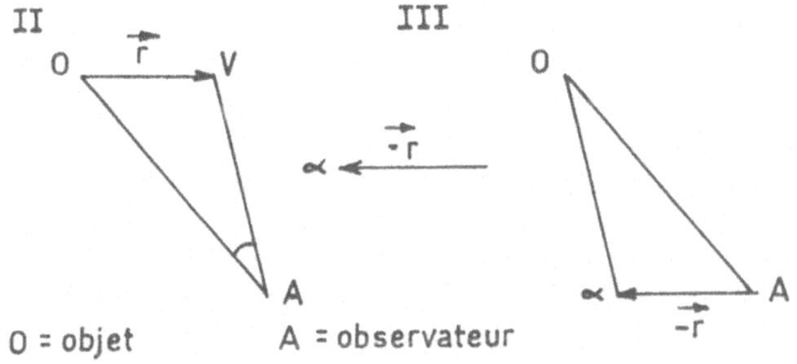

He repeats the calculation with another method based essentially on an analysis of the composition of velocities. He then finds a result that differs from the first, even though the difference is only of the second order in v/c. To explain this contradiction, he says:

> 23. A look at the work shows that an error has been made in the first deductive process. Since – while reducing the third case to the second – we applied to the whole system (in which both the object O and the observer A are) a speed r towards direction A . we should have mentioned if the same movement had to be applied also to the medium in which the rays propagate. In fact, if the medium stays still (like in our case), the case that has been reduced will be totally different from the case that we propose, since in the case that we propose the medium is at rest together with the object, while in the case that has been obtained by transformation, the medium and the observer are at rest. The difference between the two cases is thus evident and it clearly shows why the reduction is inappropriate. The speed r, towards direction Aα should apply also to the medium itself, and if the movement applies to the rays too, consequently the sinus of the angle OAV will be mr/c, like in the other case.[4]
>
> 24. But since the latter deductive process is consistent with the truth, and since the angle by which the true locus differs from the apparent locus is such that its sine is equal to mr/c, and its tangent is not equal to $\frac{mr}{c-\mu r}$, it is

[4] 23. *Hoc dum perpendemus, mox intelligemus in priore ratiocinio vitium esse commissum. Cum enim reductione tertii casus ad secundum toti systemati, in quo cum objectum O tum spectator A versantur, motum secundum directionem Aα celeritate r tribuissemus, definiri debuisset, utrum similis motus media, per quod radii propagantur, simul sit impressus an non. Namque si, uti fecimus, medium in quiete relinquatur, casus, ad quem reductio est facta, omnino erit diversus a casu proposito, quia in casu proposito medium una cum obiecto quiescebat, in casu autem mutato medium habebatur quiescens cum spectatore : ex quibus dissimilitudo casuum, ac proinde illegitima reductio clare apparet. Ipsum itaque medium in directione Aα simul promoveri debuisset celeritate r, qui motus si pariter in radio transferatur, prodibit prorsus ut altero modo anguli OAV sinus* $= \frac{mr}{c}$.

> clear that the object at rest will appear in some way to an observer in motion, while the object in motion will appear in some other way to an observer at rest, even though the second motion is equal (in absolute size) but opposite. This difference is due to the fact that we supposed that light and sound propagate in waves. However, the motion of the object that sends out rays does not alter the rays speed. In fact, if the medium is in motion, its own movement combines with the motion of the rays. It speeds up or slows down the propagation of the waves depending on whether the medium motion is in the same direction, or opposite to the movement of the rays. It thus becomes clear to everybody that what we just discussed about light is also true for sound.[5]

Euler thus discovered a fundamental difference between the wave and corpuscular theories, a difference which, in principle, would have permitted him to settle the question, if the effect had not been too small to be measured.

This difference is related to what we today call "the principle of relativity", which is the basis of all mechanics. This is valid in motion of bodies but not for the motion of waves, because there the ether defines an absolute frame of reference. Euler says further:

> 25. The second and third cases would be perfectly similar if light did not propagate in waves, but really came from an emitting source. In fact if we suppose that rays composed of light particles are sent out by a luminous resting body, at the speed c that we assign to light, then, if this same body moves, it will add its own motion to the movement of the rays. But whether or not the medium moves, this won't affect the ray motion, since the medium may be considered as void. From this hypothesis comes the other correction to the second case about apparent locus. If the luminous body moves at speed s along a direction OV, this motion will affect the natural speed c of the rays: the alteration being due to both the direction in which the rays are sent out, and their own speed which will be increased or decreased.[6]

---

[5] 24. *Cum igitur posterius ratiocinium cum vertitate conspiret, atque anguli, quo locus obiecti verus ab apparente discrepat, sinus sit* $\frac{mr}{c}$, *non autem eius tangens sit* $= \frac{mr}{c-\mu r}$, *perspicuum est aliter obiectum quiescens spectatori moto esse appariturum, aliter obiectum motum spectatori quiescenti, etiamsi motus posterior priori sit aequalis et oppositus. Ratio huius discriminis in eo latet, quod lucem instar soni per motum undalatorium propagari posuimus, quo pacto motus obiecti radios emittentis celeritatem radiorum non afficit ; verum medium, si moveatur, eundem motum cum motu radiorum miscebit ; ac propagationem undularum vel accelerabit vel retardabit, prout motus medii motui radiorum vel sit secundus vel adversus. Cuique haec plana fient, si, quae hactenus de luce sumus commentati, ad sonum atque auditum accommodentur* [Euler 1750a: 55].

[6] 25. *Perfecta autem similitudo inter casum secundum et tertium conservaretur, si lux non motu undulatorio, sed actuali eiaculatione ex carrare lucente emittatur. Si enim ponamus particulas lucis, quae radios constituunt, ex cornore lucido quiescente actu explodi celeritate c, quam luci tribuimus, idem corpus, si moveatur, suum motum cum motu radiorum coniunget : neque medium, sive moveatur sive quiescat, quicquam motum radiorum afficiet ; instar vacui enim considerari poterit.*

and a little later:

> 28. We will thus go on with this powerful hypothesis according to which the light rays are supposed to have been sent out from the luminous source by an explosion, and we also suppose that the rays coming from the sun are sent out at very high speed from this sun, reaching us in approximately 8 minutes. Although this hypothesis is less proper than the other one according to which light propagates like sound, it is however better appropriate to our purpose and makes a *movement composition* possible, which could not be achieved with the other hypothesis.[7]

These remarks constitute a first contribution to the discussion that will give way to the experience of Michelson and the ideas of Lorentz, Poincaré and Einstein.

In the next passage one can see the importance that Euler attributed to the principle of relativity and its repercussion on the whole of physics:

> 35. Following the correct laws and rules of mechanics, we must apply to a certain (composed) system a uniform linear motion, so that none of the phenomena will be (apparently) altered, but it is *not possible* to apply a *non uniform or curvilinear* motion.[8]

The result shows how far the investigations of the dynamics of light still had to go.

This first work of optics certainly influenced Euler's later research in mechanics. In his *Mechanica* [1736] he had already presented and examined in detail the principle of relativity and its consequences. He will come back to it in the *Anleitung zur Naturlehre* [1862c], written between 1745 and 1750. But here in Euler's first work a new element, the observer (*der Zuschauer*) makes his entry into physics. In technical language, Euler discovers "the passive aspect", as Eugene Wigner called it, of the principles of symmetry that he develops in this paper with his usual care.

In the *Anleitung* he says:

> 80 ... namely because we can never see the true motion of a body but rather our senses always show us directly only the apparent motion of bodies, we regard the apparent motion as a true one and study whether is necessitated

---

Ex hac autem hypothesi alia reperietur correctio situs apparentis in casu secundo. Si enim obiectum lucidum moveatur in directione OV celeritates, hoc ipso motu celeritas radio OA naturalis c afficietur, tam ratione directionis, secundum quam eiiciuntur, quam ratione celeritatis, quae vel augebitur vel diminuitur [Euler 1750a: 55].

[7] 28. *Persequemur igitur hic potissimum eam hypothesin, qua radii lucis ex obiecto lucido actu explodi ponuntur, atque assumamus radios, qui ex sole ad nos perveniunt, summa celeritate ex ipso sole esse eiaculatos, unde tempore 8 circiter minutorum ad nos pertigerint. Quamvis enim haec hypothesis minus sit probabilis quam altera, qua lumen instarsoni propagari statuitur, tamen magis est accomodata ad nostrum institutum atque motus compositionem recipit, cui us altera hypothesis minus est capax* [Euler 1750a: 56].

[8] 35... *licet enim salvis legibus et regulis mechanicis uni verso cuidam systemati corporum, motum aequabilem in directum imprimere, ita, ut nil in phaenomenis mutetur, at vel* motum ineaequabilem vel curvilineum tribuere minime licet [1750a: 60].

> by the conservation of those same forces or not. Afterwards when we have verified through other circumstances whether forces affect the body or not and how broadly these agree with those we have found, then we can conclude from this how much of the apparent motion is distinguishable from the true [motion].[9]
>
> 81. If the spectator moves away equally swiftly in a straight line and right after parallel lines estimates the areas, that is, then the maintenance of the apparent motion becomes to the so, how much the same of it maintains may be distinguished, just those strengths necessitates, as to the maintenance of the true motion.[10]

Doubtless Euler was led to make this discovery through practical investigations of the kind required by astronomy, and especially through his investigation of the aberration. However, while composing two velocities is easy enough, doing the inverse – that is, breaking down an observed movement into its components, one of which is the observer's movement – is another matter. In fact, no one seems to have predicted the aberration. Indeed, after Bradley first observed it, it still took him more than two years to explain it.[11]

The principles of wave optics

The question, "What is the mechanism underlying optic phenomena?" has been answered in essentially two ways: a luminous ray is composed of corpuscles, which is the answer given by Newton and approved by his followers, and a luminous ray is a wave, which is often associated with Huygens.

It seems that Euler opted without hesitation for the wave theory, which he sets forth in the *Nova theoria lucis and colorum* [1746]. Before expressing the new ideas that lie at its base, we must make some historic remarks.

Euler is often labeled by historians of optics as a Cartesian and an adherent to the wave optics of Huygens. It seems to me that these two characterizations lead us astray. Even though Euler always opposed the corpuscular theory of Newton, a simple look at Euler's

---

[9] 80 ...*Weil wir nämlich niemals die wahre Bewegung eines Körpers sehen können, sondern unsere Wahrnehmungen uns immer unmittelbar nur die scheinbare Bewegung der Körper anzeigen, so sehen wir die scheinbare Bewegung als eine wahre an und untersuchen, ob zur Erhaltung derselben Kräfte erfordert werden oder nicht. Wenn wir hernach durch andere Umstande vergewissert werden, ob Kräfte auf den Körper wirken oder nicht und in wie weit dieselben mit den gefundenen Übereinkommen, so können wir hernach daraus den Schluss machen, wie viel die scheinbare Bewegung von der wahren unterschieden sei* [Euler 1862c].

[10] 81. *Wenn der Zuschauer gleichgeschwind in einer graden Linie fortrückt und die Gegenden richtig, das ist, nach gleichlaufenden Linien schatzet, so werden zur Unterhaltung der scheinbaren Bewegung, wie sehr dieselbe auch von der wahren unterschieden sein mag, eben diejenigen Kräfte erfordert, als zur Unterhaltung der wahren Bewegung* [Euler 1862c].

[11] For details about Bradley's discovery, see [Hoskin 1978] and [Grant 1852: 338]. Additional – linguistic – evidence pointing out this connection is the fact that in the *Anleitung* Euler speaks of the *Zurschauer* instead of the *Beobachter*, in the very same way that he had spoken of the "spectator" in the "Explicio", and not of the "observer" as we would expect.

work permits us to discover that Newton had a profound influence on Euler's works in optics. In fact, Newton is mentioned there more often than anyone else, not only according to the general custom of the times, where it was only usual to mention one's adversaries, but also because Euler had a profound respect for Newton; he once called him *Summus Neutonus*.

Huygens's ideas, presented in the *Traité de la Lumiére* [1690], correspond more to geometrical optics, in the form that Hamilton gave it, than to wave optics as we intend them today, Actually, Hamilton's optics contains a "principle of Huygens," while, strictly speaking, the wave optics in the sense of Fresnel and Maxwell does not.[12] However, it is the analogy with acoustics that constitutes the main guide for both Euler and Huygens. Sound is transmitted by air and light by the ether, a "subtle fluid" of very low density but possessing an enormous elasticity, both of which can be estimated thanks to the relation of Newton that correlates them to the speed of light.

It seems to us that the fundamental notion of wave optics is that of *periodicity*. Newton recognized that the differences in the refraction of the colors can be explained by differences of periodicity; however, this periodicity has its origin in the bodies ("fits of easy reflexion" and "fits of easy transmission" [Newton 1704: Prop. XIII]) and is described in an adequate manner by means of the wavelength. But wavelength is not an invariant characteristic of the ray of light; it will be different, for example, after refraction. In contrast, in the *Nova theoria*, Euler characterizes a ray by its *frequency* and writes:

> In addition to the direction of the rays which stimulates our sense of vision, we must also pay attention to the impulsion frequency, since the nature of each ray is defined by these two characteristics. In fact, while the impulsions hit and touch the organ of vision, and sense is stimulated by the hits of the impulsions, sensorial perception will depend on two aspects: direction first, and then the way these hits occur, namely the number of hits received by the eye in a certain lapse of time. It is clear that, wherever the eye is, it will receive as many hits as the body (located) in A will send out vibrations in the same time frame.[13]

In effect, the frequency of a ray remained constant along its whole path; quantum mechanics linked this constancy to the conservation of its energy.[14]

---

[12] It is, however, interesting to note that a "principle of Huygens" is present in wave mechanics since it is the first order derivative in relation to time.

[13] 65. *Praeter directionem radiorum autem, secundum quam sensus visus excitatur, in ipsis imprimis pulsuum frequentia est consideranda, quippe quibus duabus rebus natura cuiusque radii continetur. Ubi enim pulsus appellunt atque organum visus afficiunt, queniam sensus ab ictibus pulsuum excitatur, perceptio a duabus rebus pendebit: primum scilicet a directione, secundum quam isti ictus perficiuntur, tum vera a numero ictuum, qui dato tempore oculum percutiunt. Manifestum autem est, ubicunque, oculus fuerit constitutus, totidem ictus ad eum pervenire, quot vibrationes corpus in A (= emitter) eodem tempore absolverit* [1746b; *Opera* III 5, p. 23].

[14] P. Costabel informed me that Malebranche had already had an idea about the frequency, but at the time I thought that it was hardly likely that Euler knew these works. However, I later found Malebranche mentioned in the mémoire of Johann Bernoulli "Recherches physiques et géométriques sur la question: comment se fait la propagation de la lumière ; pièce qui a remporté le prix de

Euler later treated the question of whether of the frequency of the red or violet light is the largest in several of his works. At first he oscillated between the two possible answers and then finally settled on the right one.[15]

Another fundamental result of this program is the statement of the present "superposition principle" presented by Euler in the following manner:

> But to me it seems not at all explicable that two or more rays that meet with an incredible speed don't disrupt each other's movement. Indeed, when several rays combine mutually, are introduced through a tiny hole into a dark cavity, or are focused by means of a mirror or a lens, no modification of the direction of the different rays is observed, even though it is still impossible that very frequent and very intense collisions don't occur.[16]

He gives a set of other, more or less valid arguments against the corpuscular theory of Newton. However, the absence of one of them, that is, the deflection of light under the influence of gravity, is quite striking. It is even more astonishing that Newton himself never considered this phenomenon. According to Whittaker this effect, characteristic of a theory of corpuscular optics, was calculated by J. Soldner in 1801 [Whittaker 1951, 2: 180]. Soldner obtained half of the value furnished by the general theory of relativity. Later, Laplace calculated further the minimal mass of a star that no longer emits light. As we know today, such a phenomenon, known as "black holes," is accounted for in the general theory of relativity, but for different reasons.[i]

Some applications to astrophysics

One of Euler's works, "Réflections sur les divers degrés de lumière du soleil et des autres corps célestes" [1752c], dedicated to what we would classify today as questions of astrophysics. Bouguer had effectively measured the intensity of the radiance of the Sun, the planets and the fixed stars. His results stimulated Euler to undertake some calculations in order to get more information on these various bodies. It should be noted in passing that he concludes that the fixed stars had to be about two light years away; in his *Lettres* [Euler

---

l'Académie Royale des Sciences" [1736]. According to the mémoire, Johann was influenced by Malabranche, and so was Jacob Bernoulli. This work (discussed in [Whittaker 1951: 95-96]) permits us to measure the progress that Euler had made; Johann arrived at the notion of frequency (though he doesn't use the term), but he doesn't consider it to be the fundamental notion that characterizes colors. For him, light propagates more quickly in an arbitrary environment than in an empty environment, and the violet rays are the quickest and seem to have a smaller frequency than red rays, since the masses associated with the red are smaller.

[15] See "Euler's Writings on Optics, Electricity and Magnetism" in this volume, pp. 211-223.

[16] 19. *...sed hoc nullo modo mihi explicabile videtur, quomodo duo pluresve radii ex diversis regionibus taro incredibili celeritate sibi occurrentes se mutuo in motu non perturbent. Quando enim plurimi radii sive per minimum foramen in conclave obscuratum introgrediuntur, sive ope speculi vel vitri caustici in focum collecti se mutuo decussant, nulla prorsus alteratio in singulorum directione percipitur, cum tamen omnino fieri non possit, ut non frequentissimae simulque vehementissimae collisiones eveniant. Hocque argumentum maximam vim habere videtur ad istam opinionem funditus evertendam* [1752c: 7].

1768-72: letter XXI] he will speak of a distance superior to sixty or more light years. But the most interesting contribution concerns the Sun. Applying the law of Kepler and introducing the result of Bouguer, he concludes:

> 22 Mr. Bouguer established a table of the reduction of the light of the stars according to their height, from which we can see that sunlight, as it would shine if its rays did not fade, being at an height of 31°, is to light as 2 is to 3. From this observation we can determine that the real brightness of the Sun is to the light of a candle as 175000 is to 1. And if the rays of the Sun had lost something while crossing the space before reaching to us, in addition to their natural decrease, proportional to the square of the distance, the brightness of the Sun would be even greater.[17]

First, note the proviso:

> And if the rays of the Sun had lost something while crossing the space before reaching to us, in addition to their natural decrease, ...

In fact, at the beginning of his work, Euler explicitly hypothesizes that light is not absorbed by the ether!

Euler deserves credit for have formulated the first "enigma of the sun" as a scientific problem. What is this? Kepler's law, according to which the intensity of light decreases and is inversely proportional to the square of the distance, immediately derives from a corpuscular theory. In the case of a wave theory, it is the *energy* that is propagated. In fact we see emerge in this work a hint of the concept of energy of a field (Euler uses the word "force"). Here we can see a parallel with what we observe in his works of mechanics (cf. [Truesdell 1960a: 173-175]. It is illustrated by the following passage:

> 4. I won't discuss the explanation of the theory further here: I shall only point out that the force of the rays that are sent out by a body depends on the strength of the vibration that applies to even the least particles of this body. I am not talking here of the frequency of the vibrations, nor of their number in a certain time frame, because I showed that the color of the rays depends on this frequency; but whatever this frequency is, we can conceive that all the particles of the body, similarly to a stretched rope, may move according to either long or short oscillations. The stronger will be the rays, the brighter will be the light: and this is the first characteristics of the luminous bodies on which my further studies will focus, that I shall name the splendor of the

---

[17] 22 ... *Mr. Bouguer a donné une table pour la diminution de la lumière des astres à chaque hauteur, d'où l'on voit que la lumière du Soleil étant à la hauteur de 31° est à la lumière, dont il éclaireroit, si ses rayons ne perdoient rien de leur force, comme 2 à 3. Ayant donc égard a cette circonstance, on trouvera que le véritable éclat du Soleil est a celui d'une bougie comme 175 000 est a 1. Et si les rayons du Soleil avoient perdu quelque chose en parcourant l'espace de l'éther jusqu'a nous, outre la diminution ordinaire, qui suit la raison des quarrés des distances, l'éclat du Soleil deviendroit encore plus grand* [1752c: 138].

> luminous body, the quality of which is proportional to the force of the rays, or to the force of the vibration.[18]

The next idea, just as audacious, illustrates from another angle this desire to introduce the principles of dynamics into optics. Thanks to his research in hydrodynamics and acoustics, Euler had realized that wave phenomena, such as sound, could exercise a *pressure* due to nonlinear terms in the equations [Euler 1777]. He used this idea to explain phosphorescence, already studied by the English physicist Wilson. We know today that this is really a nonlinear effect, but a complete understanding had to await quantum mechanics.

Then, according to Whittaker, Euler made a big discovery regarding the tail of the comets.[19] While following an idea of Kepler, he proposed an explanation of the tail of the comets and the fact that it is always in a direction opposite to that of the Sun through the *pressure* exercised by solar radiation on the matter that composes the tail.[ii] Later this statement would be reformulated and elaborated by Maxwell.

Influences of Euler's work of on his successors

It might seem surprising that Euler never tried to apply to optics the equation that he had developed and used in acoustics. The reason is certainly the following. The wave equation, the history of which is also described by Truesdell [1960a] was first proposed by d'Alembert in 1748 for a one-dimensional continuum. By the time this was finally established in 1759 for three-dimensional continua, Euler was interested mainly in other subjects, and unfortunately never returned again to physical optics.

In any case, we know that he exercised both a direct and an indirect influence on Young, Fresnel and especially on Cauchy, who would become, according to Truesdell, the leading exponent of the theory of elasticity.

In order to be able to trace the influence of Euler's ideas and their ramifications in all of physics, we must also speak of electricity and magnetism. According to Euler, these two phenomena are also demonstrations of the elasticity of the ether. This is not the appropriate place to go into this question; it is sufficient to cite Whittaker:

---

[18] 4. *Je ne m'arrêterai pas ici d'avantage à l'explication de cette Théorie: je remarquerai seulement que la force des rayons, qu'un corps lance, vient de la véhémence du mouvement de vibration, dont les moindres particules du corps sont agitées. Je ne parle pas ici de la fréquence des vibrations, ou de leur nombre qui s'acheve dans un certain tems car j'ai fait voir, que la couleur des rayons dépend de cette fréquence: mais quelle que soit cette fréquence, on conçoit que les parcelles du corps, tout comme une corde tendue, peuvent faire dans leur mouvement des excursions plus grandes ou plus petites. Et c'est de là que les rayons seront d'autant plus forts ou plus faibles, plus ces excursions seront grandes ou petites. Or plus les rayons seront forts, plus la lumiere sera brillante: et de là découle la premiere qualité des corps lumineux, 'qui entrera dans mes recherches, et que je nommerai la splendeur du Corps lumineux, dont la qualité est proportionnelle à la force des rayons, ou bien à la force du mouvement de vibration* [Euler 1752, *Opera omnia* III, 5: 131].

[19] Cf. [1951, 1: 274] where Whittaker mentions Euler's "Recherches physiques sur la cause de la queüe des comètes, de la lumière boréale..." [1746c].

> [Euler] anticipated Maxwell in asserting that the source of all electrical phenomena is the same ether that propagates light: electricity is nothing but a derangement of the equilibrium of the ether [Whittaker 1951, 1: 98].

In his E 109 "Dissertatio de magnete" [1749b], Euler first assigned magnetic phenomena to the flux "of the rarest parts of the ether", but in the *Letters* he introduces a second fluid even rarer than the ether [Euler 1768-72, II: Letter CLXXVII; Eng. trans. letter LXII, p. 215].

His exposition of his theories of electricity and magnetism in the *Letters* found an important reader – perhaps the most important reader possible – in Faraday, who could not fail to be influenced. As this influence has not yet received the attention it deserves, I take the liberty here of putting together some quotes. In his work "On the Physical Character of the Lines of Magnetic Force" [1852], Faraday wrote:

> To acknowledge the action in curved lines, seems to me to imply at once that the lines have a physical existence. It may be a vibration of the hypothetical aether, or a state or tension of that aether equivalent to either a dynamic or a static condition; or it may be some other state, which though difficult to conceive, may be equally distinct from the supposed non-existence of the line of gravitating force, and the independent and separate existence of the line of radiant force (3251)*. Still the existence of the state does not appear to me to be mere assumption or hypothesis, but to follow in some degree as a consequence of the known condition of the force concerned, and the facts dependent on it [1852: 410].

and we find in the note

> *See Euler's views of the disposition of the magnetic force; also of the magnetic fluid, or aether and its streams. *Letters*, vol. ii, letters 62, 63.

Euler is mentioned several times in Faraday's paper "On some Points of Magnetic Philosophy":

> 3301. There are at present two, or rather three general hypotheses of the physical nature of magnetic action. First, that of aethers, carrying with it the idea of fluxes or currents, and this Euler has set forth in a simple manner to the unmathematical philosopher in his Letters*; in that hypothesis the magnetic fluid or aether is supposed to move in streams through magnets, and also the space and substances around them. Then there is the hypothesis of two magnetic fluids, which being present in all magnetic bodies, and accumulated at the poles of a magnet, exert attractions and repulsions upon portions of both fluids at a distance, and so cause the attractions and repulsions of the distant bodies containing them. Lastly, there is the hypothesis of Ampère, which assumes the existence of electrical currents round the particles of magnets, which currents, acting at a distance upon other particles having like currents, arranges them in the masses to which they belong, and so renders such masses subject to the magnetic action. Each of these ideas is varied more or less by different philosophers, but the three distinct expressions of them which I have just given will

> suffice for my present purpose. My physico-hypothetical notion does not go so far in assumption as the second and third of these ideas, for it does not profess to say how the magnetic force is originated or sustained in a magnet; it falls in rather with the first view, yet does not assume so much [1855a: 81-82].

and where he has indicated the note with an asterisk, he gives the reference:

> *Euler's *Letters,* translated, 1802, vol.i. p. 214; vol. ii. p. 240, 242, 244.

Further on we find:

> 3305... "For if there be such physical lines of magnetic force as correspond (in having a real existence) to the rays of light, it does not seem so very impossible for experimenters to touch *them*; and it must be very important to obtain an answer to the inquiry respecting their existence, especially as the answer is likely enough to be in the affirmative. I therefore purpose, without asserting anything regarding the physical hypothesis of the magnet more strongly than before (3299.), to call the attention of experimenters, in a somewhat desultory manner, to the subject again, both as respects the deficiency of the present physical views and the possible existence of lines of physical force, concentrating the observations I may have to make about a few points – as *polarity*, *duality*, &c., as occasion may best serve; and I am encouraged to make this endeavour by the following considerations. 1. The confirmation by mathematicians of the truthfulness of the abstract lines of force in representing the direction and amount of the magnetic power; -2. My own personal advantageous use of the lines on numerous occasions (3174.); -3. The close analogy of the magnetic force and the other dual powers, either in the static or dynamic state, and especially of the magnet with the voltaic battery or any other sustaining source of an electric current; -4. Euler's idea of magnetic aethers or circulating fluids; -5. The strong conviction expressed by Sir Isaac Newton, that even gravity cannot be carried on to produce a distant effect except by some interposed agent fulfilling the conditions of a physical line of force; -6. The example of the conflict and final experimenta1 settlement of the two theories of light [1855a: 84-85].

and still further:

> 3317...It is assumed that the space is in a state of magnetic darkness (3305.), an assumption so large, considering the knowledge we have of natural powers, and especially of dual forces, that there is none larger in any part of magnetic or electric science, and is the very point which of all others should be held in doubt and pursued by experimental investigation. It is as if one should say, there is no light or form of light in the space between the sun and the earth because that space is invisible to the eye. Newton himself durst not make a like assumption even in the case of gravitation (3304.), but most carefully guards himself and warns others against it, and Euler[1] seems to follow him in this matter [1855a: 92].
>
> [1]*Letters*, &c. translated, Letter LXVIII., or p. 260-262.

Some references are also found in Faraday's diary [1932-36].

> 13444. The hypothesis is not so much mine as one renewed from old times. Look at Euler's *Letters* and what he says.

And four days later:

> 13471. Perhaps some communications to the Phil. Mag. in aid of the development of the theory I hold (which is Euler's or older) of the following kind.
> 13472. 1. The opening out of the ends of a long wire magnet (13450), and the proofs derived from and connected with it, of the necessity of the surrounding medium.
> 13473. 2. Polarity (of conductors) not in the body only, but in it jointly with and dependent on the like condition of the surrounding medium. This is not the polarity of the magnet [Faraday 1932-36].

Conclusion

I hope to have illustrated the moment in the history of physics when, following the discovery of Rømer, optics is for the first time compared to mechanics. It is also the moment when the notions of mechanics developed by the "geometers", as they called themselves, enter into all domains of physics.

With Young, Fraunhofer, Fresnel and Cauchy, a new era of physical optics will begin. R. Kohlrausch and H. Weber will discover in 1856 that the critical speed of a discharge of a capacitor is more or less equal to the speed of light, and after Maxwell's unification of optics, electricity and magnetism, in the twentieth century the speed of light will become the fundamental physical constant of space-time that marks *the frontier* between temporal relations and spatial relations, between energy and impulse [Whittaker 1951, 1: 232 ff.].

Acknowledgments

I was very grateful to M.R. Taton and M.J. Lévy for the invitation to the conference that they organized. I took advantage of the occasion to have discussions with many of the participants, among whom I only mention W. Hartner, J .-P. Verdet, M. Hoskin and K. Pedersen. I am also pleased to Mrs. P. Radelet-De Grave (Louvain), Mr. M. G. Barton (Brighton), as well as J.P. Antoine and Cl. Leroy (Louvain) for their suggestions and help in the completion of this manuscript.

Translated from the French by Sandro Caparrini and Kim Williams

Euler's Writings on Optics, Electricity and Magnetism[1]

We now come to one of the most remarkable men known to history.
Andreas Speiser

Introduction

Writing about Euler's way of doing research, Andreas Speiser said:

> It is therefore not at all, as with Leibniz, a juxtaposition of different sciences, but rather from a main area, mathematics, he made expeditions to more distant areas; he covers one whole range, from the *mundus intelligibilis* to matter, from the philosophy of mathematics to technology, [while] he consciously stays away from other ways [Speiser 1939: 39].

Today, Euler is really thought of as a mathematician above all, despite the fact that his writings on mechanics and physics, including astronomy, account for something more than half of his works. But in the time since plans for the complete edition of his works were made, mechanics became broadly severed from actual physics, and began to be regarded as a sort of middle area between mathematics and physics, since it was considered closer to geometry than to physics. The theory of relativity and quantum theory have taught us better, one point that Speiser strongly emphasized when he lectured on the theory of general relativity.

We know from Riemann [1892: 507] that in an earlier time Euler had been read in a different light, but in more recent decades Euler's originality and significance for mechanics have been once again seen in their true light, thanks to Truesdell's *Introductions* to three volumes of Euler's *Opera omnia* [1954; 1956; 1960a]. Today we can clearly see Euler's endeavor to understand the world *more mechanico*, an idea that was key in Euler's work, one that he held on to his whole life through, and that likewise constantly accompanied his purely mathematical work as well.

I shall come back to the relationship between mathematics and natural science at the end of this essay. It is clear that the program of understanding the world through mechanics, specially the new areas that had opened in physics – including optics,

[1] Originally published in *Leonhard Euler 1707-1783 : Beiträge zu Leben und Werk*. J.J. Burckhardt, E.A. Fellman and W. Habicht, eds. Basel : Birkhäuser, 1983, pp. 215-228.

electricity and magnetism – had to be validated.[2] The goal of this essay is to show how Euler planned to realize his program and what strides he took towards achieving that aim.

Physical optics

Euler wrote about as many works on questions of physical optics[3] as he did on the other two areas treated in this essay put together. This account must therefore necessarily be brief, and for further details, I refer the reader to my preface in the volume *Opera omnia* III, 5.

Euler formulated his program for optics, from which he would later hardly deviate at all, in his *Nova theoria lucis et colorum* [1746]. The work addresses all the phenomena known at that time: on the one hand those of light – vision in general, the origin and propagation of light, successions of pulsations and light rays, reflection and refraction; and on the other hand on the optical qualities of matter: luminosity, reflection, light refraction, and opaque bodies.

In particular, Euler gives serious consideration to Newton's corpuscular optics, which he now opposes with his own wave theory. His *own* wave theory? His *wave theory*? Do we not owe the wave theory to Huygens? Huygens's ideas, as is often remarked, were adequately formulated in Hamilton's geometrical optics, and it is there, not in wave optics, that Huygens's principle receives its full due. But with what right do we now say Euler's *wave theory?* Everybody knows that a wave equation was formulated only much later for acoustics, which was groundbreaking for Euler in optics, by Lagrange and by Euler himself, and for light only in the nineteenth century.

The answer is that optical phenomena are linked to the theory above all through the concept of *periodicity*. Such a concept is not found in Huygens but is, however, found in Hooke and in Newton. However, according to Newton, in some optical phenomena, the periodicity is more conditioned by the structure of the different materials than it is by light itself. What that involves, however, is a *spatial* periodicity, i.e., the *wavelength,* if you will, is the appropriate concept. However, monochrome light waves are characterized by their *frequency,* which remains constant along with their path (trajectory), and Euler, in his *Nova theoria*, was probably the first to state this clearly. This realization also allowed him, at least in principle, to compare theory with the observations. Therefore, one can rightly speak of his having a wave theory.

Of his numerous objections to Newton's theory, here I want to mention only one because it lies at the heart of the matter: Light rays can cross without interfering with each other.[4] In other words, Euler makes the superposition principle the basis of his optics, as Dirac would do later for quantum mechanics.

Now, it was shortly after that, in the context of the great discussion about the solution of d'Alembert's wave equation, that the principle of superposition for the small oscillations of mechanical systems was given in a completely generalized form by Daniel

[2] For the writings on optics, cf. *Opera omnia* III, 5; for those on electricity and magnetism see *Opera omnia* III, 10.

[3] Here I deal with questions of wave optics, those of geometric optics only, in so far as its application to physics lies at the heart of the question.

[4] See "Euler's Work on Physical Optics" in this volume, pp. 197-209.

Bernoulli. On that occasion, Bernoulli mentions Newton's *Opticks* [1704] in passing, but he fails to mention Euler's theory. Further, he never again treated a question of optics. On the other hand, Euler found that Bernoulli's solution of the wave equation through trigonometric series not general enough, and missed seeing at that time how it could be applied to optics. One therefore must conclude that both, each to his own damage, spoke at cross purposes!

But what is the medium in which light propagates as sound propagates through the air? According to Euler, light waves move in the *ether,* an extremely rarified, but extremely elastic medium. It is therefore necessary – again according to Euler – to develop a mechanics of the light ether. However, since this was not yet available, the analogy with sound had to serve, as far as it could.

Euler knew from his investigations in acoustics that not only can a wave field have an energy density but that it can also exert a pressure. From this he concluded the existence of a light pressure and used it to explain two extremely different phenomena. Euler reasoned that a corpuscular theory of optics can never explain phosphorescence:

> You will now see what effect the violet rays must produce on this same red phosphorescent body; first, it is obvious that it doesn't bring its smallest particles to a motion of vibration because of the incompatibility that reigns between the vibrations of the violet rays and those that the body's own particles are disposed to receive. For this reason the total effect of these violet rays will amount to inducing the particles of the body to a certain degree of tension without transmitting to them an actual movement. Therefore as soon as the body is removed from the action of these rays, its smallest particles will begin to rid themselves of their state of tension and will receive the same movement of vibration that is proper to their nature, and hence the red rays that they emit will be even that much stronger, because of the high degree of tension, than if the same body had been exposed to the red rays.[5]

Quantum theory was the first to explain phosphorescence properly, but the essence of Euler's idea is correct.

[5] *Voyons à présent, quel effet les rayons violets doivent produire sur ce même corps phosphorique rouge, et d'abord il est évident qu'ils ne sauroient porter ses moindres particules à un mouvement de vibration à cause de la contrariété qui regne entre les vibrations des rayons violet s& celles que les propres particules du corps sont disposés à recevoir. Far celle raison tout l'effet de ces rayons violets se réduira à pousser les particules du corps à un certain degré de tension sans leur imprimer un mouvement actuel. Donc aussitôt que le corps sera retiré de l'action de ces rayons, ses moindres particules commenceront à se dégager de leur état de tension & recevront le même mouvement de vibration qui est propre à leur nature,&t partant elles repandront des rayons rouges qui seront même d'autant plus forts, à cause du haut degré de tension, que si le même corps avoit été exposé aux rayons rouges* [1777: 76-77].

The other, more famous application is his explanation of Kepler's observation that a comet's tail is always turned away from the sun because of the light pressure [E.67 and E.68; *Opera Omnia* II, 31]. This discovery is even commemorated on the Swiss ten-franc note, although the artist improved the original drawing "aesthetically" and made a mess of it.

How did he fit this theory of a light ether into the general theory of mechanics? Already in one of his first and foremost works relating to this, the "Explicatio phaenomenorum quae a motu lucis successivo oriuntur" [1750a], written in 1739, we find Euler, seemingly unawares, addressing this question. He tries to calculate the aberration, which Bradley had discovered some years earlier, using his wave theory. At first, he treats the simpler case – a stationary observer sees a moving light source – and then progresses to the more difficult one – the moving observer (earth) sees a stationary light source (a fixed star) – in order to lead back to the first by means of a simple transformation. However, it now turns out that the light no longer propagates with the same speed, as the hypothesis of the stationary light ether requires. Where does this contradiction come from? According to Euler, it is simply that the applied transformation would be only permitted, if one could give the ether a speed as well; this, however, is impossible. The ether, as we would say today, distinguishes an absolute frame of reference. On the other hand, the principle of relativity is *valid* in a corpuscular theory. The calculations show that the aberrations differ according to the underlying assumption: The experiment therefore makes it possible to distinguish between the two assumptions! On the other hand, we must add "in principle", because the difference is on the order of $v^2/c^2 \sim 10^{-8}$ and therefore eludes observation. But nevertheless, because of this the question of the nature of light is intimately connected with the principles of mechanics.

Almost half – to be precise, seven out of fifteen – of Euler's writings on optics are dedicated to questions about dispersion. However, I would like to deal with only two points here. First, the question of whether red or violet light has the greater frequency occupies him again and again, and he changed his opinion about it three times, each time on the basis of a theoretical consideration to which he had been led upon hearing about a new experiment. In the *Nova theoria* he still attributes the greatest frequency to red; in two later works he corrects this opinion on the grounds of his theory of the observations of thin layers of colors, among other reasons. Then, however, he is led back to the first, wrong, opinion by a contemplation of the elasticity of thin plates, and then he finally comes back to the right one. One can see how much more narrow and more experimental were the foundations that one built on at that time and how uncertain the end results were. One also sees, however, the constancy with which Euler pursued the questions of the nature of light.

I can't refrain from pointing out a special work of Euler's here, the "Examen d'une controverse sur la loi de réfraction des rayons de différentes couleurs par rapport à la diversité des milieux transparens par lesquels ils sont transmis" [1755b]. It is admittedly not a work on physics but is rather one on mathematics, to which he had been led by an optical hypothesis (one, however, which he himself would later show to be wrong). Euler was convinced that a universal law of dispersion applies to optics, which can be briefly stated thus: we know the refraction indexes r and v respectively for red and violet light at the transition from medium a to medium b, as well as the refractive index R for red light at the transition of medium A to medium B, so from this we can calculate the refractive index V for violet light at this same transition. According to Euler, in manufacturing his instruments, Dollond followed Newton's (?) law:

$$\frac{r-1}{v-1}=\frac{R-1}{V-1}.$$

Euler now shows that one meets with a contradiction if one puts in for R and V the values r^{-1} and v^{-1} that correspond to the transition $b \to a$.

One could objects that the law is valid only for transitions from a rarer to a denser medium, and that for the transition from the denser to the rarer medium one must put in the reciprocal values. However, if we combine the formulas for two transitions $a \to b$ and $b \to c$, and compare the result with the formula for the direct transition $a \to c$, we meet again with a contradiction! Euler remarks:

> And if one pays only a bit of attention to the conditions that the nature of refraction requires, one can easily see that no other proportion except the logarithmic one can satisfy them.[6]

In other words: Euler explicitly formulates two of the axioms for groups, as we say today, and uses them to prove the uniqueness of the following law:[7]

$$\frac{\log r}{\log v}=\frac{\log R}{\log V}.$$

Finally, a consideration must be made about Euler as an *experimentalist.* In old age, time and health no longer allowed him to perform experiments himself, as he had apparently done in his youth; as we know, he first lost one eye, and later the second one.

It is likely, however, that he invented machines and optical instruments, and he devised an instrument for determining the exact measurements of the refraction index of liquids (see Euler's "Expériences pour déterminer la réfraction de toutes sortes de liqueurs

---

[6] *Et pour peu qu'on fasse attention a ces conditions, que la nature de la réfraction exige, on s'appercevra aisément qu'aucune autre proportion que celle des logarithmes ne sauroit leur satisfaire* [1755b : § IV].

[7] For details, I refer to my discussion in *Opera Omnia* [III, 5: XL ff.], where I note where individual points are formulated differently today.

transparentes" [1758] and his son Johann Albrecht's "Sur la réfraction des fluides" [1769]). The device was simple enough: the liquid was poured into the lens-shaped cavity between two glass lenses. Then Euler calculates the conditions with extreme care and discusses the conditions under which the desired refraction becomes as large as possible. A thorough knowledge of the geometry of this lens system then allowed a rather precise measurement of the refractive index. Using this device, his son, Johann Albrecht, then executed measurements.

If one reads through the series of the five works of Johann Albrecht Euler [1769], one can't help being extremely perplexed at first: he worked in part with indefinable liquids such as "infusion of nutshells or peach leaves" or simply "French wine," and so forth! We should not forget that exact chemistry was then only beginning; indeed, the works show that there was hardly any substance beside water that was even only moderately well-defined at the time. Besides water, alcohol (*esprit de vin*), was still by far the best. So what else could they measure?

It is a wonder that in spite of this Johann Albrecht (or his father?) nevertheless devised meaningful measurements. He measures how the refractive index of a salt solution changes when one increases the concentration of the salt, as well as how the refractive index of water changes as it is heated. That is admittedly only modest and hardly exact, but it is still a significant beginning for a field of research that would become immensely important in the nineteenth century. Euler's device was still used as late as 1900 (see, for example, [Pulfrich 1927]).

Electricity

Apart from a summary [2004] [i] of a paper by Aepinus [1761], Euler has left no written works on the theory of electricity, but he did comment on it extensively in the *Lettres à une princesse d'Allemagne*. The introduction to the new field at the beginning of the seventeen letters on electricity gives not only a good and often surprising picture of the impression that the numerous new discoveries made on researchers at the time, but also shows the importance that Euler attributed to these questions.[ii] In letter XXIII he says:

> The subject which I am now going to recommend to your attention almost terrifies me. The variety it presents is immense, and the enumeration of facts serves rather to confound than to inform. The subject I mean is electricity, which for some time past has become an object of such importance in physics that everyone is supposed to be acquainted with its effects.
>
> You must undoubtedly have frequently heard it mentioned in conversation; but I know not whether you have ever witnessed any of the experiments. Natural philosophers of modern times prosecute the study of it with ardour, and are almost every day discovering new phenomena, the description of which would employ many hundreds of letters; nay, perhaps I should have

> never done. ... I could not bear to think of letting you remain unacquainted with a branch of natural philosophy so essential; ...[8]

And already in the following letter, he introduces his central idea:

> There is no room to doubt that we must look for the source of all the phenomena of electricity only in a certain fluid and subtile matter; but we have no need to go to the regions of imagination in quest of it. That subtile matter denominated *ether*, whose reality I have already endeavoured to demonstrate, is sufficient very naturally to explain all the surprising effects which electricity presents. I hope I shall be able to set this in so clear a light, that you shall be able to account for every electrical phenomenon, however strange an appearance it may assume. The great requisite is to have a thorough knowledge of the nature of ether.[9]

In other words, the same ether that light waves move in, is responsible for the electrical phenomena as well: optical and the electric properties are thus closely linked. Whittaker says:

> [Euler] anticipated Maxwell in asserting that the source of all electrical phenomena is the same aether that propagates light: electricity is nothing but a derangement of the equilibrium of the aether [Whittaker 1951-53: v. 1, p. 98].

How did Euler conceive of the interaction between this ether and ordinary matter? Here a brief warning is in order, because one can easily read Euler from the wrong point of view. He even says in letter CXXXIX (XXIV):

> The greatest part of natural philosophers acknowledge their ignorance in this respect. They appear to be so dazzled by the endless variety of phenomena which every day present themselves, and by the singularly marvellous

---

[8] [Eng. Trans. Euler 1833, 2 : 79-80]. *La matiere sur laquelle je voudrois à présent entretenir V. A., me fait presque peur. La variété en est si surprenante et le dénombrement des faits sert, plutôt à nous éblouir qu'à nous éclairer. C'est de l'Electricité dont je parle, et qui depuis quelque tems est devenue un article si important dans la Physique, qu'il n'est presque plus permis à personne d'en ignorer les effets.*
Je ne doute pas que V.A. n'en ait déjà entendu parler très-souvent, quoique je ne sache pas si Elle en a aussi vu faire les expériences. Tous les Physiciens en parlent aujourd'hui avec le plus grand empressement, et on y découvre presque tous les jours de nouveaux phénomenes, dont la seule description rempliroit plusieurs centaines de lettres; et peut-être ne finirois-je jamais ... Je ne voudrois pas laisser ignorer à V.A. une partie si essentielle de la Physique; .... [Euler 1722, letter CXXXVIII].

[9] [Engl. Trans. Euler 1833, 2: 82]. *Il n'y a aucun doute qu'il ne faille chercher la source de tous les phénomenes de l'électricité dans une certaine matiere fluide et subtile, mais nous n'avons pas besoin d'en feindre une dans notre imagination. Cette même matiere subtile qu'on nomme l'*Ether*, et dont j'ai déjà eu l'honneur de prouver la réalité à V.A., est suffisante pour expliquer très naturellement tous les effets étranges que nous observons dans l'électricité. J'espère mettre V.A. si bien au fait de cette matiere, qu'il ne restera plus aucun phénomene électrique, quelque bizarre qu'il puisse paroître, sur l'explication duquel Elle puisse être embarrassée. Il ne s'agit que de bien connaître la nature de l'éther* [Euler 1722, letter CXXIX].

> circumstances which accompany these phenomena, that they are discouraged from attempting an investigation of the true cause of them. They readily admit the existence of a subtile matter, which is the primary agent in the production of the phenomena, and which they denominate the *electric fluid*; but they are so embarrassed about determining its nature and properties, that this important branch of physics is rendered only more perplexed by their researches.[10]

For the most part this "electric matter", was simply proposed as a new, different substance, that could be explained by means of the experiments that were performed with charged bodies, often, however, without the author taking into consideration the mechanical laws that these had to obey. The most important of these explanations was probably that of Franklin, to whom we owe the discovery of the conservation of charge. Franklin's writings, his letters to Peter Collinson [1751], were known to Euler through Aepinus, who set forth a detailed theory of electricity and magnetism in his book *Tentamen theoriae electricitatis et magnetismi* [1759]. Euler's principal aim here was to understand electrical forces in the context of the theory of continuum mechanics that he had created, more precisely, hydrodynamics and acoustics. One must always keep this in mind when Euler speaks about the ether.

Concerning the interaction of ether with matter, Euler says that all bodies are traversed in all directions by pores, which under normal circumstances are filled with ether. The electric qualities of a body depend, for example, on how easy or how difficult the pores are to open. In conductors – metals and such – the ether can easily flow in and out; in this case, the pores are always open, so to speak. With other materials a certain strength is required to open the pores through friction: these are the insulators. Then the ether will either be pressed out of or into the pores; these are positively and negatively charged bodies.

Euler is among those who rejected the idea that there were two types of electricity. As later development showed, he was the only one among these few who wasn't primarily interested in this seemingly fundamental question. Expressed in modern terms, he was primarily interested in the electric field, not in the charge. Since his ether was a compressible fluid, it automatically fulfilled, as Euler knew, a continuity equation and, with it, a principle of conservation. From these basic ideas, he explains qualitatively a whole series of appearances and experiments, including, among others, the Leyden jar discovered by Musschenbroek.

---

[10] [Eng. Trans. Euler 1833, 2 : 81-82]. *La plupart des Physiciens avouent là-dessus leur ignorance. Ils paroissent si éblouis de la variété infinie qu'ils découvrent tous les jours, et par les circonstances tout-à-fait merveilleuses qui accompagnent ces phénomenes, qu'ils perdent tout le courage d'en oser approfondir la véritable cause. Ils y reconnoissent bien une matiere subtile, qui en est le premier agent, et qu'ils nomment la matiere électrique, mais ils sont si embarrassés d'en déterminer la nature et les propriétés, que cette grande partie de la Physique en devient plutôt embrouillée qu'éclaircie* [Euler 1722, letter CXXXIX].

Magnetism

On questions of magnetism, Euler wrote a whole series of works, and no fewer than nineteen letters in the *Lettres à une Princesse d'Allemagne* are dedicated to it. He devotes about a twelfth of the letters to this subject, two more than for electricity.

If we survey the history of magnetism in Europe since its beginnings in the thirteenth century, we can see that of the elements of the discussion that present-day physics deals with, the oldest is not force but the field lines. Actually, we can trace these back as far as to Pierre le Pélerin de Maricourt (1269) [Radelet-De Grave and Speiser 1975].

In reading Euler's ideas about magnetism in the *Lettres*, the first thing we notice is that he introduces the topic through a practical application, namely as a means of determining the longitude of a ship using a compass, an idea that had first been proposed by Henry Oldenburg [1667]. Actually, the science of magnetism was clearly divided into two branches at that time: on the one hand the science of geomagnetism, whose goal was the most exact possible description of the earth's magnetic field, as we say today; and on the other hand, the actual theory of magnetism itself.

The discoveries in geomagnetism of the declination, temporal fluctuations and the inclination led to the preparation of magnetic maps of the world. A climax of this development was Edmund Halley's famous magnetic chart, on which lines are drawn where the declination has the same value ("isogons") [Halley 1683]. All of this is explained by Euler to his royal pupil in his well-known lucid style, which reveals a detailed knowledge of the technical problems. The task that Halley faced was in fact a complicated one: from individual values of the declination, which, however, are not evenly distributed over the whole earth, he wanted to find the "isodeclinations" and determine indirectly from those, the nature of earth's magnetism, that is, the number and location of its poles. Halley believed that the assumption of a simple magnetic dipole is not enough to explain its curves, and proposed assuming four movable poles. This is where Euler took up the subject, and he dedicated several works to this question.

To Euler, Halley's suggestion seemed unnecessarily complicated and rash. He showed – exactly like physicists who are seeking to interpret the measurements critically in order to extract laws from them still do today – that one can do with only two poles, if one assumes that they are sufficiently distant from the geographical poles and are movable. He calculated the magnetic field that results from this assumption, and gave the field lines. In the area of magnetism, this is the first example of a mathematically defined field.

The first (non-trivial) mathematical formulation (that is, in accordance with Truesdell, after the first steps were taken by Euler himself) is owed to d'Alembert [1752]. Euler then developed and set forth the field concept in full generality in his writings on hydrodynamics. This is an example of how his ideas about the nature of magnetism developed into those of hydrodynamics. To paraphrase Wolfgang Pauli, one always tries to work with ideas with which one had an early success.

Once again we see that a fluid stands at the center of Euler's thinking, and we now must turn to his ideas about the dynamics of magnetism. Externally, Euler's theories absolutely fit into the traditional Cartesian tradition, which Huygens had developed and which was continued by Daniel Bernoulli, among others. The ether penetrates and surrounds the earth in a big vortex. Specifically, it can penetrate into the pores of iron, and if these pores are directionally parallel, then the iron becomes a magnet. From these basic assumptions, he then also succeeds in explaining the observations qualitatively.

Euler has in general, as Mrs. P. Radelet has shown [1982], merely refined these theories; he did however go much deeper into one essential point. Like Daniel Bernoulli, he too is interested in the question in general of how the vortex had been formed. In modern terms, we would say that Euler clearly saw that the ether can have kinetic as well as elastic energy.

Two further points that are important because of their effect of his later theories, to which I will come back, merit comment here. First, in contrast to his predecessors, Euler's views were obviously formed with reference to the hydrodynamics that he himself had formulated. He constantly strove to formulate a "dynamics of magnetism" following the example given by Newton's mechanics, and the discoverer of hydrodynamics knew better than anyone else just what this meant.

The second point regards the ether. Originally – and this is found in the *Anleitung zur Naturlehre* [1862c] as well – in Euler's opinion the same ether was responsible for gravity, magnetism, electricity and light. He seems never to have had any doubt that the light ether is responsible for the electric phenomena as well. He was undecided about what magnetism involved. In the *Lettres,* he is of the opinion that the elastic medium that transfers magnetic forces must be less dense than the light ether. Apparently, the results of new experiments, as we saw with the question of the frequency of light rays, led him to alter his opinion. Permanently? This I am not able to answer; the dating of the *Anleitung* is uncertain, as we know,[11] and this prevents us from following Euler's ideas chronologically. However, an examination would be worthwhile.

It seems that of the three areas discussed here, Euler's ideas on magnetism produced the longest and most remarkable echoes. I therefore allow myself to correct an error in my Introduction to the *Commentationes opticae* in Euler's *Opera omnia* [D. Speiser 1962]. Concerning Faraday, I had written: "His physical views are indisputably closely related to those of Euler, but it seems unlikely that Euler had an influence on Faraday" [1962: LV]. When I wrote that I had in mind Euler's scientific works, but I had forgotten about the *Lettres à une Princesse d'Allemagne*! By 1872 the *Lettres* had been translated into nine languages and published in about a hundred and ten editions. Faraday knew the *Lettres* and repeatedly mentions them.[12]

Closing remarks

This discussion of Euler's works in optics, electricity and magnetism must still be complemented in two directions.

First, as is always the case with Euler, the picture would be incomplete without mention of his intense interest in technical questions. In the case of magnetism, I have

---

[11] On the dating, see [Truesdell 1960a: C], where Truesdell gives a series of arguments for a later dating of the *Anleitung*. In regards to this, I would like to point out the relationship between the statements in this work with analogous ones in the *Theoria motus corporum solidorum seu rigidorum* [1765a].

[12] See "Euler's Work in Physical Optics" in this volume, especially pp. 207-209.

already pointed it out, but just as it is with regards to mechanics and geometrical optics, it is valid as well for all the areas discussed here. This interest is evident in almost every single one of his letters to Princess d'Anhalt Dessau. Euler could not have been farther from our idea of a mathematician living only in the world of formulae. The letters convey an impressive picture of his extensive technical knowledge, and they inspired many mathematical examinations.

We are led to the other point by asking what effect Euler had on the physics that developed after him. This is not the place to go into the individual effects. One would have to look to the influence of the *Lettres* once again; as Faraday's example shows, this was extremely far-reaching. But it is worthwhile to make a few observations.

Only a small number of researchers, among them Johann II Bernoulli (1710-1790), did *not* align themselves with Newtonian corpuscular theory of optics. Even Lambert, who admittedly preferred Euler's theory (see, for example, his *Photometria* [1760]), didn't want to commit himself. The decision in favor of the wave theory came only later, as a result of the discoveries of Young and Fresnel, both of whom were familiar with Euler's works. But in the new century, experimental research then took over in this area.

As concerns electricity and magnetism, through the work of Daniel Bernoulli and Franz Ulrich Theodor Aepinus and then decisively by Coulomb, the elastic ether was soon superseded by the concept of forces at a distance. Sommerfeld and Reiff recorded this development [1902], and its classic formulation, which probably determined the thoughts, at least of the physicists in Germany, is found in Helmholtz's *Über der Erhaltung der Kraft* [1847].[13]

In all three areas, the development at first turned away from Euler's ideas and then and only later, at the time of Faraday and Riemann, in a certain sense returned to him. Truesdell determined almost the same thing as regards the aftereffects of Euler's works on mechanics. How did this happen?

If today – when physics has assumed a very different character from what it had in the nineteenth century and indeed in many ways has more in common with the physics of the eighteenth century – we seek to try to evaluate Euler's place in history, we can see more clearly the developments of the last 300 years. What then was Euler's ether, which had a central place in his ideas about physics? It was, as we said, not just another of many other substances that were so quickly proposed and just as quickly invalidated. Euler's ether is the carrier of the *force field*. The *field concept,* the formulation of which goes back to Euler and d'Alembert, is the big step forward that we owe to the eighteenth century. The laws of point mechanics describe infinitely small alterations during an infinitely small interval of time, as Newton had taught. The laws of extended bodies describe infinitely small alterations during an infinitely small interval of time *and* of a portion of space. Euler had carried this over above all in his great works on hydrodynamics. From Euler the field concept goes first to the French school and then – probably through Cauchy – reaches

[13] Just at the time I was correcting the galleys of this paper for its original publication, I received the article by Mr. R. Halleux (Liege) entitled "Huygens et la théorie de la matière", pp.187-196 in *Huygens et la France*, Vrin, Paris, 1981 [Halleux 1981]. I can see in this paper that Huygens stated similar ideas, which presumably influenced Euler. An examination of the particular leap made by Euler's ideas will have to wait for another paper.

Stokes and Maxwell. Be that as it may, for the work of individuals, the program that was connected with the concept of the central forces, which the physics of the nineteenth century had already mastered – and which had led to splendid results – from Cauchy, Faraday and Maxwell on would slowly but surely be replaced by the program of field physics, to which even Newton secretly paid homage. However, until the beginning of this century the physical bearer of the mathematical field remained the *elastic ether*.

Here, finally, we must mention Euler's posthumously published *Anleitung zur Naturlehre* [1862b]. This is, on the one hand, a detailed and critical representation of the whole foundation of mechanics, its most important concepts and results, above all the ("classical") principle of relativity, and on the other hand, a bold speculation, whose goal it is to sketch a program for an interrelated, unified theory of natural science.

Euler believes that all appearances can be traced back to the interactions of only two kinds of matter, *die grobe*, coarse or gross, and *die subtile*, subtle or fine. The gross matter appears to us thanks to the differences in the possible geometrical formal structures of the various materials. Presumably, Euler left their exploration to chemistry, one of the areas that he consciously stayed away from. The subtle matter – more precisely, the ether – is responsible for gravity, electricity, magnetism and optics. In the *Anleitung,* we find in systematic form the program that he tried to work out concretely in his individual works and the *Lettres*.

In the nineteenth century such speculations were viewed as strange and even irritating, and were hardly received with any approbation. Even so, it is possible that Riemann knew the *Anleitung*, that he studied it and was influenced by it. Be that as it may, in the twentieth century we have learned to think differently. Speculations with the goal of basing physics on an interrelated system of equations, a "unified field theory", as we say, now appears a completely legitimate, even necessary area of research. Today we can better understand Euler.

With the help of the ether, Faraday and Maxwell united optics, electricity and magnetism, and today the field concept, which superseded the ether, is at the center of physics. Present-day attempts to find a unified field theory are not so radically different; thanks to quantum theory, we understand some things better today, while as regards others we come up against essentially the same puzzles even today.

Let's go back to the beginning. Euler's work, as Andreas Speiser noted, was undoubtedly driven by mathematics. It was the source of his imagination and the driving force for all of his diverse examinations. Euler apparently had an unlimited faith in the power of the mathematical way of thinking. This doesn't mean that the knowledge of nature was of secondary importance for him; on the contrary, nature was constantly the object of his reflections and research. A large part – perhaps the predominant part – of his results in the area of analysis is the product of his efforts to recognize natural laws and formulate them adequately.

How and where do the mathematical deductions intersect with the empirical observations and experiments of physics? How does geometry interact with the mechanics, and mechanics with physics? Euler's answer is that geometry, that is, knowledge of the space in which we live, is a part of mechanics, and this is a part of physics. His theories agree with those of Newton, Riemann and Einstein.

Acknowledgments

I am indebted to P. Radelet-de Grave for discussions and her help in the preparation of the references. I also thank most cordially my wife and Miss H. Speiser aided me in the preparation of the manuscript.

Translated from the German by Kim Williams

The Distance of the Fixed Stars and the Riddle of the Sun's Radiation[1]

Prelude

In presenting the history of science, we rightly underline the final synthesis of a discovery which was arrived at only after a long period of development. But the roots of this development reach often far back, often to a time long before the new branch began to evolve on a broad scale. They appear to us as part of a prehistory, and to study them is particularly enticing.

When I. Bernard Cohen, A. Koyré's collaborator in the critical Edition of Newton's *Principia* [1927] invited me to contribute to the *Mélanges* collected in honour of his friend, I accepted with the greatest pleasure. To express my gratitude I choose a subject connected to one of Koyré's main investigations, the article of his that sums up the most important results of Euler's paper "Réflections sur les divers degrees de lunière du soleil et des autres corps célestes" [Euler 1752]. Written in 1750, Euler's paper belongs among those that opened doors to new branches of science, in this case astrophysics.

It is surprising to find a piece of work that penetrates far into various uncharted regions and pointing out empirical facts, which were puzzles at the time and would only be solved much later. One of these concerns the spatial extension of the part of the universe, which is not only accessible to our senses, but to our scientific investigations as well. Based on empirical data combined with a calculation and a critical estimation Euler shows that this extension must be larger than the distance of Saturn – at that time outermost planet – from the sun, by at least a factor of 60! Indeed Euler predicts the distance of the fixed stars to be at least half the value of the distance of Alpha Centauri, the closest fixed star known today. By comparison, the next empirical discovery, Herschel's discovery of the planet Uranus in 1778, enlarged the planetary system by a factor of only 2.

Euler's other result is even more striking. Based on the same empirical facts, Euler shows that the sun's radiation – or more precisely, its energy output – is of a magnitude such that the sun's state and condition cannot be compared to anything observed on our earth. Euler would be proved right only more than 150 years later, through the discovery of the nuclear forces!

---

[1] Originally published in *Mélanges Alexandre Koyré* Paris: Hermann, 1964, pp. 541-551.

Introduction

A. Koyré, in his "Baltimore Lectures", has described the long and hard process of thought that led "from the closed world to the infinite universe" [Koyré 1957]. This process not only replaced one view of the world with another: it replaced a method of research with another, more systematic one, since the new science was able to connect mathematics with empirical data. This fateful discovery divides the history of science into two parts. Historically, however, there was no sharp division. The meaning of the new advance had to be absorbed by small steps.

This note depicts one of these steps, at a stage which was no longer cosmography, yet was only a beginning of one branch of modern astrophysics: Euler's research on the brightness of celestial bodies, "Reflections sur les divers degres de lumiere du soleil et des autres corps celestes" [1752]. The subjects treated here may appear very inhomogeneous, and loosely connected, but they are presented here together, not only because they were discussed in the same work, but also because they are all the outcome of investigations applying Newton's conception, according to which the phenomena of nature had to be understood as processes between *forces*. They are thus intimately tied together.

It is a pleasure to be able express my gratitude to Alexandre Koyré, whose books, and even more so, whose personality, stimulated the author's interest in the history of physics.

The brightness of the sun

The stimulus for Euler's investigations was the *Essai d'optique sur la gradation de la lumière* by the French physicist Pierre Bouguer [1729]. This rather small volume was later enlarged by Bouguer and printed again in 1760, two years after his death. Curiously enough, for the benefit of the German reader, it was translated seven years later by Joachim Richtenburg – into Latin!

The goal which Bouguer set for himself was to define a standard for the brightness of light. Newton had not touched upon this side of optics, but Huygens had in his *Cosmotheoros* [1698]. The problem is analogous to the one Réaumur – with whom Bouguer was in contact – solved four years later for the measurement of temperature. In both cases a law of physics was used to reduce the scale of the investigated quantity to the measurement of a length, a procedure which has been applied repeatedly since then. Bouguer's idea was simple. We can never compare sources of light which have different brightnesses, that is, we can only say that two light sources are equal or unequal, but we can never say how much stronger one light is than the other. However, with the help of the law according to which the apparent luminosity decreases with the square of the distance, we can overcome this defect. To compare, for example, two candles, it is sufficient to place them at different distances from the observer so that both appear equally bright. Then, according to Kepler's law, the strength of the two lights is proportional to the squares of their respective distances. If two light sources are not visible simultaneously (for instance, the sun and the full moon), we must compare both with a third one, which serves as a standard: the "Bouguer candle". To measure the strength of the sun, which allows neither direct observation nor any comparison, we must "dilute" its rays with help of a lens, the focal length of which is well known.

Thus Euler obtains his main experimental result: the light of the sun is approximately 300,000 times brighter than the light of the full moon. This result was the goal of Euler's investigations.

Theory of light

For a physicist the question now became how these notions could be built into a theory of light. A physical theory had to conceive light as a dynamical process, one which could be described in space and time, and it had to fit these notions into the framework established by Newton's axioms. This had been imperative at least since Rømer's discovery that light, as Galileo had already conjectured, propagates with a finite velocity. Newton's corpuscular optics, which could answer most questions in a qualitative way, was the first attempt in this direction. In the *Opticks* Rømer's discovery is stated as proposition XI in part III of Book Two, before the description of the phenomena of thin plates. Perhaps its appearance at this point of the *Opticks* is somewhat unexpected, but of course Newton had good reasons for inserting it here, where the topic under investigation is the interaction of light with matter. The notions with which he depicts this interaction refer to *motion*. The intensity of a light beam on the other hand is clearly proportional to the number of corpuscles, and Kepler's law, that the intensity decreases with the inverse square of the distance, is almost obvious.

Euler's inquiry followed a different path. The insight into physics that he had gained from his work on problems of acoustics had convinced him that light, like sound, was a wave motion. As sound is propagated through the air, so light is propagated through the ether. He had formulated his ideas for the first time in 1746 in his treatise *Nova theoria lucis et colorum*. This paper was the optical counterpart to his *Dissertatio de sono* [1727].[2]

The *nova theoria* was for the most part but a program: only in 1759 did he discover the wave equation. But the program was formulated clearly. His principal argument against Newton was that in a corpuscular theory the superposition principle, as we call it today, could not hold. Already in his first work he stated, what might be called the fundamental theorem of wave optics: light is a periodic motion of the ether, and the colour of the light is determined by the *frequency* of the wave (rather than by the wave length). The frequency, as he shows, remains constant through whatever medium the light might pass.[3] Newton, who (together with Hooke) was the first to recognize the importance of the notion of periodicity for optical phenomena, seems rather to have favoured the notion of *wave length* as the fundamental quantity.[4] But as wave length is not an invariant of the propagation of light, it characterizes a colour only for a single medium. It is, however, the key property for such phenomena as the colour of thin plates, and in these matters Newton's hand was luckier than Euler's.

[2] Cf. also [Truesdell 1956 : xxiv] and [Euler 1746, *Opera* III, 5: 1].

[3] Cf. e. g. [Newton 1704: Book II, Part III, Prop. XVI-XX].

[4] The R. P. Costabel kindly informed me that the notion of frequency and a similar statement appear already in the work of Malebranche; cf. [1958-1978, vol. 3]. See also "Euler's Work in Physical Optics" in this volume, p. 203, note 14.

But what determines the *intensity* of a light ray? Clearly, the amplitude of the wave, that is, the strength of the oscillation of the smallest parts of the ether. Later it would be recognized that the amplitude is intimately connected with the energy of the wave, but before the wave equation was found, it was impossible to establish this connection. However, as we shall see in the work which will be described here, a rudimentary form of the energy theorem for a radiation field begins to emerge.

The prehistory of astrophysics

In addition to the mathematical formulation of wave optics, Euler was always interested in the connection of optics with other branches of science. Indeed, the synthesis of all parts of physics was one of the central stimuli for his investigations. To this goal he devoted much of his activity, and his efforts to reveal links with neighbouring fields run through all his works in optics and electricity and magnetism.

Thus, when he saw Bouguer's work, he asked at once whether these results could not, when combined with his own ideas about the emission of light, form the starting point for new investigations.

Was it possible to infer properties of the stars from their apparent brightness? So far the law of gravitation was the only tool which the physicist had for attacking such problems, but he could apply it only to the sun and to the planets! Was it now possible to use also the laws of optics for investigating the fixed stars, which hitherto had escaped every physical description? This would clearly be a new way of attacking the problems which the celestial bodies present to us. The paper, therefore, belongs to the prehistory of modern astrophysics; perhaps it is even the first of mathematical astrophysics.

Luminosity of the fixed stars

What was, indeed, the problem? To begin with, a law had to be found which stated the connection between the apparent luminosity of a star and its specific source intensity (as Euler called it, *éclat*). Yet in order to uncover such a connection, Euler had to clarify Bouguer's notions and to introduce supplementary ones. Euler states the law connecting these quantities. He believes that the apparent luminosity **J** was proportional to the inverse square of the distance R (Kepler), to the specific source intensity E, to the area $d\sigma$ and to the cosine of the angle i under which the ray hits a surface, but that it is independent of the angle of emission:

$$\mathbf{J} = \int \frac{E \cos i}{R^2} d\sigma$$

He errs! According to this formula the sun would appear brighter on the border than in the center. Later, Lambert, who was in turn stimulated by this work to write his famous *Photometria*, corrected Euler's error. The apparent luminosity, he claimed, depends also upon the angle under which the ray is emitted from the surface, whereas Euler believed it did not. One might feel that the mistake was hardly excusable, but since Euler was interested only in relative intensities, it had no severe consequences for his results.

In his *Photometria,* Lambert said about this paper:

> If I am not mistaken, Euler is the only one, who, scrutinizing exactly and perspicaciously the essence of light, also took this circumstance [the direction of emission] into consideration and made it the object of a calculation, when he investigated the brightness of the planets [1760: 29].

Luminosity of the moon and the planets

Let us now turn to Euler's applications of his formula to the moon and to the planets, that is, to bodies which only return alien light. The luminosity of the moon and the inner planets naturally depends upon the magnitude of the illuminated part which we see; Euler easily mastered this problem of stereometry. But this was not enough. The solution could only be determined by applying a principle of physics as well, stated by him somewhat crudely, "a body cannot reflect more light than falls on it". He calls a body which reflects all light "white" (Lambert was later to say that it has an *albedo* of 1). In a corpuscular theory this simply meant that only incident corpuscles were reflected, but how could one formulate this in a wave theory? In wave optics, light is not a propagation of matter, but a propagation of energy. In 1750 Euler could not yet formulate this mathematically, but the underlying physical idea was clear to him; he says:

> The stronger the rays are, the more brilliant is the light, and from this follows the first quality of luminous bodies, which enters into my investigations, and which I shall call the splendour of luminous bodies, whose quantity is proportional to the force of the rays, or the *force of the vibrating movement.*

One must, in addition, compare this with:

> One conceives that the particles of the body, entirely like a stretched rope, can make bigger or smaller elongations in their movement. And it is from this that the rays will be stronger or weaker according to whether these elongations of the body's movement are bigger or smaller .

That means that the behaviour of the ray is linked to the mechanical behaviour of the radiating body, to be precise, to the strength of its internal vibrations. Here the significance of the energy is already clear. We need not pursue his investigations of the planets further. Essentially he tried to calculate how much light was reemitted and how much was absorbed. From this he drew conclusions about the planet's surfaces. But of course the observational data at his disposal were too meagre to be effective.

Distance of the fixed stars

Now to the fixed stars! Euler showed how – in principle at least – it was possible to calculate the distance of a star by measuring its apparent luminosity. But as the apparent luminosity was not yet measured at his time, he had to rely on crude estimates. In addition to the distance, the specific intensity of radiation and the size of the star entered into his formula. It was reasonable then to assume that in this respect the stars may be comparable to the sun, although today we know that this holds only for a small minority of them.

Furthermore, since the apparent radius of a star had escaped any measurement, Euler estimated a quarter of a second to be its upper limit. From these data, and by using Bouguer's estimate of the sun's brightness, he then calculated that a fixed star is at least 4000 times further away than the sun. However, a superficial comparison of the brightness of a star with that of the moon already showed that this estimate of the star's apparent radius was much too high (he guessed by at least a factor of 30). Thus the star must be at least 120,000 times further away than the sun.

Today we know that this value, even for the closest star, is two times too low, but Euler approached the right order of magnitude. For understanding the impression these numbers might have made on Euler and his contemporaries, and the feelings they may have created, we might compare these investigations with those of present day physics about the age of the universe. But this result, as Euler observed, was valid only under an hypothesis: the $1/r^2$ law is true only if light is not absorbed in the ether! For this there was no direct evidence. But it was known that all parallaxes must be very small and certainly could not exceed 3″ since no parallax had been found.[5] This purely geometrical argument confirmed that the stars must be very distant; what is more, it helped to sustain the hypothesis that light was not absorbed in the ether. That Euler points out this last mentioned hypothesis is surprising indeed: it illustrates how empirically minded he was as a physicist, in spite of what sometimes has been alleged about his tendency to *a priori* reasoning.

The whole discussion is, of course, very simple, according to the meager empirical evidence of his time. But clearly, compared with earlier efforts, it heralded a new epoch. Needless to say, the program he outlined has been worked out in more recent times.

Brightness of the sun

Euler next applied his equation to the sun by making use of Bouguer's measurement of the sun's brightness for his calculations. He found a surprisingly high value for the total emitted radiation. From this he concluded that *the matter of the sun must be entirely different from every other combustible matter on earth, and it must be in a state of ignition (i.e., temperature!) which no body on earth is ever in.* These facts appear even more curious, since no change in the state of the sun can be traced for at least several thousand years. The idea that the sun is an ordinary glowing body *therefore provides no explanation of those facts.*[6] Modern physics, as is well known, has confirmed this bold conclusion. The

---

[5] The first parallax was measured 1837 by F. G. W. Struve.

[6] *Le Soleil étant donc un tel feu, dont la force ou l'éclat surpasse plusieurs mille fois toutes les espèces de feu, qu'on est capable de produire sur la terre, il faut absolument, que la matiere, dont ce corps lumineux est formé, soit d'une nature tout a fait différente des matieres, qui se trouvent sur la terre. Et en effet, de quelque matiere qu'on voudroit s'imaginer, que le Soleil fut composé, et a quelque degré de chaleur, qu'elle eut été portêe, ou elle auroit été bientot réduite en cendre, ou le degré de chaleur y auroit bientot diminué. Donc, puisqu'on ne remarque aucun changement dans la substance du Soleil depuis quelques milliers d'années, et que cela même est un des plus grands sujets de notre admiration, nous en devons être encore plus surpris, après avoir reconnu cette excessive force de lumière, qui y est ramassée, et qui surpasse tant de fois les feux les plus brillants, que les hommes sont en état d'exciter* [1752 in *Opera* III, 5 : 139]. Before Bouguer, James Gregory had

nuclear forces, which in ordinary life are not manifest, are responsible for the sun's energy supply.[7] Euler was probably the first who recognized and stated clearly that the radiation of the sun was an enigma of the first order of magnitude for classical physics, and who was able to prove this with a numeric estimate.

Before concluding this note, we permit ourselves a small digression in order to give a hint of the peculiar power of Euler's endeavor: to extend Newton's notion of force to all physical phenomena.

From Euler to present-day physics

The atomistic structure of matter had always been the subject of philosophical and scientific speculation, long before it became an heuristic principle of modern science, let alone before the existence of the atom was really proven. The history of the nucleus presents a striking contrast to this. No vestige of any speculation about something like a nucleus of the atom seems to have occurred – with, perhaps, one possible exception. In the last of his 31 queries at the end of his *Opticks*, Newton says:

> Now the smallest Particles of Matter may cohere by the strongest Attractions, and compose bigger Particles of weaker Virtue, and many of these may cohere and compose bigger Particles whose Virtue is still weaker, and so on for divers Successions, until the Progression end in the biggest Particles on which the Operations in Chymistry, and the Colours of natural Bodies depend, and which by cohering, compose Bodies of a sensible Magnitude [1704: Book III, query 31].

This, of course, is not to say that Newton foresaw the nucleus and the nuclear forces ("The smallest Particles may cohere by the strongest Attractions"), but it demonstrates the penetrating eye of one who, through experience gained by his experiments and through his attempts to formulate theories, had acquired a deep insight into how nature might possibly work out its inner mechanisms. This absence of such a speculation is not surprising: the laws that atoms obey govern the chemical processes that we experience every day, but the world of the nucleus is extremely remote from daily life. That the only allusion in this direction, the one just quoted, comes from the discoverer of the concept of force, the founder of modern dynamics, demonstrates that for once we really are faced with the exception which confirms the rule!

The existence of a nucleus becomes manifest primarily through its dynamic (rather than geometric) properties, and this aspect also provided the guiding principle through which nuclear physics had oriented itself at its beginning. Thus it is not accidental that we find in the same work of Euler some of the first indications of the energetic aspect of light and the first investigation on the sun's radiation, and in particular his forceful emphasis of the magnitude of the riddle it poses for us!

---

already made such measurements, from which Newton drew some conclusions. I am grateful to M. Hoskin for information about this.

[7] This was conjectured as soon as radioactivity was discovered. The state of science at the beginning of this century is described in [Pringsheim 1910]. A satisfactory explanation was however given only by F. G. Houtermans and R. d'E. Atkinson in 1929, and by H. A. Bethe in 1939.

Yet two of the most important experiences in our daily life were explained through nuclear physics. One is the abundance of the different elements which was explained by the shell structure of the nucleus. The other one, no less important since it makes possible all activity on earth, is the sun's radiation. Since this is one of the oldest riddles of nature, one which perhaps occupied man's imagination more than all others, it seemed worthwhile to describe how it was attacked for the first time by the methods of modern physics. The investigation which has received attention here belongs to the epoch of physics in which the first fundamentals of modern science had been consolidated, and from which another generation started to explore new grounds. In assuming a few thousand years as the age of the universe, Euler adopted a relic of earlier epochs. On the other hand, the conclusions that he drew from his theories with respect to the stars, and even more with respect to the sun, point to present day physics.

Acknowledgment

The author is indebted to Drs. J. Murdoch and J. L. Olsen for helpful advice.

Johann Heinrich Lambert[1]

Prelude

The first task of the historian of science must be to make accessible to readers from various disciplines the works that marked the development of science, because in history of science, as in other fields of history, nothing can take the place of the original documents and nothing is as interesting and stimulating as reading them. Here is why I was delighted with the publication of Johann Heinrich Lambert's *Les propriétes remarquables de la route de la lumière par les airs et en general par plusieurs milieux réfringens, sphériques et concentriques*, courageously undertaken by Mr Alain Brieux. In this preface our only aim is to make the reading of this work of Lambert easier and to describe briefly its historical context, and not to attempt a critical edition.

The life of Lambert

The eighteenth century discovered the cult of genius. Lambert, beginning from a meager education to say the least, and arrived at the acme of science, was considered as the very example of genius. As an autodidact, Lambert had few equals because he have to overcome obstacles that would have stopped even those who were more gifted, and this shows in his work.

The life of Lambert has often been told and several of his contemporaries left firsthand accounts. Astronomer Johann III Bernoulli wrote an obituary for him in the *Nouvelles littéraires*. There is also a text in praise of Lambert by Johann Heinrich Samuel Formey, Perpetual Secretary of the Academy of Berlin presented on 29 January 1778 [1778], as well as an article by the philosopher Lichtenberg in the *Teutscher Merkur* and the *Souvenirs de 20 ans de séjour à Berlin* by Dieudonné Thiébault. Formey and Thiébault devoted parts of their articles to *petite histoire* and funny anecdotes, especially when they regarded laughable situations. However, the fact that the authors knew Lambert personally lends credence to their information, even though these might not be perfectly accurate. Among the most recent biographies we should mention the one of D. Huber and that of the astronomer R. Wolf.

---

[1] The present text is the union of two papers on Lambert and optics. The first is the Introduction to the facsimile edition of *Les propriétes remarquables de la route de la lumière* (Paris: Alain Brieux, 1977). The second was originally published as "L'oeuvre de J. H. Lambert dans le domaine de l'optique," pp. 207-224 in *Actes du Colloque international et interdisciplinaire J. H. Lambert* (Mulhouse 1977), Paris: Orphys, 1979.

Johann Heinrich Lambert

Johann Heinrich Lambert was born at Mulhouse on 28 August 1728. His family, probably of Huguenot origin, took refuge there because at that time Mulhouse was associated with the former Swiss Confederation. His father was a tailor and Johann Heinrich was the eldest of a large family that lived in very difficult conditions. Thus, his parents could give him only an elementary education until he was twelve years old, and at a very young age he had to work to help support his family. It is said he made small pictures to sell for small change and that with this money, he bought candles so that he could study at night. The obstacles that he had to overcome were certainly large ones.

Upon learning that he had been refused a grant to study theology, his parents forced him, in spite of his protests, to be an apprentice tailor in his father's shop and, during his few free hours, to take care of his younger brothers and sisters. However, he was still able to do some other work. Thus it came about, a singular happenstance, that he copied some notes for J.J. Rousseau.

One day one of his comrades lent him a mathematics book and he found himself irresistibly attracted to this science. Then good fortune came to his aid: a mason who worked in his father's house, surprised by the boy's intelligence and by his pertinent remarks, lent him another book on geometry and it was in this way, by studying on his own, that he learned mathematics.

When he was fifteen years old, his parents made him enter as an employee into a business in Montbéliard, in the Upper Alsace, where he could learn French. But in 1745, the professor of law J.R. Iselin, who was also a member of the Academy of Berlin, encouraged him to go to Basel in order to assist in the publication of a newspaper, and so that he could begin to study mathematics and philosophy. Presumably he came into contact with the Bernoulli family, and perhaps even with the elderly Johann I, at this time at the height of his fame and generally recognized as the first mathematician on the Continent, but Johann III Bernoulli, who knew him, doesn't say anything about it in his obituary.

Be that as it may, in 1748 he was able to forget his material worries because Count Pierre de Salis, former ambassador in London, took him to Chur in the canton of Graubünden to educate his grandsons. Salis possessed a large library and Lambert had the time and the means to study as he wished; Formey tells of his lifelong friendship and the devotion to Salis. All his biographers agree that it was one of the happiest periods of his life. It was only in 1755 that he published his first works in the *Acta Helvetica*, but he began a large number of works that were subsequently published.

Although his main interest was always oriented toward the sciences and philosophy, he knew German, French, Italian and Latin and, according to Formey, he was capable of composing verses in all of these languages. He also undertook serious studies of law, to such a point that he was even able to make the case in defense of the city of Chur in a suit between the city and its bishop. Later, he was as member of a commission that was to rectify the border between Graubünden and Lombardy.

He lacked only one thing: a mentor to guide him. During all of his years of study he was his own sole teacher.

In 1756, together with his pupils he made with customary Grand Tour that took him, among other places, to Göttingen, Utrecht, Leyden, the Hague (where he met Musschenbroek), Paris (where he made the acquaintance of d'Alembert), Marseille, and Turin. It was during this journey that he took to print in Holland the book *Les propriétes remarquables de la route de la lumière*, which would appear in the Hague in 1758. He also began the *Novum Organon* and the *Perspective*.

After his return, he would remain another in Chur. Then in 1759, he obtained a position in the Electoral Academy of Bavaria, founded that same year by the Elector Max Joseph III , and he lived in Augsburg as a professor. It was in Augsburg that the *Photometria* [1760], his second major work appeared. There one finds treatments of the problems discussed earlier in the preface of the *Propriétes remarquables*.

Lambert, not satisfied with this situation, decided to go to St Petersburg in order to find a more interesting position there. However, when he passed through Berlin, many of the members of the academy were as impressed by his extraordinary personality as they were by his knowledge. Led by the philosopher Sulzer and supported by Euler, they decided to try to make him a member of the Berlin Academy. This was not easy, as it required the authorization of the king and, according to Formey, a "conspiracy" was formed to persuade Frederick II. Frederick, however, asked to see Lambert in person. The conversation that followed is legend: the king, who wasn't capable recognizing Lambert's value, refused to make "the biggest fool that I've ever seen" a member of his Academy. But in the end those who supported Lambert prevailed, especially after one of them had the idea of telling the king that Lambert was "a man like La Fontaine", a comparison that, according to Formey, was often repeated. Evidently, the king, for whom everything French was exemplary, could not resist the flattering idea of having another La Fontaine in his court. Besides, the rumor had spread that the Petersburg Academy wanted Lambert. In any case, Lambert was nominated. Credit must be given to Frederick the Great: after the battle of Kunersberg and the fall of Pitt, and before the death of Elisabeth II, the king went through the most difficult phase of the Seven Years' War, indeed, of his entire reign, but not once during this critical period did he drop the Academy; on the contrary, with the nomination of Lambert he made it stronger. From that time on, Lambert quit worrying about material security and, for the rest of his life, he was able to work quietly in Berlin.

During the first year, he was in close contact with Euler and, after Euler's departure for St. Petersburg in 1766, they maintained an important scientific correspondence. The king increasingly appreciated Lambert and subsequently covered him with honors. He was named member of several commissions and was even named the superior adviser to the department of buildings; this nomination brought him material benefits as well. However, Lambert didn't put his prodigious faculties of calculation at the king's disposal as much as Euler had, even though the application of science to practical questions interested him.

It is during his stay in Berlin that the greatest number of his works – certainly containing the results of previous studies – appeared. This fact can be explained by the possibility that was offered to him to publish in the *Mémoires* of the Academy, which meant that he didn't have to look for publishers.

As Formey says,

> During these twelve years, that were truly spent as if in a dream, Mr Lambert, in his element, never ceased working for the betterment of science and the public good. He accomplished a great number of excellent works and he published innumerable pieces, all worthy of him, in our *Mémoires*, in the *Ephémérides* of Berlin and in many other collections. All of his writings offer the two great characteristics of universality and originality.[2]

---

[2] *Pendant ces douze années qui se sont véritablement écoulées comme un songe, M. Lambert, dans son élément, n'a cessé de travailler à l'accroissement des Sciences et au bien public. Il a mis au jour quantité d'excellens Ouvrages et a répandu des Pièces sans nombre, toutes dignes de lui, dans nos*

Formey tells us about the cause of his death and his last moments. His illness started in 1775 with a cold. Lambert, an autodidact to the end, refused to consult some physicians and treated himself according to his own ideas, which were, in this case, unhappy. In Formey's words, *C'est ainsi que, voulant toujours être inventif, il l'a été à ses propres dépens: artifex periit arte sua* (This was how, wanting to be always inventive, he did so at his own expense: the artificer died by his own art). He lost his robust constitution: *On l'a vu fondre comme la cire au feu* (One saw it melting as wax by fire). However he remained an optimist to the end worked until the day of his death, which arrived suddenly on 25 September 1777.

Even this brief account of his life is sufficient to give an idea of Lambert's extraordinary personality, which was striking to all those who knew him. He was, in all senses of the word, original. Proud of being self-taught and of having learned everything all by himself, as his conversation with Frederick II shows, he tackled all problems he studied in his own way, and measured himself against them.

The originality of his way of proceeding is always evident when his works are studied. It goes without saying that such ambition had to go hand-in-hand with extraordinary energy and tenacity to arrive at such results. But, if the originality led him to results that were unexpected and uncommon, it was also the reason why his influence was weak and that he is not often recognized by posterity, as he properly deserves to be. His method isn't easily inserted in the ordinary current of science, because he followed his own paths, starting from zero.

It also goes without saying that Lambert was the epitome of the absent-minded professor and was entirely enamored of his ideas, making him forget all that surrounded him. Dieudonné Thiébault, the French man of letters and confidant of Frederick II regarding everything that concerned the Academy, described Lambert in these words:

> Once he had begun a discussion, whatever it was, it was no longer possible to interrupt or stop him : it was sure from the very beginning that he saw so clearly the plan to be followed and he followed it so assiduously that nothing could dissuade him from it. His ideas were always perfectly regular and ordered; if an objection was raised, he stopped only long enough for the objection to be voiced. But he never answered it: he took the thread of his reasoning up again as though he had never been interrupted, since the objection that was made had a proper time and place, and that the discussion would only have lost had it been displaced from the plan that he had laid out from the very beginning.[3]

Mémoires, dans les Ephémérides de Berlin et dans plusieurs autres Recueils. Tous ses Ecrits offrent les deux grands caractères de l'universalité et de l'originalité [Thiébault 1813: iv, 28].

[3] *Lorsqu'une fois il avoit entamé une discussion, quelle qu'elle fût, il n'étoit plus possible de l'arrêter ou de l'interrompre : on étoit sûr que, dès le début, il voyait si bien le plan qu'il avoit à suivre, et y étoit si fidèle, que rien ne pouvoit l'en détourner. L'ordre de ses idées étoit toujours régulier et parfait ; si on lui faisoit quelques objections, il ne s'arrêtoit qu'autant qu'il falloit pour laisser dire ce que l'on vouloit : mais jamais il n'y répondoit : il reprenoit la suite de son*

He was conscious of having become what he was through his own efforts and this procured for him a self-assurance that sometimes shocked those who didn't know him, and often made those who did know him laugh. Innumerable anecdotes were told about him. I take the liberty of telling one of the lesser-known ones, recounted by Thiébault:

> The new member of the Academy occupied himself first thing with his acceptance speech, and decided to use it to resolve an important question regarding the reflection of light : but he had some experiments still to verify, for which he needed a large mirror ; but he only had among his own things a small pocket mirror that was barely big enough to allow him to put his wig on. For this reason he decided to go to a café, the best one in town, located in front the castle at the corner of the main street.
>
> Entering the room on the ground floor, he greeted the people who were there, that is, five or six officers and some gentlemen who were playing tarot : he greeted them in his own manner, without looking at them, and tilting his head from left to right, and he placed himself in front of a large mirror that enjoyed beautiful light : here he took out his sword, he thrust it forward, retracted it, thrust forward again, menacing the mirror now with a third and now with a fourth, he stopped, meditating on all that he did and saw, and beginning again the same sequence he went on for a good half-hour, without noticing that all these persons who didn't know him and didn't know what to make of this exercise took him for a fool and had surrounded him and were ready to grab and disarm him should it be necessary. When he had finished his trials and his reflections, he put his sword back into its scabbard, he glanced indifferent at those who surrounded him, said goodbye to them in the same way that he had greeted them, and left to write an essay that was worthy of the admiration of the scholars.[4]

---

raisonnement comme si on ne l'eût pas interrompu, parce que l'objection qu'on lui avoit faite, devoit se trouver dans un moment et dans un ordre plus convenable, et que la discussion n'auroit : eu qu'à perdre à s'écarter du plan qu'il s'étoit tracé d'abord [Thiébault 1813: iv, 30].

[4] *Le nouvel académicien s'occupa d'abord de son discours de réception, et décida d'y résoudre une question importante sur la réflexion de la lumière : mais il lui restoit quelques expériences à vérifier, pour lesquelles il avoit besoin d'une grande glace ; et il n'avoit dans tout son mobilier qu'un très petit miroir de poche, qui suffisoit à peine pour mettre sa perruque. Ainsi, il prit le parti de se rendre pour ses expériences, à un café, le premier de la ville, placé en face du château, au coin de la grande rue* [Thiébault 1813: iv, 28].

En entrant dans la salle du premier étage, il salua les personnes qui s'y trouvoient, c'est-à-dire, cinq ou six officiers et quelques bourgeois qui jouoient aux tarots: il les salua à sa manière, sans les regarder, et en jetant diagonalement sa tête de gauche à droite, et se plaça devant une grande glace qui étoit exposée à un très beau jour : là, il tira son épée, la porta en avant, recula, se rapprocha, menaçant cette glace, tantôt par une tierce, tantôt par une quarte, s'arrêtant, méditant profondément sur tout ce qu'il faisoit ou voyoit, et recommençant le même jeu pendant une bonne demi-heure, sans s'apercevoir que tous ces messieurs qui ne le connoissant pas et ne sachant que penser de cet exercice, l'avoient pris pour un fou, et l'entouroient disposés à le saisir et à le désarmer s'il en étoit besoin. Quand il eut bien fait ses essais et ses réflexions, il remit tranquillement son épée dans le

Se non è vero ... (if it is not true...) it is at least well embellished!

But it would be wrong to stop at such a superficial level. I will cite also what Formey says in praise of Lambert:

> His straightforwardness was born of his rectitude, and sometimes led to inflexibility. One had to get out of his way, otherwise he would knock one down, run one over, without regard or distinction for anyone. Rather than neglecting worldly customs, he ignored them.[5]

According to Johann Rudolf Wolf, he was a difficult colleague to deal with, but he improved with the years and in spite of his weaknesses, he was universally respected.

Lambert's work

Two features are particularly characteristic of Lambert's work: the universality of his interests and the originality of the questions that he asked. We not only possess his mathematical works, but also those of physics, astronomy, meteorology, technology, in short, covering the whole gamut of the natural and mathematical sciences, as well as a large number of philosophical works. And, as we saw, he also knew law.

A detailed examination reinforces this impression: in mathematics he works on analysis, algebra, geometry and prepares some notes on logarithms. In physics he deals with mechanics, acoustics, optics, magnetism; only elasticity seems to escape him, as it did Newton.[6] His works in astronomy, the most important group of his work, range from numeric tables to practical use to cosmological speculations, passing through the calculations of the orbits of comets.

A close look at the list of his publications shows that the works of the various domains are distributed throughout their author's whole life; it would appear that that he dealt with all these questions for his whole life. Only for astronomy does the period of Berlin distinctly signal a center of gravity, made evident by the number of his publications; during these years, he published more than fifty works dealing with astronomy.

However, even in this domain his first book goes back to as early as 1761: these are the *Insigniores orbitae cometarum proprietates*. According to Johann III Bernoulli, who was for a part of his life an astronomer in Berlin, Lambert's interest for these studies was sparked by the appearance of a comet (in 1744 or 1746) that Lambert had observed and studied when he was still young. His interest in comets is evident throughout all his works.

Here we have especially discussed the scientist, but it should not be forgotten that his glory at the time rested especially on his *Novum Organon*. Indeed, while he was alive

---

fourreau, jeta un coup d'oeil indifférent sur ceux qui l'environnoient, leur donna le même salut qu'à son arrivée, et s'en alla rédiger un Mémoire digne de l'admiration des savans [Thiébault 1813: iv, 28].

[5] *De sa droiture naissoit sa fermeté, poussée souvent jusqu'à l'inflexibilité. Il falloit s'ôter de son passage; autrement il heurtoit, il reversoit, sans égard, distinction, ni acception de personnes. Il négligeoit les usages du monde plutôt qu'il ne les ignoroit* [Formey 1778: 87].

[6] It is difficult to categorize Lambert's works. This problem, already difficult to resolve in the case of his contemporaries, is made even more complex by the large number of his works that cross disciplinary borders.

Lambert was as famous as a philosopher as he was as a scholar. In Berlin he was friends with Moses Mendelsohn and his work would also influence Kant, with whom he exchanged some letters. Kant wanted to dedicate his *Critique of Pure Reason* to Lambert, but was prevented by doing so by Lambert's death.

It is sufficient to mention some of the works of Lambert to illustrate the originality of his thought. In mathematics he proved that e, π and other numbers are irrational. The very idea that it should be necessary to prove this, as well as the method that he used, was new. He calls these numbers "transcendent quantities" and this term has remained, although today it designates a more precise notion (a frequent occurrence in the evolution of scientific language). In any case, his work opened a new line of research that reached a first culminating point at the beginning of the second half of the nineteenth century and that is still today a fascinating branch of mathematics.

Lambert also treated the famous problem of parallels. He is among those who came the closest to non-Euclidean geometry. Even if he is not actually numbered among those who discovered the non-Euclidean geometry he is, along with Girolamo Saccheri, one of their most important precursors.

To illustrate his originality as a physicist we can mention his contributions to studies of the resistance of the fluids and friction as well as his work on the fluidity of sand, topics that physicists and mathematical physicists of the 1760s mostly (and happily) left aside. Several of his works are dedicated to meteorology, including a project for a world organization to supervise the weather of the whole earth by means of twenty observation stations.

The most popular of his books was undoubtedly the *Cosmologische Briefe* published in 1761. These letters were translated into French with the title of *Systèmes du monde* and published at least three times. After his death they were also published in English with the title *Systems of the World.* This book, in which he considers the universe as being filled with stellar systems, is still interesting reading today, if for no other reason than to see how much he widened the horizon of the universe. In this book, we find for the first time a clear concept of the Milky Way as well as a description of it. Whereas at that scientists had attributed to the universe a measure of a few light years, Lambert is the first to speak of millions!

The *Photometria* is another example of how diversified his interests were. In it he deals with a number of questions that range from physical optics to psychophysics, and treats nearly all of them in an original way. He was the most important precursor of the psychophysics, which began to develop in the nineteenth century with Gustav Fechner and others. One could say the same thing about his writings concerning perspective.

Finally, when one browses the list of his works concerning problems of applied physics and technology, one is immediately struck by the large number among them that are dedicated to problems of scale and equipment for measuring. Besides the *Photometria* we find some works on the micrometry, on scales, hygrometry, pyrometry, pendulums, barometers, and more, and to these works, one can add another dealing with mixture of colors starting with white and the three primary colors, as well as a philosophical work on *taxéométrie* (i.e., the measure of order). This preoccupation, like the problem of the definition of scale and the equipment for measuring, run through Lambert's works like a red thread and must be considered the counterpart of his philosophical speculations.

LES

PROPRIETÉS REMARQUABLES

DE LA

ROUTE DE LA LUMIÈRE,

PAR LES AIRS

ET EN GENERAL

PAR

PLUSIEURS MILIEUX REFRINGENS SPHE-RIQUES ET CONCENTRIQUES, AVEC LA SOLUTION

DES

PROBLÈMES,

QUI Y ONT DU RAPPORT, COMME SONT LES REFRACTIONS ASTRONOMIQUES ET TERRESTRES, ET CE QUI EN DEPEND.

PAR

J. H. LAMBERT.

A LA HAYE

Chez H. SCHEURLEER, F. Z. & Compagnie, M. DCC. LVIII.

Frontispiece of *Les Proprietés Remarquables*

Optics before Lambert

Eighteenth-century optics was dominated by Newton's *Opticks* [1704], in which he set out his corpuscular theory: light is composed of corpuscles that emanate radiant bodies. However, this work is largely dedicated to physical optics. Only the first twenty pages contain some remarks on geometric optics proper, while the problem of the passage of light through the atmosphere, that is, through a medium whose refractive index varies continuously, is not mentioned at all. Huygens's *Traité de la lumière* [1690a] should be mentioned as well. This work is often cited as the origin of the wave theory of optics, but it is actually contains the foundations of geometric optics: the notion of periodicity, fundamental for a wave theory, is not found here.

The ideas that are the basis for such a theory were developed by Euler starting in 1746 in a set of works published in vol. 5, series III of the *Opera*. However, until the end of the eighteenth century, no one followed the Euler's ideas. Lambert himself was undecided between corpuscular and wave optics, while saying in the *Photometria* that wave optics seemed likelier to him, but that it had not yet succeeded in predicting new effects. His results in *Les propriétés remarquables* are due to geometric optics and are not at all dependent on a hypothesis about the nature of light.

Before Lambert, great progress had been made in geometric optics, especially in the work of Clairaut and Euler. Euler's first works are published in the vol. 6, series III of the *Opera*, and Lambert certainly knew a part of it.

As regards astronomical refraction, to which *Les propriétés remarquables* is dedicated, its history goes back up to Tycho Brahe and Kepler. At the time of Lambert, astronomers used Cassini's refraction tables. It was in 1729 that P. Bouguer submitted to the academy of Paris a mémoire entitled "*Essai d'optique sur la gradation de la lumière*", in which he expresses "Bouguer's law". The first physical theories elaborated in this domain are due to Thomas Simpson, who calculated, in 1743, the refraction according to the density of air, and to Edmund Halley, who calculated it according to the pressure.

Lambert's work in optics

The principal part Lambert's work is constituted by two books based on the research that he had been able to carry out in Chur and had written at the beginning of his career. These books are *Les propriétés remarquables de la route de la lumière par les airs,* which appeared in 1758, and the *Photometria sive de mensura et gradibus luminis, colorum et umbrae*, which appeared in 1760. The six or seven small works he later wrote on this subject can be considered as appendices.

Les propriétés remarquables de la route de la lumière par les airs (1758)

This first book of Lambert's sometimes shows some weaknesses, but it can be considered a characteristic example of an eighteenth-century book of science, not only because of its objectives but because the specific manner of treating the subject used in the "age of reason".

According to the *Bibliographia Lambertiana* by Max Steck [1943], Lambert began this work in 1756 at the latest, making use of previous studies; the final redaction of the manuscript dates from March 1758. The work appeared in The Hague in June of the same year. It enjoyed an immediate success, was translated into German by Tempelhoff and published twice in Berlin, in 1772 and 1773. The original French edition is very rare nowadays, but fortunately the book has been reprinted.

It is in view of practical applications that Lambert undertook this work; indeed, his theory of the refraction of light through the atmosphere was meant to correct the astronomical observations and the geodesic measures, and the organization of the book is governed by this goal.

Another characteristic feature is the presentation of the theory in a deductive form *more geometrico* in the Euclidean style: hypothesis, thesis, proof, corollary.

Lambert limits himself to the use of only two "experiences" (experiments) as the starting point for his geometric reckonings. He also refuses to adopt a physical relation between refraction and density, declaring: "All that I would have won by it, would have been to have had one differential equation rather than another, without having made other progress".[7] Indeed at the time this relation was still unknown and Lambert succeeds in avoiding its use by assimilating the luminous ray to an arc of a circle of which he determines the radius from the observations.

The measurement of astronomical refraction goes back to Tycho Brahe and to Kepler, but Lambert was especially inspired by three works that he mentions in his preface:

- Pierre Bouguer's *Essai d'optique sur la gradation de la lumière* [1729];[8]
- Euler's article on the different degrees of the light of the stars [1752c];[9]
- Kästner's translation of the treatise on optics by Smith.[10]

We don't know if Lambert already knew Euler's work, "De la réfraction de la lumière en passant par l'atmosphère selon les divers degrés tant de la chaleur que de l'élasticité de l'air" (E219), as the similarity between Lambert's fig. 3 in the *Propriétés remarquables* and Euler's fig. 1 in "De la réfraction de la lumière…" suggests (figs. 1 and 2). It might also be that both used the same source, or that the similarity is fortuitous (but we don't think so).

But Lambert's approach to this problem differs entirely from Euler's.[11] Euler attacks the physical problem as a whole and also studies the hypotheses on the relations between

[7] *Tout ce que j'y aurois gagné, ce seroit d'avoir eu une equation différentielle pour une autre, sans en avoir eté plus avancé* [Lambert 1758 : 4].

[8] In this work Bouguer laid the empirical foundations permitting to measure the intensity of light and tried to define a scale of brightness.

[9] Bouguer's book stimulated Euler to write his paper "Réflections sur les divers degrés de lumière du soleil et des autres corps célestes" [1752c]. In this work, one of the first in the domain of astrophysics, Euler approaches the notion of energy of a field. See [David Speiser 1962] and also the article "The Distance of the Fixed Stars and the Riddle of the Sun's Radiation" in this volume, pp. 225-232.

[10] The original English edition is *A Compleat System of Opticks*, by Robert Smith [1738]. It was translated into German by Abraham Gotthelf Kästner in 1755 and into French by Nicolas-Claude Duval-Le-Roy in 1767. According to E. Anding, the German translation is the one that Lambert used and that he studied in depth.

the temperature, the pressure and the density. Lambert, on the other hand, abstains purposely from making such hypotheses and his merit is precisely to show how far the nature of a very particular problem permits avoiding a physical hypothesis, because, at the time he was writing, there was no thorough understanding of thermodynamics.

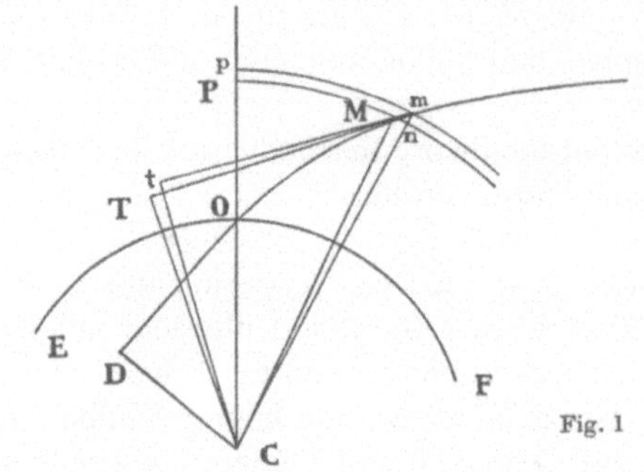

Fig. 1 (above) Fig. 1 from Euler's "De la réfraction de la lumière..."

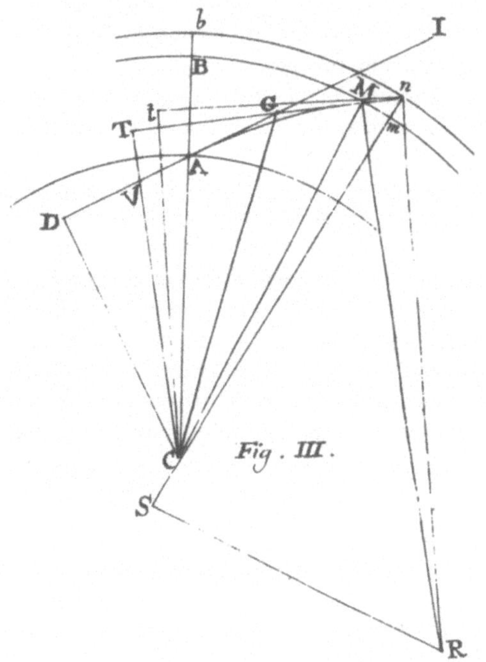

Fig. 2 (right) Fig. III from Lambert's *Propriétés remarquables*

To make Lambert's idea clear, it is necessary for us to say something about geometrical optics. At its base is Snell's law, which says that the ratio between the sine of the angle of incidence α and the sine of the angle of refraction β is independent of the angle of incidence:

$$\frac{\sin\alpha}{\sin\beta} = \text{const} = \frac{n_\alpha}{n_\beta},$$

where n_α and n_β are the indices of refraction (expressed, for example, in relation to the vacuum). If the ray runs through a sequence of layers of different densities, it is necessary to make use of a local system of coordinates at the boundary between every two pairs of layers (fig. 3).

[11] Cf. also on this topic the article of A.P. Youschkevitch [1979].

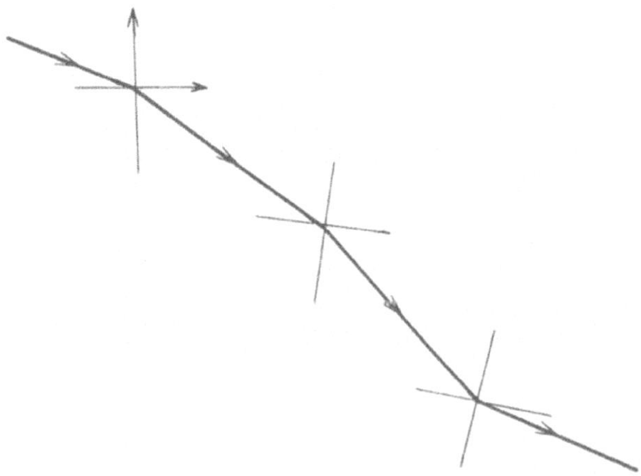

Fig. 3

It can be seen that if the optical density varies continuously one is led to a problem of integration that is in general very complicated.

But in some cases one can do better. Let us consider for example a succession of parallel layers (fig. 2). To simplify the argument, we suppose that the first and the last layers are vacuums.

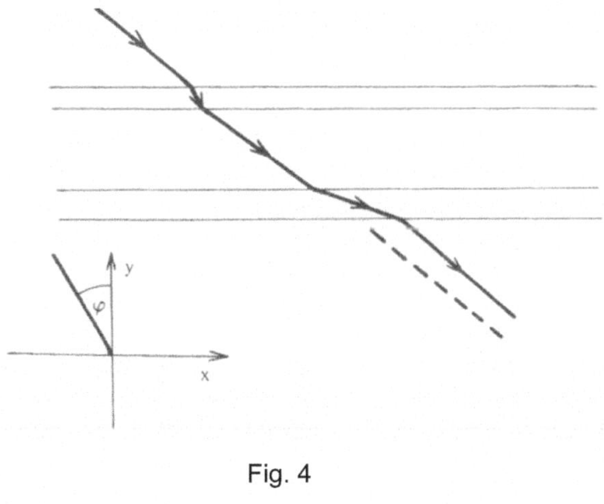

Fig. 4

In this case we can say that whatever the density of each layer (and it can also vary continuously), the direction of the ray after it passes through the sequence will be the same as before. This apparently trivial fact is a familiar one in daily life.

We can now introduce a global system of coordinates and write

$$n_i \sin \alpha_i = \text{const} ,$$

where n_i and α_i are respectively the indices of refraction and the angle of incidence of the ith layer

An analogous simplification is possible in the non-trivial case of a system of concentric spherical layers in which Lambert is interested (fig. 3).

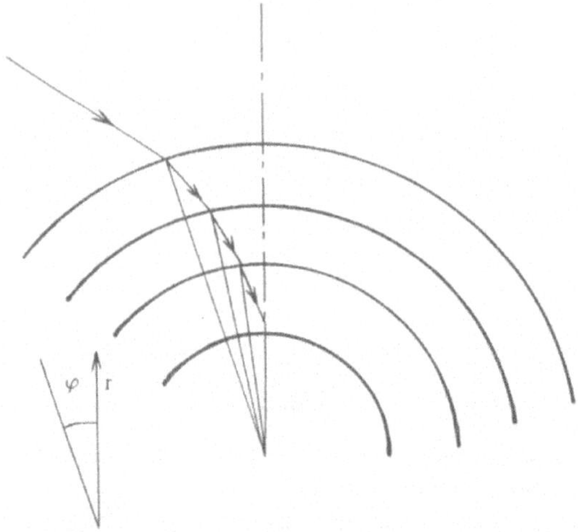

Fig. 5

In this case we can also introduce a global system of coordinates – this time of polar coordinates – and we can formulate the law:

$$r_i n_i \sin \alpha_i = \text{const}$$

along the ray of light. In short, along the trajectory the product of ray r_i , the index of refraction n_i and the sine of the angle of refraction α_i is a constant (r being the distance from the point considered to the center of the layers).

In the case of the parallel layers, the simplification is the result of translational symmetry, in this case the law of refraction has the spherical symmetry of the problem. This theorem is, in the context of the optics of Hamilton-Jacobi, analogous to the conservation of the angular momentum.[12]

Obviously, Lambert doesn't formulate the law in this way. He says in Theorem 7 that for each pair of points along the ray, the ratio between the normals that go from the center of the sphere to the tangent of the ray is constant whatever the angle of incidence is. This result is now known as "Bouguer's law", and from it he deduces theorem 16. He establishes that the locus of the centers of curvature of the trajectory of the ray lie on a straight line orthogonal to the straight line that passes through the center of the spheres and through the point where the ray penetrates the non uniform media (where the angle of incidence is defined); this theorem is used constantly in what follows.

[12] One can find an elegant derivation of this law in the context of the optics of Hamilton-Jacobi in the classic *Principles of Optics* by Born and Wolf [1970: 121].

We don't know if Lambert had independently discovered Bouguer's law or if he knew it from Bouguer's mémoire. In Bouguer, the derivation of the law is once again linked to a physical hypothesis, that is, that the refractive index depends only on the density; this is approximately true, but not strictly exact for the atmosphere. In any case, Lambert deserves credit for having clearly set forth the geometric character of this law.

Lambert then shows that one can apply this theorem to the atmosphere if one disregards the variation of the separation between the two media at the price of a minor error that decreases with the angle of incidence. To do this he bases his argument on the theorem 16 described above.

While reading the discussion of the theorem, in which he tries to classify (without being complete, contrary to what he affirms) the different types of trajectories, one cannot fail to be struck by the analogy presented by this discussion to the one concerning central forces in mechanics. This mathematical analogy between geometrical optics and mechanics was discovered in 1696 by Johann I Bernoulli and permitted him, in particular, to solve the famous problem of the brachistochrone. Later, in the hands of Hamilton and Jacobi, this analogy would be developed completely and would become a fruitful heuristic principle for rational mechanics. However in Lambert we find no allusion to this analogy.

But it is the applications that constitute the object of the second and third parts of the work. They are presented in the form of problems and to solve them Lambert once again resorts to the geometry of the circle. The starting point of his reasoning is the assimilation of the curve of the light rays to a circular arc, which is a hypothesis permitted in the particular case of the atmospheric refraction. The discussion is especially detailed and provides us with a vivid illustration of the scientific activities of the eighteenth century.

The book testifies to the spirit of its time, using the geometric methods to the very end, with the aim of reducing or eliminating the hypotheses. It would be false to say that Lambert wanted to reduce the physics to geometry, but geometry guides and dominate his thought process. It would also be false to say that the *Propriétés remarquables* lacks originality, but all things considered, it is the pursuit of *classic* ideas through to their very last consequences that characterizes the book. As we shall see, it will be quite different in the *Photometria*.

Interlude

There exists a very interesting exchange of letters between Lambert and Daniel Bernoulli that begins with Lambert sending Daniel one of his first drafts of the *Propriétés remarquables*, and also contains long passages concerning the *Photometria*.[13] In the first half of the century Daniel Bernoulli was the best experimental scientist among the great mathematicians (*géomètres*); Lambert assumed that role for the second half.

***Photometria sive de mensura et gradibus luminis, colorum et umbrae* (1760)**

The *Photometria*, written in Latin, is Lambert's fundamental work on optics and had already been announced as such in his first book. The richness and diversity of the book

[13] The correspondence between Daniel Bernoulli and Lambert, edited by P. Radelet-De Grave and V. Scheuber, was published on the occasion of the conference in which part of the present paper was given [Daniel Bernoulli 1979]. The letters related to optics are those numbered from 7 to 12.

make it difficult to discuss. To illustrate this, it is sufficient to list the qualities that the editor (and alas, also the historian who writes about it!) of this work must possess. He should know geometrical optics, Newton's corpuscular optics, Euler's wave theory of light, the technology of optical systems; he should be an astrophysicist, and should also know the anatomy of the eye and its physiology. On top of all that, he should know the psychology of vision as it currently stands *and* as Goethe intended it, and should be well versed in the science of geometric perspective and the perspective of colors, sciences that Lambert enriched.[14] And above all, he should be a specialist of the history of experimentation and the equipment used for it, because Lambert was an inventor and ingenious constructor. Fortunately we have at our disposition an edition of the most important parts of this work, translated into German by E. Anding in an exemplary manner and enriched with explanatory notes which are still of great value today [Lambert 1760, trans. Anding 1892].

The book is composed of seven parts:

- an introduction of the topic and the fundamentals: direct light, its various manifestations and its intensity; clarity and illumination;
- the reduction of the intensity of light during the passage through transparent bodies, especially through glass (that is, an optical system);
- the reflection of light by the opaque bodies;
- the perception of light and subjective clarity;
- the diffusion of light during the passage through transparent bodies and especially through the earth's atmosphere;
- a theory of the illumination of the planetary system;
- the colors and the shadows.

It is in this work that all of Lambert's originality appears. He tackles questions that had hardly been asked before him. Systematic research had advanced in other directions, in pursuit of other ideas. Lambert clearly expresses the limit (§ 17).

Lambert says that the first part of optics deals with the propagation of light (*de motu...*). Up to this point, researchers had dedicated nearly all their efforts to this argument. But there is also a second part which deals with physical questions regarding, for example, the intensity of light and its perception. Effectively, today we can say that everything that physics accomplished up to Maxwell and Dirac is the explanation of phenomena starting from the laws of systems of propagation; no one can possibly deny its successes.

But these problems must necessarily go beyond the physical setting. The originality of the *Photometria* consists in the application of the methods and results that had been elaborated during the eighteenth century to questions that were for the most part new. To Lambert belongs the credit for applying means of calculating concrete examples down to the smallest details.

Here I restrict myself to presenting the most important and most remarkable points. Among these, in the first chapter, is Lambert's famous cosine law (§ 70), with which he corrects an error committed by Euler in his work "Réflections sur les différents degrés du

[14] Cf. the papers by H.W. Arndt [1979], C. Debru [1979], S.L. Jaki [1979] and H. Pfeiffer [1979]. *Editor's note.* Cf. also [Andersen 2006].

soleil et des autres corps célestes" [1752c],[15] a work that Lambert himself claims to have been inspired by, and which A.G. Kästner had explained in his introduction to the translation of Smith's *Opticks*. Lambert formulates it thus: "The brightness (*luminosité*) decreases in inverse proportion to the square of the distance, in proportion to the sine of the angle of incidence, and to the size of surface area of the source *turned toward the object illuminated*, and proportional to the intensity of the source."

In modern terms,

$$dL = df \cdot J \cdot \cos\varepsilon \frac{1}{r^2} \cos i \,,$$

where L denotes the luminance; df, the area of the radiating surface; J, the intensity; $\cos\varepsilon$, the cosine of the angle of emission (a factor added by Lambert that Euler had disregarded); $1/r^2$, the dependence of the distance according to the law of Kepler; $\cos i$, the cosine of the angle of incidence.

Here we find Lambert approaching the law later discovered by Fechner, of whom Lambert is the most important precursor, but he seems to miss it because of measurements that are not sufficiently exact (§§ 265-270).[16] These investigations on the attenuation of light in optical systems are probably among those that have had the greatest influence on his successors.

In one of his letters Lambert writes to Bernoulli about the *Photometria*:

> The first chapter sets forth the principles and the difficulties, and the third subjects them to the most rigid examination that I could think of. The second part contains the experiences that I made with glass. I will be delighted, Sir, if you would take the trouble of examining them, and your feelings regarding them would be extremely valuable. You will find the order and the relationships [between topics] changed throughout, and completely different of the one of the first draft, that I would have destined for a different object had I remained longer in Göttingen.
>
> What will appear to you to be the thing of greatest novelty is the method for determining the degree of the whiteness of every body, which is found in *Theorem 35*, and its consequences, of which I had communicated some results in my previous letter to you, but at that time those were still merely drafts.[17]

---

[15] Cf. [Lambert 1760, trans. Anding, III, p. 75]. According to Anding, this form of the formula is due to Beer.

[16] Cf. the works of Pfeiffer [1979] and Debru [1979] on this topic.

[17] *Le premier chapitre expose les principes et les difficultés, et le troisième les assujettit à un examen aussi rigide, que j'ai pu l'imaginer. La seconde partie contient les expériences que j'ai faites par le verre. Je serai charmé, Monsieur, si vous voulez vous donner la peine de les examiner, et Vos Sentiments là-dessus ne pourront que m'être extrêmement estimables. Vous en trouverez l'ordre et la liaison changée de fond en comble, et toute différente de celle de la première ébauche, que j'aurois destiné à tout autre but si j'etois resté plus longtemps à Goettingue.*

Ce qui y paroitra de plus nouveau, c'est la méthode de déterminer le degré de la blancheur de chaque corps, qui se trouve Théorème 35 et ses conséquences, dont je vous avois communiqué

In the text, he underlines once more the importance of this formula that permits the determination of the *albedo* of a surface. The notion of *albedo* due to Lambert is the one remained in use.

The fourth chapter is almost totally dedicated to geometric questions relative to the lens and on physiological questions. His main result is perhaps the proof that the opening of the pupil is not determined by the size of the object but by the quantity of light that falls on the retina. On the other hand, this chapter deals very little with the subjective aspect of vision.

The fifth chapter contains a description – this one quantitative as well – of the phenomena of the atmosphere. According to Anding, the explanation of the twilight represents the first significant progress in this field since since Ibn al-Haytham (generally known under the Latinized form of his given name, Alhazen).

The applications of his formulas to the planets and to the fixed stars, contained in the sixth chapter, extend the work of Euler mentioned earlier. The results of Lambert obtained at the very beginning of modern astrophysics have contributed to our understanding of celestial phenomena, an understanding that decidedly broadens our horizons. Further, the views and speculations contained in his *Cosmologische Briefe* are based in part on these experimental results.

It is the last part, which deals with colors and shadows,[18] which today seems the most interesting, and might be the most original. In 1892, the very attentive Anding noted that this whole branch of science was even at his time still just at its beginning, concluded by mentioning the names of Fraunhofer, Purkinje and Helmholtz.[19] Even those who accept and even admire Goethe's theory of color must regret that he didn't study the work of Lambert, which he mentions in passing, more closely; with this study he would have avoided a series of errors and misunderstandings regarding Newton. The antecedents of these investigations by Lambert should be studied again: one finds much of it in Goethe's *Farbenlehre* [1949] but Goethe never faces the mathematical questions.

A few years ago the scientific world was surprised by the experiments of the American Edwin Land concerning the perception of the colors. To the best of my knowledge, these experiments have never been explained in a satisfactory manner. It is very probable that the works of Lambert contain valid – perhaps even crucial – indications.

But for the historian of the sciences, the *Photometria* besides being a scientific document of great value, is of interest for yet another reason: it contains the assessment of the science of the time, because Lambert was an excellent judge of his predecessors and his contemporaries and he presents his opinion with frankness. There is a very good description of the situation of the optics of the time, that is, of the corpuscular theory of Newton that was dominant at the time and of the wave theory that goes back to Descartes and Huygens but is due mainly to Euler. Lambert credits Euler with having entirely transformed the system of his predecessors and having adapted it to most phenomena [Lambert 1760: § 18]. Lambert leans towards a wave theory, and with the exception of Euler, he was almost the only one of his epoch to do so. However he is unable to accept it

quelques résultats dans ma précédente, mais alors ils n'etoient encore que simplement ébauchés (Lambert, April 1760, letter no. 9a in [Daniel Bernoulli 1979].

[18] See also [Pfeiffer 1979].

[19] [Lambert 1760, trans. Anding 1892, III : 168].

definitely, because, as he says, "it has not yet passed the ultimate test, that is, to predict new effects [Lambert 1760: § 18]. In this regard it is necessary to underline once again the great value of the commentaries of E. Anding for the history of science of the eighteenth century. It is very much to be hoped that in a complete edition of the works of Lambert, Anding's work will be given a place next to the books that they served so well.

Postlude: the other works

In addition to the two books discussed here, we possess a half-dozen works of Lambert about which I will make just a few remarks.

One of these works, "Anmerkungen über Strahlenbrechung" [1779], is a continuation of the investigations presented in the *Propriétés remarquables*: it shows that the results published by Tobias Mayer can be easily obtained by means of the results that had been presented in the *Propriétés.*

The *Photometria* finds a continuation in "Vom Glanze der Venus" [1780], a charming and amusing work to read because the narration of his observations shows us how Lambert worked when he was most at home in a problem.

Two papers on perspective, "Sur la perspective aérienne" [1774] published in Berlin in 1774, and "Mémoires sur la partie photométrique de tout l'art de peindre" [1768], are remarkable among other reasons for their discussion of the writings of Leonardo da Vinci.

"Sur les porte-lumières appliqués a la Lampe" [1770] shows Lambert as the inventor, or in any case one of the inventors, of beacons. It should not be forgotten that Lambert was indeed an inventor and constructor of devices. It was the analogy with the megaphone that gave him the idea to place a metallic cone behind candles or oil lamps in order to concentrate the light emitted by a source while multiplying the clarity by a factor of 10. Thus Lambert can be considered as one of the inventors, perhaps the first, of automobile headlights.

On 28 June 1760 Daniel Bernoulli wrote to Lambert:

> Dear Sir,
> I received your excellent *Photometria* and it is impossible for me to demonstrate all of the gratitude that such a beautiful work merits. I beg you in any case to believe that I am sensible to the whole value of it and everything that renders it precious to me in particular. I therefore have the honor of thanking you a thousand times for it and make you as many again the most sincere compliments on the great successes with which your truly immortal work must meet.[20]

Posterity has confirmed Daniel Bernoulli's judgment and has extended it to all of his work.

---

[20] *Monsieur, J'ai reçu votre Excellente Photometria et je ne sauroy témoigner toute la reconnoissance, qu'un si beau present merite. Je vous prie cependant de croire, que j'en sens tout le prix et tout ce qui me le doit rendre pretieux à moi en particulier. J'ai donc l'honneur de Vous en faire mille et mille remerciemens et autant de complimens des plus sinceres sur les grands succès, qui doivent rendre votre ouvrage véritablement immortel* (Lambert, 28 June 1760, letter no. 10 in [Daniel Bernoulli 1979].

In a recent letter, an eminent mathematician wrote to me that nowadays it was truly a *scandal* that the works of d'Alembert are still unpublished. Perhaps scandal is not the right word, but it is infinitely deplorable that there is no complete edition of the works of Lambert. Andreas Speiser, who edited some of the analytic, algebraic and arithmetic works of Lambert [1946-48], said frequently that it is an absolute rule that people whose complete works are not published are never cited. In conclusion, I would like to express my desire that Lambert's complete works will be forthcoming.

Acknowledgments

I am very grateful to Mr L. Bossy, who helped me to understand the work of Lambert, as well as Mr V. Scheuber, who helped in the compilation of the references. I especially thank Mrs. P. Radelet-de Grave for the care with which she edited Lambert's text, verified the literature and assisted me in writing of the Introduction to the facsimile edition of the *Propriétés remarquables*.

Pierre Louis Moreau de Maupertuis

Pierre Louis Moreau de Maupertuis[1]

Introduction

The immediate occasion for the present reflections was the preparation for the publication of the Bernoulli correspondence with Fritz Nagel, especially that between Johann II Bernoulli and Maupertuis, de la Condamine, Mme du Châtelet, Voltaire and Bernhard Merian, the philosopher from Basel in Berlin.

The spread of Newton's mechanics into science

Maupertuis's works as a physicist are part of the history of the penetration of Newton mechanics into science on the Continent. As everyone knows, Newton derived Kepler's laws from his own laws of motion in his great work, the *Principia*, but the book contains much more. Table 1 shows how Newton's ideas penetrated of into the world of science.

Book	The first, most important successors
De motu corporum I	Jacob Hermann Johann I Bernoulli Leonhard Euler later: Analytic celestial mechanics Alexis Claude Clairaut Jean d'Alembert Leonhard Euler Joseph-Louis Lagrange Pierre Simon de Laplace
De motu corporum II	Jacopo Riccati Daniel Bernoulli Johann I Bernoulli Jean d'Alembert Leonhard Euler
De mundi systemate	Pierre Louis Moreau de Maupertuis Daniel Bernoulli Charles Marie de La Condamine

[1] Originally published as "Pierre Louis Moreau de Maupertuis (1698-1759) in *Pierre Louis de Maupertuis. Eine Bilanz nach 300 Jahren*. Ed. Hartmut Hecht. Berlin: Verlag Arno Spitz GmbH, 1999, pp. 341-362.

Table 1. The first, most important successors to Newton on the Continent

Of the two books on motion, which regarded what we today call "general mechanics", the first is dedicated to point mechanics, for which Newton created the valid basis, while the second is dedicated above all to hydrodynamics. The first noteworthy echoes on the Continent came from Jacob Hermann and Johann I Bernoulli in the year 1710; twenty-four years later appeared the significant follow-up with Euler's *Mechanica*, and with Alexis Claude Clairaut came the breakthrough in higher celestial mechanics.

The reaction to the second book appeared more slowly: it would stimulate important continuations by Jacopo Riccati. The works of Daniel and Johann I Bernoulli were decisive, and around the time that Maupertuis's stay in Berlin came to an end appeared the important works of d'Alembert and Euler.

The history of the more clearly written third book, *De mundi systemate* (On the system of the world), in which Newton set forth his theory of universal gravity,[i] proceeded quite differently. Everyone knows what this is: the planetary system is held together by the attraction of each body to every other. The attraction decreases with the square of the distance: $F \propto 1/r^2$, and the Keplerian laws derive from these assumptions. In the opinion of most – as well as of Newton himself – the theory had a gap: it could not establish any mechanism that conveyed the gravity. Because of this, the Cartesians above all fought the theory. According to Descartes's theory, a vortex drove the planets about the sun, which explained why the planes of their orbits almost coincide. Johann Bernoulli had also tried to devise such a mechanism for Newton's theory.

The dispute between the two theories lasted fifty years, from 1687, the year of the publication of the *Principia*, until approximately 1737. The decision in favor of Newton was the result of a long battle that was led by a group of mostly young French scholars. Newton's book had been translated into French by Mme du Châtelet in 1759. Mme du Châtelet was the center of a circle which also included Voltaire, as well as two young Swiss, Johann I Bernoulli and Samuel König from Bern, about whom we will hear more. But the most active members were Maupertuis and his friend La Condamine. Outside France, Newton's theories were defended above all by Daniel Bernoulli.

This was the state of affairs around 1730.

Maupertuis's training in Basel

Maupertuis was enrolled as a student of Johann Bernoulli's in Basel on 30 September 1729. Under Johann, Maupertuis acquired a firm foundation in the newest mathematics and mechanics of the time, meaning above all Cartesian geometry, as well as the elementary parts of the new infinitesimal calculus, which had been developed by Leibniz and the Bernoulli brothers. Presumably the lessons were not very different from those that Bernoulli had given more than thirty years before to the marquis de l'Hôpital. The mechanics taught were probably made up of parts of Huygens's *Horologium*, Newton's *Principia* and from works by Bernoulli himself. Some of these works dealt with Newton's new mechanics, which Maupertuis would come to accept, with its foundations and its most important application, the theory of gravity. Although his teacher had objections to these, Maupertuis became more and more convinced. Admittedly, Maupertuis was not a first-class mathematician, but he acquired a good understanding of both sciences. Above all, he understood what it meant to subject a theory to experimentation, because Maupertuis seemed to have a natural instinct for measurements and their evaluation, at a time when this was still barely understood. This would come to serve him well.

The years between his residence in Basel and the great expedition

I can be brief about what happened during the years that followed. Maupertuis was in various places, including Montpellier, where he occupied himself with biology, among other things. He dealt with questions of heredity as well as evolution, concerning which he was in contact with Buffon, but I won't enter into these. I will, however, allow myself to note that it would be worthwhile to pursue the question of whether Maupertuis was among those who first recognized that the earth is much older than reported in the Bible, which Newton firmly believed. In any case, one perceives that Maupertuis was a very versatile and inventive scientist, open to all new theories. He must therefore be forgiven if his critical sense was not strongly developed and sometimes deserted him.

His other main interest was the shape of the earth. He wrote a book about this question, *Sur la figure de la terre* [1738]. The great question of these years was whether the earth was flattened or if it is an extended ellipsoid. The first theory was predicted by Huygens's theory of centrifugal force and from Newton's theory of gravitation; the second was supported by the measurements of the French astronomer Cassini. Who was right? For many, the answer depended on the validity of Newton's new mechanics.

The Paris Académie was mostly on the side of its member, Cassini, but a group of *jeunes loups* who didn't trust these very difficult measurements fought against the *vieux lions*. Maupertuis was their representative. The Académie then decided to send an expedition to the equator in Peru and Ecuador, but rather than Maupertuis, his friend La Condamine was chosen to lead it. However, this first expedition didn't produce any immediate result and for this and other reasons it was decided to send a second expedition to Lapland in the extreme north. This time, Maupertuis was put in charge.

The priority dispute between Newton and Leibniz

In order to understand the explosive spread of the affair that later brought Maupertuis to grief, we do well to remember a similar affair that had occurred at the beginning of the century.

Newton had taken the final, crucial step in the calculus in the year 1666, but at first hardly published anything. In 1674 Leibniz discovered his differential calculus in Paris, stimulated by the reading of a work of Pascal. He too was to publish his discovery only some years later, which then stimulated the Bernoulli brothers to amplify this new science. Leibniz never claimed the priority of his discovery, though he did claim its independence. By the end of the eighteenth century practically all mathematicians used his calculus and that of the Bernoullis.

But then Newton conceived the that Leibniz had stolen it from him. A dispute developed, the emotions of which can still be found today in the literature. Leibniz was defended by his friend Johann I Bernoulli, but in vain, since the Royal Society condemned

Leibniz as a plagiarist. In Germany, where Leibniz was regarded as the greatest scientist on the Continent and as the greatest philosopher of his age, the outrage was huge, and one suspects all the more so since just shortly before the "perfidious Albion" had opposed Louis XIV in the War of the Spanish Succession. Be that as it may, Newton's new physics were not popular in Germany, and on the other hand, the mere suspicion of plagiarism was not seen as a threat. We must keep both in mind in what follows.

The Berlin Academy of the Sciences

The Berlin academy was founded by Frederick I in 1701, one year after the establishment of the Prussian kingdom. Leibniz had organized it, and it began promisingly, but Frederick's successor, the soldier king Frederick Wilhelm I, was only interested in his rule and his army. However, his son, Frederick the Great, was of the same mind as his grandfather. Immediately after he assumed power, he announced his desire to restore the academy to its original greatness.

It appears that Frederick first wished to have as president the philosopher Wolff, who was regarded as the heir to Leibniz. But Wolff declined the invitation, and the king turned to France, the country, whose *esprit* he admired above all others. When the king inquired about an officer, he often asked, *Hat er Fortüne?* Not in the sense of fortune in terms of wealth, but in terms of luck, and even better, in the sense of "Does he know how to grab luck with both hands?" No doubt Maupertuis's dynamic expedition impressed him. Also, Maupertuis belonged to the aristocracy and had become known as *homme de letters*, something that the king valued highly.

Undoubtedly Maupertuis also possessed a knack for administration and organization, and these were now in demand. Because the king's striving for glory in battle was even greater than his striving for scientific fame, Maupertuis took part in the Silesian wars. Maupertuis's task therefore was not easy, but even under normal circumstances one would have to have spoken of great success. In 1745, he was nominated president.

The first goal was naturally to bring a series of first-class scholars to Berlin, and in this he succeeded. Albrecht von Haller, the founder of the Göttingen academy, declined, and Daniel and Johann II Bernoulli wanted to join merely as foreign members. As Adolf von Harnack writes in his history of the academy [1900], Maupertuis apparently had a preference for the Swiss, in fond memory of his time in Basel. So he also appointed Bernhard Merian and J. G. Sulzer from Winterthur as professors of philosophy. But more than Maupertuis, it was the king who succeeded in winning, from St. Petersburg, the man who would be the academy's most important member for the next twenty-six years: Leonhard Euler.

Euler wrote over 300 works during his Berlin years, including eight books, many of which dealt with Newton's mechanics. He was *directeur de la classe scientifique*, and since Maupertuis didn't speak German, a large part of the administration work fell to Euler.

It is worthwhile to reflect for a moment. The man who had led Newton's theories to victory in France now presided in Berlin. At his side worked the leading expert of the new science. Next to both of these we must consider as well the foreign members Daniel and Johann II Bernoulli. This means that the center of the new Newtonian mechanics, at that time the dominant science, was now in Berlin, that is, in a country that until this point had been the territory of Leibniz and Wolff. Euler's express goal was to extended Newton's foundations of mechanics to other branches of science as well, and he declared again and again that the monadology of Leibniz and Wolff was insufficient for laying the foundations of science. Sooner or later there were bound to be repercussions, but for ten years all was well.

Euler's *Methodus Inveniendi Lineas Curvas* and Maupertuis's principle of least action

In 1744 in Lausanne was published Euler's *Methodus inveniendi lineas curvas...*, (*A Method for Finding Curves Enjoying Some Properties of Maximum or Minimum*), concerning the shortest curve between two given points on some surface, or the curve with a given length that encloses the largest area, and other extremal properties in geometry and mechanics. This was the first detailed representation of a new branch of analysis founded by Jacob Bernoulli. According to Constantin Carathéodory, the editor of the volume of the *Opera omnia* dedicated to the *Methodus inveniendi*, Euler had already begun to work on such problems in St. Petersburg. He published his first work in this area there, and discussed it with his colleague from Basel, Jacob Hermann, a student of Jacob Bernoulli, a significant circumstance, as we shall see.

The book, according to Carathéodory one of the most beautiful ever written in mathematics, was already completed in St. Petersburg. The first applications were in the area of differential geometry, but in 1743 in Berlin Euler wrote two important appendixes, *Additamenta*. In the first, he developed the theory of the elastic curve, which goes back to Jacob Bernoulli; in the second, he examined the principle of least action. For long time, it was *the* standard work, and it also contains the Euler-Lagrange equation of calculus of variations, which is still used by mathematicians and physicists today.

Euler had been directed towards this question through his correspondence with Maupertuis. In modern terms, the principle is:

$$\delta A = \delta \int (E - V)^{1/2} \mathrm{d}s = 0 ,$$

First, what is the meaning of this equation? The answer and an explanation are provided in an appendix at the end of this paper.

Since a priority question is also involved here, I must go into some detail.

Euler's *Additamentum II is* the first publication in which the principle of least action is found. If, however, one reads carefully through it, it can be seen that Euler in fact reacts to a suggestion and examines it. He proceeds in such a way that the result, which follows from the new principle, is compared to that produced by what he calls the "direct method". The "direct method" that Euler always used is none other than that of Newton. The reasoning behind it is that in mechanics the principle applies to forces, as we say today, which receive energy. It therefore also applies to gravity, for example, but not to friction and viscosity. The result is that the principle, though of great practical value, is not universally valid. Euler doesn't comment on its validity in optics, and he himself used the principle only rarely.

The three works of Maupertuis are less weighty in comparison to the *Additamentum.* In the first, entitled "Loi du repos des corps" [1742], he shows that the laws of impact can be derived from a minimum principle in the case of elastic impact, and impact was fundamental according to Cartesian mechanics. In the second, entitled "Accord de différentes loix de la Nature, qui avoient jusqu'ici paru incompatibles" [1748a], he shows the same for optics, if one accepts (in agreement with Descartes, Newton and Leibniz, but in disagreement with Fermat, Huygens and Euler) that in a vacuum light expands more slowly rather than faster than in water or air. The laws of two different areas of the physics therefore follow from the same basic principle.

He tried in the third work, "Les lox du mouvement et du repos deduites d'un principe metaphysique" [1748b] to derive this basic principle of physics from the principles of metaphysics. Such an action seems strange to us today, and it would appear to be a conservative, not to say a reactionary, attitude in the nineteenth century. It definitely appeared that way to many. But metaphysical or theological observations are found in the works of Kepler, Galileo, Descartes, Newton and Leibniz as well; in Voltaire's book on Newton's physics, where one least of all would expect this, the first hundred pages, that is, one fifth, deal with metaphysical questions. I mention this since one can often read that the dispute between Maupertuis and Euler was motivated by religion, or at least philosophy. That is wrong. I would like to quote the end of *Additamentum II.* Here Euler, with a sidelong glance at Maupertuis and his numerous followers, says

> Although the power of this consideration is not yet sufficiently understood, so I am not sure that it agrees with the truth, nevertheless I doubt that a sounder metaphysics can lead to greater evidence; I leave to others this business of which metaphysics to teach.[2]

The French mathematician Jean Dhombres, one of the editors of the Bernoulli edition, emphasized that metaphysics and science finally part ways in the Maupertuis affair: this is perhaps one of the most significant outcomes of this story. As the Neo-Kantian Ernst Cassirer as well as Andreas Speiser have pointed out, even Kant, a philosopher who never spoke of the "demon natural science", has learned from Euler.

So to whom is the principle of least action to be attributed? Euler always has spoken of it as a principle of Maupertuis. He is often praised on the one hand for his generosity and condemned on the other for his servile behavior towards his president. I myself always spoke in my lectures of a single principle of Euler and Maupertuis. However, the reading shows that Euler was right and dealt with it correctly. In order to see this, one need only know what was meant by a "principle" in the eighteenth century. It must be admitted that Maupertuis's mathematics was completely inadequate, and that the principle's fruitfulness came from Euler, but the idea that all areas of physics are held together by *one* principle, which was apparently what Euler was saying, came from Maupertuis. In the present work I will speak of the principle of Maupertuis and Euler.

We now come to the affair that lead to Maupertuis's downfall.

---

[2] *Cujus ratiocinii vis, etiamsi nondum satis perspiciatur, tamen, quia cum veritate congruit, non dubito, qui ope principiorum sanioris Metaphysicae ad majorem evidentiam evehi queat; quod negotium aliis, qui Metaphysicam profitentur, relinquo* [Euler 1744: *Additamentum II*, § 16].

The work of Samuel König

Fate seems to have predestined Johann Samuel König for the role that he played in the Maupertuis affair. He was born in Büdingen in Germany, the son of a pastor who had had to leave Bern for religious reasons, and for part of his life he too was an exile. It isn't seemly to pass judgment on a man who had often had a difficult life, but historians obviously must do their job, so I can't avoid it.

The young König was supposed to have returned to Bern in order to study law, but he felt more drawn to the sciences and went to Paris, where he was a *mémorialiste* at the Académie.

Back in Bern, he became part of the circle of Samuel Henzi who, with others, sought to push through political reforms in order to put the government on a broader footing. The attempt failed: Henzi and two others were beheaded, and König banished for ten years.

The Henzi conspiracy caused a big stir in Germany as well. Lessing even wanted to hold turn it into a dramatic play (according to Michael Stettler, it was the more conservative Albrecht von Haller who dissuaded him from this project). But Haller also successfully recommended König for the Franconian University in Westfriesland in Holland. Henzi himself apparently remained a hero for all who desired reform, and apparently a part of this fame was later reflected onto König.

König, however, was in no way only the upright fighter for truth and right against all authority that he is often made out to be. The historic truth is somewhat different, and in order to understand the scandal that he caused, as well as the reactions of Maupertuis and Euler, I must now mention some earlier events.

The correspondence of Mme du Châtelet with Johann II Bernoulli casts a shadow on König's character, even if the marquise limits herself to allusions. A witness of these events is more explicit, writing in 1740 to Johann Bernoulli:

> I am not at all sure that M. König has the most respectful recognition for the goodness of Madame du Chastelet [sic]. Not only was she in charge of his brother's keep and treated him at Cirey as one of the family, but when she directed M. König to Paris, she made his brother come too, she lodged both of them in an apartment that she rented [...]. She maintained a lackey for them, and gave orders to her headwaiter to provide them without restriction with whatever they asked for … .[3]

He then describes further how both of them grossly abused the trust of the marquise. Even taking into consideration the perhaps insufficient maturity of a 28-year old, the incident casts a shadow on König.

---

[3] *Je suis bien sûr que tout autre que M. König seroit de la plus respectueuse reconnaissance pour Les bontez de madame du Chastelet. Non seulement elle se chargea de L'entretien de son frère et le fit traitter à Cirey comme le fils de la maison, mais lors qu'elle mena M. König à Paris, elle fit encore venir ce frère, elle les logea tout deux dans un apartement qu'elle Leur loua [...]. Elle leur entretint un laquais, et donna ordre a son maitre d'hôtel de leur faire fournir sans aucune restriction tout ce qu'ils demanderoient....* Voltaire, letter to Johann Bernoulli, 30 January 1740, in [Voltaire 1965: 255].

Who was this witness? It was none other than Voltaire, the same Voltaire who would later defend König. One may object to Voltaire's well-known *médisance* (malicious gossip) but one cannot forget this testimony when it comes to judging later events.

I now come to König's famous publication in the *Acta Eruditorum* in 1751, entitled "De universali principio aequilibrii & motus in vi viva reperto, deque nexu inter vim vivam & actionem, utriusque minimo, dissertation" (Dissertation on the universal principle of equilibrium and motion found in live force, and on the relationship between live force and action, which is a minimum) [König 1751].

Historians have concentrated almost exclusively on the short appendix at the end, but the main part is historically informative as well. König's goal is to show that Maupertuis's principle is wrong. To this end, after an introduction consisting of definitions, principles, theorems etc., he cites seven examples to show this. These seven are:

– the law of the lever	Archimedes, Galileo
– the equilibrium of moments of forces	Varignon, Mariotte
– the parallelogram of forces	Newton
– lifing a heavy rod	Galileo
– the equilibrium for liquids in communicating vessels	Stevin
– the regulation of the axis of rotation of a rod that carries two unequal masses	Huygens
– the elastic impact of two bodies	Wren, Huygens

One needn't be a physicist in order to see two things immediately. First, of the seven examples, five belong to statics, and therefore have nothing to do with the principle. Second, the solutions to the seven problems were found long before König.

How does König proceed? He is, if you will, an orthodox Leibnizian of the purest kind, who would base the whole of mechanics on the *vis viva*, and therefore on kinetic energy. This is a detour for problems of statics, and doesn't lead far for the other two either, as everyone would have known by then. One gets the impression of a man who had been out of contact with research for years and had only lived in his own world of thought, who in this case merely circles about the *vis viva*. This sentence is characteristic: "However, all of those who follow Galileo in estimating the equilibrium as the product of mass times acceleration certainly find themselves in grave error".[4] However, Huygens, Newton, Jacob and Johann Bernoulli and Euler, among others, belong to these successors of Galileo. At best one could exclude Daniel Bernoulli: however, I will come back to this.

Refuting a proposed new principle for mechanics requires a careful, logically consistent procedure. Even with the best intentions, what we find in König cannot be described as other than confused. Definitions, axioms, hypotheses appear in colorful, disorganized sequence. The work therefore was written in a hurry, and is disordered, and careless. Pierre Costabel, the editor of two volumes of the correspondence of Johann Bernoulli with Varignon and of one volume of the Euler correspondence, arrived at a similar judgment by another route.

[4] Illi autem omnes, qui duce Galilaeo aequilibrium ex facto velocitatis iniziali in massam aestimare iusserunt, in magno certe errore fuerunt [König 1752: *L. Euleri Opera omnia,* s. II, v. 5, p. 317].

An example must suffice to show what concepts are involved. The whole work is about Maupertuis's action, but this doesn't hinder König from suddenly using this concept in Newton's sense of force. Only there it is about something quite different, namely, about a force, a vector that is a directed quantity. However, since König did not consider the acceleration, there is only a weak link. All leading scientists at that time knew that in order to solve a problem of n bodies, $3n$ equations are needed, for example, Newton's. The *vis viva*, and this is the weakness of this approach to mechanics, only provides one equation, and both of König's examples are of this kind, with, as we say today, only one degree of freedom. But Maupertuis's principle can come straight to the rescue: Euler's equations following from it are generalized Newtonian equations. König needed only to have read the *Additamentum II* in order to see this. But the irony is that he doesn't want to know anything about it, and this renders this paper sterile.

The consequences triggered by this astounding work are disappointing indeed, and one can understand Euler's anger when later many agreed with König's opinion on this question, precisely since he understood nothing at all about it. This shows clearly the essence of the affair: at the time there were not even a dozen people in all Europe who were capable of well-founded judgment on this question; the situation was similar to that of the theory of general relativity at the beginning of the twentieth century.

Now, one might object that it was in this very work that König stated the principle that made him immortal.

Since I too, like many other colleagues, was led astray and wrote that, here (in 1729) Daniel Bernoulli had formulated a statement, which König later stated in complete generality [Radelet-de Grave and Speiser 1987a: 30, 138]. I must cry, *mea culpa*! In modern terms, König proves for two bodies the theorem according to which the kinetic energy of a system of two masses is equal to the kinetic energy of the system with respect to the centre of mass (that is, with respect to a system of moving axes whose origin is at the centre of mass) plus the kinetic energy of the centre of mass if the total mass were concentrated in it (see [König 1752]). But we find this theorem seventy years earlier with Wren. However, König then adds in a scholion that the same is generally true no matter how many bodies there are. But there is no formulation in three dimensions and no proof, especially in the case of curved paths. In this case we find more in Daniel Bernoulli. König discusses the principle only very vaguely. It is useless to look for the fundamental connection with the Galilean principle of relativity, since König could not correctly grasp this principle without Newton's concept of force. This perhaps is not his fault; at the time, only Euler understood this. However, a year earlier in the *Mémoires* of the Berlin academy, Daniel Bernoulli had written on the *vis viva* of the planetary system in full generality [1750], and König would only have had to apply his discovery to it and prove it with its help in full generality.

These are not the only inconsistencies of the work; to show the most important, we must read the end. König quotes the end of a letter from Leibniz to Jacob Hermann, in which he writes:

> Action is not what you think, the consideration of the times there between; it is as the product of the mass by the time, or of the time by the *vis viva*. I have remarked that in the modifications of motions it ordinarily becomes a

> Maximum or a Minimum. One can deduct several propositions of great consequence from it; its power to serve has determined the curves that the bodies describe attract have one or several centers. I wanted to treat these things among others in the second part of my Dynamics, but I held off; the bad reception to the first caused by prejudice having disgusted me.[5]

In the next twenty pages or so, König tried to demonstrate the absurdity of Maupertuis's idea, and then closes with a quotation from Leibniz, his hero, who thus supposedly already knew this principle! The work concludes with this about-face, and it is this quotation that caused the scandal.

The action of the Academy and its judgment

Maupertuis was indignant over this attribution of priority. It isn't clear whether he was most concerned about defending his priority or fearful of an accusation of plagiary. In any case, he turned the case over to the Academy.

The action of the Academy and its judgment are recorded in an official document written by Euler, "Exposé concernant l'examen de la lettre de M. de Leibnitz..." [1752a]. The report includes a criticism of König's understanding of Maupertuis's principle, and an enumeration of all the steps taken by König and the Academy to verify the authenticity of the letter, since no trace of such an idea has been found in Leibniz's work and letters, nor in those of Johann I Bernoulli. On 28 May Maupertuis asked König about whole letter, and received a reply on 8 June saying that it came from Samuel Henzi. Maupertuis then asked the French ambassador in Bern provide information on other letters. On 7 October, Formey, the secretary of the Academy, wrote an official letter to König, which was delivered by the ambassador, but he received no answer until 10 December. At König's request the letter was sought, first in Bern and then in Basel in Hermann's estate, but without success.

Euler closes, "it is definitely manifest that his accusation is of the worst and that this piece was forged…".[6]

It must be emphasized that König himself is *not* called a forger, as has been claimed again and again; surely this was deliberate. All documents were submitted to the Academy during a plenary session; with all the members present, three-quarters of the total reached an unanimous decision by a secret vote. König then resigned as a corresponding member of the Academy.

---

[5] *L'Action n'est point ce que vous pensés, la considération du tems y entre; elle est comme le produit de la masse par le tems, ou du tems par la force vive. J'ai remarqué que dans les modifications des mouvemens elle devient ordinairement un Maximum ou un Minimum. On en peut déduire plusieurs propositions de grande consequence; elle pourroit servir à déterminer les courbes que decrivent les corps attirés à un ou plusieurs centres. Je voulois traiter de ces choses entr'autres dans la seconde partie de ma Dynamique, que j'ai supprimée; le mauvais accueil, que le prejugé a fait à la premiere, m'ayant degouté* [König 1752: *L. Euleri Opera omnia,* s. II, v. 5, p. 323].

[6] *…il est assurément manifeste que sa cause est des plus mauvaises et que ce fragment a été forge ...* [Euler 1752a: *Opera,* s. II, v. 5, p. 72].

I must comment here. In the literature, König is often praised for his integrity; I myself have serious reservations about this. A scientific publication should increase our knowledge, or at least provide clarification. König hardly did either. His arguments, as I have tried to show, are far off the mark. Unfortunately, his own work is hardly taken into consideration in historical research. There are decided sectarian camps, and a former colleague's emotional pronouncement on the subject is due more to this than to merely unpleasant aggressiveness. König, who was with Johann Bernoulli in Basel just a few years after the condemnation of Leibniz by the Royal Society, had to know what such a publication might lead to, although the effect was much greater than he could have envisioned. His claim that an academy doesn't have the right to comment on priority questions would definitely not be accepted today.

The dispute would probably have passed without a great fuss, since König's own defense, *Appel au public* [1752], contained little that was new. But then the fight was joined, not by a scientist, but rather by a man of letters, and what had been a relatively modest affair quickly grew into a public scandal.

Voltaire's intervention

When Maupertuis returned from his expedition, Voltaire had covered him with praise. The relationship between the two was still warm during the first years in Berlin as well. What changed all this? What made Voltaire write *La diatribe du Docteur Akakia et du Natif de St. Malo* [1753] an angry pamphlet in which Maupertuis is portrayed as a total fool and charlatan, someone of whom the public must beware? Certainly this was not due only to the judgment of the Academy because, as we saw, Voltaire was hardly an admirer of – indeed was quite nonchalant about – König. Thus one can't get around the fact that the grounds for Voltaire's having taken König's side lay in his almost paranoid jealousy of Maupertuis's prestige, as most of those who have examined the situation have claimed.

How did he strike his target? Everyone who has read some of his short stories knows how amusing and humorous his jokes and mockery are, but who goes to the trouble of checking his claims, especially if the topic is arid science?

The French Jacques Tuffet did just that in his critical edition of the *Diatribe*, and the result is disastrous for Voltaire. I will mention three examples. The first two are taken from medicine, which at that time was an area in which everyone could cope with relatively ease.

Maupertuis writes:

> Medicine is very far from being at the point where one can deduce the treatment of the illnesses from a knowledge of the causes and effects: the best physician is the one who argues the least and observes the most.[7]

His intended meaning is clear. But in Voltaire this becomes "The best physician [...] is the one that argues the least".[8] Period.

[7] *La médecine est bien éloignée d'être au point où l'on pourroit déduire le traitement des maladies de la connoissance des causes et des effets: le meilleur médecin est celui qui raisonne le moins et qui observe le plus* [Maupertuis 1752: 347].

Then, in Voltaire, we find, "he advises us [...] to pierce [the patient's] skin needles".[9] You rightly suspect, of course, that Maupertuis speaks about acupuncture, concerning which, according to Tuffet, Maupertuis had advocated *careful* experiments [Maupertuis 1752: 206].

The third example comes from celestial mechanics. Voltaire ridicules Maupertuis for writing that there are stars that look like "millwheels" [Voltaire 1753: Édition critique de Tuffet, 6]. What the figure below shows, however, is no normal star but a galaxy.

Spiral galaxy

Its shape is due to centrifugal force and gravity, that is, precisely the forces whose combined effect Maupertuis had examined; thus his prediction was in no way a fluke. In the universe there are approximately ten billion such millwheels. This prediction of flattened galaxies recalls those of neutron stars and the black holes by Zwicky and by Baade. Zwicky too, like Maupertuis, seems to have been a very imaginative scientist, and although because of this he was less critical, it made him a very good observer.

But Voltaire's writing was successful: who doesn't like to sit with the mockers and laugh with Voltaire? And on the other hand, how many people at that time could follow

[8] *Le meilleur médecin* [...] *est celui qui raisonne le moins* [Voltaire 1753: Édition critique De Tuffet, 8].

[9] *...il nous conseille* [...] *de percer sa peau [d'un malade] avec des aiguilles* [Voltaire 1753: Édition critique de Tuffet, 7].

the arguments of Euler, Maupertuis and König? In Berlin there was maybe one, and in all of Europe perhaps a half a dozen more. Today Voltaire would not have gotten off so easily; in England or America, for example, he would probably be sued. But Maupertuis, not at all the charlatan that his jealous rival made him out to be, was only a exhausted man and had to abandon Berlin. Even today his reputation suffers from it. But he was not the only one to have to leave Berlin: Voltaire too had to go. The king had prohibited him from re-publishing the pamphlet, and when Voltaire did anyway, he had it burned publicly. Since Voltaire had also broken the laws regarding payments, he was forced to leave Prussia. He never saw the king again. But until today Maupertuis has been deprived of the credit that he deserves in science.

Maupertuis's end

After the attacks of Voltaire Maupertuis was a broken man, and his health became increasingly worse as well. He never returned to Berlin; he left the businesses of the Academy to Euler, and mainly lived in France. En route to Italy, he arrived ill in Basel, where his friend Johann II Bernoulli took him in and looked after him for several months. He died on 27 July 1759 and was buried in the church of Oberdornach, close to Basel. Otto Spiess was able to show that in all likelihood, contrary to other claims, he still lies there today.

What remains from the work of Maupertuis?

Above all, Maupertuis was the first who was able to confirm empirically, from his expedition, that Newton's theory of gravity goes beyond Kepler's laws. The new theories became generally accepted thanks to him, among others.

The ascension of Berlin as the center for science also remains to his credit. That ascension begins with Maurpertuis and Euler; indeed, the beginning was vertical! Even if today we think of these merits as historic, they are nevertheless permanent.

Some of Maupertuis's predictions were brilliant. That of the spiral galaxy I have mentioned; of other in the areas of biology and linguistics I cannot speak with authority.

But above all, he was the first to attempt to formulate a principle that would link all areas of physics, one that all laws would originate from. In classical physics, his principle was admittedly useful, but it hardly proved to be fundamentally valid. Euler's misgivings were justified: the appearances of heat hinder the principle.

It is a different case in quantum theory. Einstein was the first to point out that the first priority is always to quantify action, not energy or angular momentum, and it is the action of Maupertuis, not that of Hamilton, that underlies the theory of Bohr-Sommerfeld. But in modern field theory, principles of action or force are now increasingly proven to be fundamental, certainly since Hilbert successfully derived Einstein's general theory of relativity from such a principle. Today, they have become indispensable in subatomic particle physics through the famous principle of Emmy Nöther; this was particularly emphasized by Feynman and Dirac. However, these are not due to some metaphysical reasons, as Maupertuis thought, but because of their connection with the symmetry properties of space, time and matter, that is, particles and fields. That we look for the reason elsewhere today does not change the fact that it is often so in historic development; in physics, symmetries belong to that which is very close to philosophy. In any case, the first to speak of such a connection was Maupertuis.

Translated from the German by Kim Williams

Acknowledgments

I heartily thank Mr. Hecht of the Berlin-Brandenburg Academy of the Sciences for the invitation to speak in Berlin, as well as all the participants of the conference, with whom I could exchange ideas. I also thank my collaborators on the Bernoulli-Edition, Mme P. Radelet-De Grave, who has worked independently of on Maupertuis, Mrs. U. Wandeler, Mr. F. Nagel, and Mr. M. Mattmüller for their support and Mrs. S. Busam Golay for the revising the manuscript.

The principle of the least action

The determination of the solution of a mechanical problem is composed of two parts:

1. the law, in general a differential equation of the second order, which is always valid, independent of the circumstances;
2. the boundary conditions; their formulation is different in ordinary Newtonian mechanics and in the formulation by a variational principle.

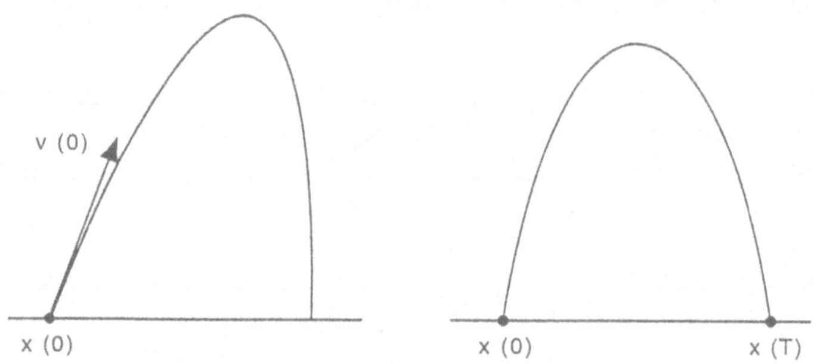

In Newton's equations, one must stipulate the initial location and the initial speed: x(0) and v(0) (figure, left).

In the case of an action principle, the initial condition for the situation becomes x(0) and the end condition x(T) (figure, right). One therefore spoke of a "teleological" rather than a "causal" principle; however, both concepts are misleading.

The Maupertuis-Euler principle corresponds to the task of the artillery soldier: he knows the position of his gun, that of his target, and the kinetic energy of the grenade due to the powder. The problem then is to find the angle of the gun, that is, the angle under which the grenade must be shot.

Editor's Notes

The Beginning of the Theory of Elasticity from Galileo to Jacob Bernoulli (pp. 1-14)

i. Galileo is only concerned with *statics*. The first non-trivial extension of similarity to *dynamics* occurs perhaps in Newton's *Principia*, Bk. II, prop. XXXII, concerning the motion of two systems of mass-points similarly disposed, of like motions and having densities in proportion to the distances. This theorem is the basis of Newton's theory of the resistance of fluids.

ii. "It is opportune at this moment to begin to investigate, according to what ratio the moment of its weight increases in relation to its resistance to breaking, in a prism or cylinder remaining parallel to the horizon, as it elongates; which moment I find to increase in a ratio duplicate of the elongation" (*Conviene ora che cominciamo a investigare secondo qual proporzione vadia crescendo il momento della propria gravità, in relazione alla propria resistenza all'essere spezzato in un prisma o cilindro, mentre, stando parallelo all'orizonte, si va allungando; il qual momento trovo andar crescendo in duplicata proporzione di quella dell'allungamento*) [1638: Second Day, Prop. III].

iii. However, recent historical studies have found an earlier example of Galileo's problem of the breaking of a horizontal beam in Bernardino Baldi's *In mechanica Aristotelis problemata exercitationes* [1621]. For a full account of the question, see [Becchi 2004].

iv. "But in my reasoning I preferred not to consider the bending of the whole beam, or rather I assumed a shape already reduced, through the prior bending caused by the weight, to the form we attributed to it... However, consideration of the bending would furnish a new and not inelegant problem" [translation by Truesdell, 1960a, p. 64]. *Ego in ratiocinationibus meis nolui considerare flexionem totius trabis, vel potius posui figuram a pondere per flexiones praecedentes jam ad eam figuram quam ei dedimus esse redactam... Flexionum autem consideratio novae adhuc nec inelegantis operae foret* [Leibniz 1850-63, 2: 16-17].

v. Euler cites Galileo's *Discorsi*, for example, in the *Mechanica* [1736: Praefatio and §§ 135, 196, 577]. The question about which authors were read by Euler is not easy to answer. For example, in the *Algebra* the problem is posed of finding two numbers whose sum is 10 and whose product is 30 [1770a: § 700]. This is the same problem that, two centuries before, had led Cardano in the *Ars magna* to introduce complex numbers [1545: Chap. XXXVII]. Does this mean that Euler had read Cardano's book?

The Second New Science in Galileo's *Discorsi* (pp. 17-38)

i. A certain "Cristiano Vurstisio" from Rostock is cited by Sagredo in the *Dialogo* as having given "two or three" lectures on Copernicus's theory at the University of Padua, presumably during the 1590s [Galileo 1638 in *Opere* VII: 154]. It was once believed that Vurstisio could be identified with Christian Wusteisen from Basel; unfortunately, this mathematician died in 1588 and never visited Padua. There was a however a Christopher Wursteisen, from Rostock, in Padua around 1595, and there are reasons to think that Galileo, who was then a young professor in Padua, first heard of the Copernican system from him. See Drake's *Galileo at Work* [1978 : chap. 3]).

ii. The standard English translation of Galileo's *Discorsi* is that of Henry Crew and Alfonso de Salvio of 1914, but it is hardly adequate. The translations given here are by the editors.

iii. On Pierre le Pélerin de Maricourt and his book on magnetism, see [Radelet and Speiser 1974; 1975].

iv. Literally, Galileo's relationship is $V_1 : V_2 = (S_1 : S_2)\cdot(T_2 : T_1)$. Indeed, in Euclidean terminology, "composed" means "equal to the product of". See what Galileo himself says in the *Discorsi* [*Opere*, VIII: 359]).

v. In this context, "moments" can be freely translated as "degrees" or "intensity". For an exhaustive discussion of the notion of moment in Galileo, see [Galluzzi 1979].

vi. See the Third Day of the *Discorsi*, [Galileo 1638 in *Opere*, VIII, p. 203]. Galileo's reasoning is clearly wrong, but the result is correct: it is impossible for a freely falling body to increase its speed proportionally to the space traversed. An analytical proof (by then, an easy exercise in elementary calculus) was given by Jacob Hermann in the *Phoronomia* [1716: § 146] and by Euler in the *Mechanica* [1736: § 135].

vii. It is known that Galileo performed some of the experiments he described. The starting point of the revision of thinking about Galileo's experiments is the paper by Thomas Settle [1961].

viii. There are some ambiguities in Galileo's use of the concept of *impetus*; for a short discussion of the problem, see [Dugas 1954: chap. IV, § 3]. However, we can safely assume that in the *Discorsi* he meant by *impetus* a quantity proportional to the velocity (like momentum). This can be seen, for example, from Prop. 4 in the Fourth Day, which teaches us how to determine the impetus of a projectile at each point in its parabolic path (*Quomodo in datae parabolae, a proiecto descriptae, punctis singulis impetus sit determinandus*). This proposition (not a theorem, for there is no real proof in the modern sense) is in effect a statement of the vectorial composition of velocities (or of momenta) for freely falling bodies.

Huygens's *Horologium Oscillatorium* and Newton's *Principia* (pp. 41-49)

i. In the Introduction to his *Essai d'une nouvelle théorie de la résistance des fluides* [1752], after a sharp but well-founded criticism of Newton's results in hydrodynamics, d'Alembert gives the following final judgment: "In spite of all these observations, we must still admire the efforts and wisdom of this great Philosopher, who after having been so fortunate as to find the truth in a large number of other questions, dared to be the first to pave the way to the solution of a problem that no one before him had attempted. Also, this solution, although not very exact, shines with all the genius of the inventor, of this fertile mind and resources, which no one possessed to a greater degree" (*Malgré toutes ces observations, nous n'en devons pas moins admirer les efforts & la sagacité de ce grand Philosophe, qui après avoir trouvé si heureusement la vérité dans un grand nombre d'autre questions, a osé se frayer le premier une route pour la solution d'un Problême que personne avant lui n'avoit tenté. Aussi cette solution, quoique peu exacte, brille par-tout de ce génie inventeur, de cet esprit fécond e ressources, que personne n'a possédé dans un plus haut degré que lui*) [1752: XX].

ii. For more information on the life of Huygens, see the very complete biography by Vollgraff published in Huygens's *Oeuvres* [1888-1950, 22: 385-771]; for his work, the best place to start is the collection of essays edited by Bos and Rudwick [1980] and the article by Bos in the *Dictionary of Scientific Biography* [1972].

iii. See, for example, the *Dialogo dei massimi sistemi*: "The other particolar is truly remarkable; it is that same pendulum makes its oscillations with the same frequency, or very little different – almost imperceptibly – whether through large arcs or very small ones along a given circumference" (*L'altro particolare, veramente maraviglioso, è che il medesimo pendolo fa le sue vibrazioni con l'istessa frequenza o pochissimo o quasi insensibilmente differente, sien elleno fatte per archi grandissimi o per piccolissimi dell'istessa circonferenza*) [1632: *Opere* VII, p. 475; trans. Stillman Drake].

Newton's *Principia* (pp. 51-69)

i. Friedrich Schiller, 1872, "Fantasie an Laura": Tilge sie vom Uhrwerk der Naturen / Trümmernd auseinander springt das All / In das Chaos donnern eure Welten / Weint, Newtone, ihren Riesenfall! (Blot it out from the mechanism of nature and the All bursts asunder in fragments; your worlds thunder into chaos; weep, Newtons, for their giant fall!)

ii. "So certain is this that we may boldly state that it is absurd for human beings even to . . . hope that perhaps some day another Newton may arise who would explain to us, in terms of natural laws unordered by intention, how even a mere blade of grass is produced." See Kant's *Kritik der Urtheilskraft* [1790: § 75].

iii. The most authoritative biography of Newton is now that of Westfall [1982]. See also [Hall 1992].

iv. As it turns out, Newton was not the originator of the 'milled edge'. The first experiments in edge-marking were made in France as early as 1558, and the final step is attributed to Peter Blondeau, who coined for the Commonwealth with a mechanical press and later received a royal patent on the concept of edge-marking to prevent counterfeiting. The technique was further developed by Jean Castaing (or Chastaing), ca. 1680. According to C.E Challis' A New History of the Royal Mint, "he (Newton) made no decisive contribution to the development of minting processes. It is possible that he never once visited the coining press room." [1992: 419]. (Thanks are due to Robert W. Hoge, Curator of North American Coins and Currency for the American Numismatic Society, and to Dr B. J. Cook, Curator of Medieval and Early Modern Coinage of the British Museum, for this information.)

v. There was indeed a dark side to Newton. "I am safest in people that are afraid of me", he once admitted (cited in [Westfall 1982: 845]).

vi. A distinction should be made between the resistance suffered by a body moving through a fluid, which Newton supposes proportional to the square of the velocity (*Principia*, Bk II, prop. XXXIII to XXXV), and the viscosity, which is supposed to be proportional to the relative velocity of the parts of the fluid (Principia, Bk II, Sect. IX). The first hypothesis was widely accepted throughout the eighteenth century, while the second was almost forgotten.

vii. Newton's experiments in mechanics are described by Bertoloni Meli [2006: § 9.4].

viii. However, Newton knew that the orbit of a comet is not a parabola: see Bk. III, Lemma IV, coroll. 3, and the section just before Bk. III, prop. XLII.

ix. For a history of the mathematical theory of comets in the eighteenth century written by one of the main actors on that stage, see [Lagrange 1780]. A readable history of the theories of comets up to the end of the century can be found in vol. 1 of Pingré's *Cométographie* [1783-84].

x. Newton's version of the classical principle of relativity appears in Bk. I of the Principia, corollary V to the Third Law.

xi. On the early ideas about the principle of inertia in the late Middle Ages, see, for example, Moody [1951], Grant [1964] and Franklin [1976]. While the modern form of the principle goes back at least to Galilei and Descartes, the use of the word "inertia" in this context is due to Newton. Newton's use of the locution "vis inertiae" to denote something which is not a force was criticized by Euler in his "Recherches sur l'origine des forces" [1752c: § 9].

xii. More probably, the third law of motion derives from the theory of elastic impact. It is significant that in the *Principia*, immediately after stating the third law, Newton discusses the laws of impact. There is also a very interesting passage in Newton's *Arithmetica universalis* (published in 1707, but based on lectures given in Cambridge during the years 1673-1680: "Problem 12. Given the sizes and motions of two spherical bodies moving in the same straight line and colliding with each other, to determine their motions after recoil. The solution of this depends on framing these stipulations: that each body shall suffer as much in reaction as it impresses in its action upon the other; and that after recoil they shall depart from each other with the same speed as before they approached" [Newton, 1707. English translation from the Latin by D. Whiteside from *Mathematical Papers* **5**, p. 149]. (I owe this citation to Prof. George E. Smith, Tufts University, Cambridge, Mass.) The rule stated by Newton must be understood in the sense that in the impact of two bodies, the momentum lost by the incoming object is equal to the momentum gained by the second object. See also the exposition of the third law given by Maclaurin in his *Account of Sir Isaac Newton's Philosophical Discoveries* [1748: chap. 2, § 5]. Newton's third law in the case of elastic impact is equivalent to the conservation of linear momentum, already known to Descartes, Huygens and Wallis. Indeed, there is a remark by Descartes in his *Le Monde, ou le Traité de la lumiere* [1664] that looks like an earlier version of the third law: "As a second rule, I suppose that when one body pushes another it cannot transmit to or subtract from it any motion without losing or acquiring at the same time an equal quantity" (*Je suppose pour seconde Regle: que, quand un corps en pousse un autre, il ne sçauroit luy donner aucun mouvement, qu'il n'en perde en mesme temps autant du sien; ny luy en oster, que le sien ne s'augmente d'autant*) [1664: 215]. Newton did not create the third law out of nothing; his merit consists in having elevated a rule for the impact between billiard balls to a fundamental axiom of mechanics.

xiii. Without going farther than the beginning of the eighteenth century, we must add to this list the copies of the *Principia* printed in Holland in 1714 and in 1723, the English translation by Andrew Motte in 1729 and the "Jesuits' edition" of Le Seur and Jacquier in 1739-42 discussed below.

xiv. It is interesting to note that this edition was reprinted at least three times before 1850. Thus we can safely assume that it was the main source for scholars interested in the *Principia* up to the nineteenth century.

xv. For the details of the Marquise du Chatelet's work on the *Principia*, see [Zinsser 2001].

xvi. There were many editions of this book during the eighteenth century. It is, in fact, one of the best example of the popular interest in science during this century. It must be noted that several copies purportedly printed in Naples in 1737 were in effect counterfeited.

xvii. According to Brewster's Memoirs of the Life, Writings, and Discoveries of Sir Isaac Newton [1855: vol. 1, chap. XII], the Newtonian philosophy was taught by David Gregory at Edinburgh and by his brother James at St. Andrews before 1690. John Keill gave lectures on Newton's theory at Oxford in 1697.

xviii. As early as 1907 Einstein suspected that a relativistic theory of gravitation could explain the motion of perihelion of Mercury; he briefly expressed this opinion in a letter of 24 December 1907 to Conrad Habicht [Einstein 1987, 5: 82]. In a manuscript of Spring 1913, Einstein tried, in collaboration with his friend Michele Besso, to calculate the effect by means of a first version of the general theory; he obtained only half of the expected value and did not publish his work [Einstein 1987, 4: 360-473]. The correct result appeared in the second of the sequence of three papers in which he gave the final version of the theory, aptly entitled "Erklärung der Perihelbewegung des Merkur aus der allgemeinen Relativitätstheorie" [1915].

xix. An exposition of some of these criticisms of Newton's presentation of mechanics is given by Dugas [1950: Part. 4, chap. 10.]. As for Mach's "operational definition" of mass, a crucial point of his formulation of the basic principles of mechanics, it should be noted that it was probably inspired by a similar discussion in the first edition of Poisson's *Traité de Mécanique* [1811: §§ 310-12]. An earlier definition of inertial mass by means of the acceleration imparted by known forces to systems of identical objects was implicitly given by Euler in the *Theoria motus corporum solidorum...* [1765: § 151].

xx. In continuum mechanics the principle of moment of momentum is equivalent (with some assumptions), to the symmetry of the stress tensor: $T^{ik} = T^{ki}$. Boltzmann [1925] observed that this symmetry can be taken as an axiom.

xxi. In the second half of the much-discussed Scholium generale appended to the second (1713) edition of the Principia, Newton seems to consider time and space as attributes of God. This surprising point of view is nonsense for pure science and quite unorthodox for religion. It was severely criticized by Leibniz in his first letter to Samuel Clarke (November 1715). For more details, see [Jammer 1954: chap. 4] and [Snobelen 2001].

xxii. In modern terms, Newton's law of cooling states that "the rate at which heat is lost by a body to its surroundings is proportional to the difference in temperature between them." The law first appeared in a paper by Newton published anonymously in the *Philosophical Transactions* under the title "Scala graduum caloris" [1701]. In the original formulation, Newton says: "The heat which hot iron loses in a certain time, is as the total heat of the iron. Therefore, if equal times of cooling be taken, the degrees of heat will be in geometrical proportion, and therefore easily found by the tables of logarithms" (*Calor quem ferrum dato tempore amittit est ut calor totus ferri. Ideoque, si tempora refrigerii sumantur aequalia calores erunt in ratione geometrica, & propterea per tabulam logarithmorum facile inveniri possunt* [Newton 1701: 827]). Truesdell [1980a: 48] asserts that Newton's "degree of heat" is a temperature, and translates this verbal formulation into the formula $d\theta /dt \propto (\theta - \theta_0)$, where θ is the temperature, t is the time and θ_0 is a constant. The law was probably discovered in the course of some experiments in chemistry around 1693. For more details, see [Ruffner 1964].

The Science of Mechanics in the Eighteenth Century (pp. 71-84)

i. Indeed, Leibniz coined the term *function* (Latin *functio*) in a manuscript dated August 1673. He used it again in articles dating from 1692 and 1694, to denote segments expressing quantities in the Cartesian plane. Details are given in Youschkevitch's classic paper on the history of the concept of function [1976-77: § 6].

ii. For the development of ideas expressed in the *Principia* during the first half of the eighteenth century, see [Truesdell 1967] and [Guicciardini 1999].

iii. This is what the physicist, philosopher and historian of science Ernst Mach asserted in his *Mechanik in ihrer Entwickelung* [1883]. However, Mach's opinion has now been completely demolished: the work of historians during the last fifty years has shown the amazing number of discoveries in fundamental science made during the eighteenth century.

iv. See Nauenberg [2001] and Wilson [2001].

v. The three-body problem was first explicitly stated by Clairaut in "Du système du Monde dans les principes de la gravitation universelle": "Given three bodies with their positions, masses and velocities, find the curves that must describe their attraction, supposed proportional to their mass and in inverse ratio to the square of the distances" (*Trois corps étant donnez avec leurs positions, leurs masses et leurs vitesses, trouver les courbes qu'ils doivent décrire par leur attraction supposée proportionelle à leur masse et en raison inverse du quarré des distances* [1749: 335]). For the history of the three body problem see Marcolongo [1918] and Wilson [1985: § 4].

vi. The outstanding reference for the history of perturbation theory in the eighteenth century is the book-length article by Wilson [1985].

vii. Euler had first stated the importance of this "new principle" in his "Recherches sur le mouvement des corps célestes en general" [1749b: § XXIII].

viii. While these equations today seem so natural that we tend to read them in Newton's *Principia*, it is a fact that for about fifty years after the publication of the *Principia* only a few particular mechanical systems could be studied by means of some *ad hoc* principles whose degree of generality was difficult to ascertain. By simply perusing the most important works on mechanics of the beginning of the century, it is easy to see that "Newton's law" does not appear at all, or is relegated to a secondary role. According to Euler, the novelty of his principle does not consist in the equations themselves, but in their generality. Thanks to his "new principle", Euler was immediately able to find the general equations for rigid bodies and, in a few years, for perfect fluids, too.

ix. There is no reasonably complete history of analytical mechanics as yet. For the principles of mechanics in Lagrange, see [Fraser 1983]; for some nineteenth-century developments [Lützen 1995]. The history of the principle of least action can be found in [Pulte 1989].

x. *Lagrange's equations* appeared for the first time in Lagrange's "Théorie de la libration de la Lune" [1782: § 11].

xi. The famous *Laplace's equation* for the gravitational potential first appeared in two dimensions [1779: § XXXVII], then in three variables in spherical coordinates [1785: § VIII] and lastly in three variables in Cartesian coordinates [1789: § II]; see also the first volume of Laplace's *Traité de Mécanique céleste* [1799: partie Ière, livre IIème, § II]. However, it had been discovered much earlier by Euler in the course of his researches on fluid mechanics, under the hypothesis of the existence of a velocity-potential [1761: § 67].

xii. The *Poisson brackets*, of fundamental importance in quantum mechanics, were introduced in [Poisson 1809]. Poisson's paper is described by Dugas [1950: part 4, chap. 5].

xiii. The *Hydrodynamica* is also dominated by the hypothesis of *parallel plane sections*, by means of which every problem is reduced to one-dimensional fluid motion. From the mathematical point of view, this means that Daniel Bernoulli could use ordinary differential equations instead of partial differential equations (not yet known at that time).

xiv. More precisely, of a three-dimensional, axially symmetric flow. This is perhaps d'Alembert's greatest conceptual triumph. However, he did not see its importance, and we owe to Euler the full development of the *field viewpoint* in mechanics, which in his hands grew to become the foundation of our mathematical understanding of nature.

xv. While for earlier authors the "pressure" was simply the weight of the fluid, Euler's *internal pressure* is a much more subtle and general idea. It is, in fact, the first three-dimensional example of a contact force in continuum mechanics. See [Truesdell 1968b] and [Casey 1992: II.9].

xvi. While the continuity equation is unquestionably due to Euler [1761: § 36 for an incompressible fluid; 1757b: § 17 for the general case], the idea of studying the change in volume through motion of an infinitesimal parallelepiped of fluid goes back to d'Alembert [1752: § 48].

xvii. Euler considers *all the forces acting on an element of the body*, not just the external forces (*...omnes vires, quibus corpus hoc tempore sollicitantur...*, § 27). In modern textbooks, the internal forces are eliminated by means of Newton's third law. This is a subtle point in regard to the foundation of mechanics. For a history of the question, see [Truesdell 1964]; some hypotheses about Euler's ideas on this question have been put forward by B.L. van der Waerden [1983]. Truesdell's assertions for the beginning of the nineteenth century have been revised in [Caparrini 1999].

xviii. The classic reference for the history of elasticity in the seventeenth and eighteenth centuries is [Truesdell 1960a]. Previous efforts, such as those of Saint-Venant [1864], Todhunter and Pearson [1886-93], Love [1892-93: *Historical introduction*, p. 1-34] and Timoshenko [1953] fall short in several ways of recognizing the achievements of these early authors.

xix. For a history of elliptic integrals and functions, see [Houzel 1978].

xx. In effect, Parent demonstrated that it is not true that the moment required to bend a beam a given amount is independent of the position of the neutral axis.

xxi. These have been mostly reprinted in [Fuss 1843] and [Eneström 1906-07]. The entire correspondence will be published in vol. 3, series IV of Euler's *Opera omnia*. Of course, Euler's general theory of the equilibrium of the elastic curve by considering the minimum of potential energy, one of the outstanding achievements of rational mechanics, is developed in the *Additamentum I* to the *Methodus inveniendi* [1744].

xxii. Euler's papers on engineering are collected in vols. XV, XVI and XVII, series II, of the *Opera*. They make abundantly clear that Condorcet's opinion [1785], often cited even today, that Euler's theories were calculations for calculations sake is simply nonsense. The *Théorie complette de la construction et de la manoeuvre des vaisseaux* [1773], a practical textbook on ships, was first published in St. Petersburg, republished in Paris and translated into English, Italian and Russian.

xxiii. The *latitude* at sea can be easily determined by calculating the height of the sun on the horizon. To determine the *longitude*, it is necessary to compare the local hour with the hour in a place whose longitude is known; thus, it would be necessary to have always at hand a perfect clock showing the hour at the point of departure. There are many books and articles on the question, but the clearest exposition is still that given by Euler in his *Lettres à une princesse d'Allemagne* [1768-72: letters CLXI-CLXIX].

xxiv. A notable exception, in tune with our subject, is the magnificent account given by James Bell of the history of experimental solid mechanics [1973].

xxv. Prof. Michel Vandyck kindly informs me that he plans to publish his research in the future.

xxvi. Taylor's and d'Alembert's methods for deriving the motion of an oscillating string, together with Johann Bernoulli's methods in hydraulics, essentially consisted in isolating a portion of the continuum and considering the forces acting on it. They showed to Euler that the fundamental law of mechanics is the "Newtonian" equation $\boldsymbol{F} = m\boldsymbol{a}$. (Historically, this fact shows that it is impossible to write a history of the fundamental principles of mechanics without considering its most advanced branches.)

xxvii. For a history of Bessel functions, see [Watson 1922].

xxviii. An English translation of Daniel Bernoulli's obscure statement of d'Alembert's principle in the case of the suspended chain can be found in [Truesdell 1960a: 160].

xxix. Daniel Bernoulli complained to Euler (letter of 20 October 1742) that his father had published results on the motion of several dynamical systems he had first obtained about ten years before, but not yet published; see [Truesdell 1960a: 182, footnote].

xxx. Here is a modern paraphrase of d'Alembert's reasoning as it appears in the *Traité*. Consider a system of mass points under the action of both external and internal forces. Because of the mutual actions, the nth point cannot take the acceleration $\mathbf{a}_n$ imparted by the external force $\mathbf{f}_n$. Intuitively, if the system were subjected only to the *lost forces* $\mathbf{f}_n - m_n\mathbf{a}_n$, it would not move at all. Thus, d'Alembert's principle asserts that *the lost forces form a system in static equilibrium*.

This description differs in several ways from d'Alembert's. First, there are no general equations (and no vector notation) in the *Traité*. Second d'Alembert does not employ the concept of force; he applies his reasoning to the "motion" of the points of the system, i.e., to the impressed velocities, which are proportional to the accelerations. While d'Alembert's principle marks a significant advancement in our knowledge of mechanics, it is possible to find its roots in several papers of Jacob, Johann and Daniel Bernoulli.

For a historical study of d'Alembert's principle, see [Truesdell 1960a: 186] and [Fraser 1985]. Since there are several interpretation of d'Alembert's principle in the literature, it is also interesting to read the discussion in Poisson's *Traité de Mécanique* [1811: vol. II, livre IIIème, chap. II] for the opinion of an expert a few decades after the discovery.

xxxi. The definitive treatment of the problem of the heavy hanging cord is due to Lagrange [1762]. Lagrange first considers a set of similar weights equally spaced, then passes to the limit.

xxxii. In a memoir written in 1744, that is, a few years before the publication of d'Alembert's work, Euler had obtained a system of partial differential equations for the *finite* motion of a string [1751b: § 32]. Moreover, a modern reader looking for the familiar wave equation in d'Alembert's original paper is likely to be disappointed, for here the partial derivatives appear only as coefficients of differential forms. Thus, for example, the equation $\partial^2 y/\partial t^2 = \partial^2 y/\partial s^2$ occurs on p. 216 in the enigmatic form $\alpha = \beta$. The modern expression (except for trifling details) first appears in Euler's "Remarques sur les mémoires précedens de M. Bernoulli" [1755b: § 22]. (Euler's comments on the new notation for partial differential equations are worth noting.) Indeed, while d'Alembert has the merit of having discovered and solved the one-dimensional wave equation, it was Euler who applied it to the propagation of sound [1766a: § 18], gave its generalization to three dimensions [1766b: § 42] and developed most of the basic mathematical apparatus for dealing with it.

xxxiii. The controversy on the vibrating string is described by Riemann [1868], Burckhardt [1901-08], Truesdell [1960a: part III] and Ravetz [1961]. In short, Euler believed that a "discontinuous function" in the eighteenth-century sense of the term (essentially, a continuous, piecewise analytic function) could be a solution of the wave equation, since clearly a plucked string is physically possible. D'Alembert objected that because of the derivative $\partial^2 u/\partial x^2$ in the equation, only "continuous" (i.e., twice differentiable) functions are allowable. This polemic marks the birth of the idea of weak solutions for partial differential equations (see [Lützen 1983]).

xxxiv. Daniel Bernoulli simply *assumed* that any solution of the wave equation could be found by superposition of eigenfunctions. Since the vibrating string can take any shape at the beginning of the motion, this idea is roughly equivalent to the representation of a "general" function by means of a Fourier series. Obviously, a serious proof lay beyond the reach of eighteenth-century mathematics.

Euler had already given a trigonometric series as a solution of d'Alembert's equation [1749b: § XXX]. However, Euler did not believe that this was the most general solution.

xxxv. This method, today taught to undergraduates, first appeared in a letter from Euler to Lagrange dated 9 November 1762 [1892b].

xxxvi. The most accessible text by Chladni is the *Akustik* [1802]. For the debate on Chladni's experiments see [Bell 1991].

xxxvii. From October 1759 to October 1762. See Lagrange's *Oeuvres*, vol. 14, or, better yet, Euler's *Opera omnia*, series IV, vol. 5. (The reader of Lagrange's correspondence, as published in the *Oeuvres*, should be warned that the spelling has been modernised by the editors. For an example of Lagrange's real style, see [Caparrini 2007]).

xxxviii. Cauchy had first given a verbal description of his main results [1823]; the mathematical details followed later [1827]. These two papers make easy and pleasant reading for a modern expert. The importance of Cauchy's discovery cannot be overemphasized; it is the culmination of many decades of development of contact forces. For a history of the concept of stress, see [Truesdell 1968b].

The Discovery of Rigid Body Mechanics (pp. 85-103)

i. While there are no rigid bodies in special relativity, it is possible to define a "rigid motion" in four-dimensional space-time; the idea is due to Born [1909]. See Pauli [1921: § 45] and Synge [1956: chap. I, § 20].

ii. Lagrange's historical introductions to the three sections of the *Méchanique analitique* on statics, dynamics and hydrodynamics should be taken with a grain of salt, for they systematically downplay the role of the Basel school of mathematical physics (consisting of the Bernoullis, Hermann and Euler). They influenced the ideas of Mach, whose *Mechanik in ihrer Entwicklung* [1883] was once considered a reliable history of mechanics.

iii. In the *De motu corporum ex percussione* [1703b]. While this work was published posthumously in 1703, the existence of a manuscript from 1656 shows that by that date Huygens had already obtained all of his results. Huygens had already published in 1669 a short note containing his main results [1669].
The *De motu*, written in implicit polemic with Descartes's erroneous theory of impact, discusses the impact of elastic bodies. It is remarkable for the axiomatization of the subject, the use of the principle of relativity for the derivation of the main results and the *proof*, from considerations of symmetry, of the conservation of momentum and kinetic energy.

iv. See [Biarnais 1988] and [Herivel 1988].

v. The concept of "center of oscillation" was definitely discarded in favor of the moment of inertia thanks to Euler; see the "Recherches sur la connoisance méchanique des corps" [1765b: § 2].

vi. This formula occurs in the form of a purely verbal statement in Huygens's *Horologium oscillatorium* [1673: Pars IV, Prop. XVIII]. It was probably discovered on October 1664. For a historical account of Huygens's derivation see [Erlichson 1996].

vii. This is a generalization of Galileo's remark in the third day of the *Discorsi*, scholium to prop. 9: a body descending along an inclined plane cannot rise on a second inclined plane higher than its point of departure [1638, *Opere* VIII: 244]. Galileo had first stated this in a letter to Paolo Sarpi dated 16 October 1604 (*Opere* X, p. 114). This result was generalized by Newton in the *Principia*, (Bk I, § 8, Prop XL): for a body sliding without friction on a curve in a vertical plane, the speed depends only on the height.

viii. This principle had in effect been clearly stated by Galileo in the *Discorsi*: see the experiment on the simple pendulum described by Salviati during the third day, a few pages before the first theorem of the section on uniformly accelerated motion [Galileo 1638 in *Opere* VIII: 206].

ix. The general history of the principle of energy in classical mechanics has yet to be written. The mathematical statement and proof of the conservation of energy for a system of mass points appears, perhaps for the first time, in Lagrange's "Théorie de la libration de la Lune" [1782: § 12].

xi. Jacob's solution to the problem of a pendulum with two equal weights is also notable for its foreshadowing of *d'Alembert's principle*, first stated in the *Traité de Dynamique* [1743]. By means of the law of the lever, Jacob considers the equilibrium of the velocities lost by the two weights because of the constraints. (Since the velocities acquired in a given infinitesimal time are proportional to the accelerations, Jacob's method is equivalent to the equilibrium of lost moments.)

xii. In a letter to Lagrange of 23 October 1759 [1892-a], Euler says he had finished writing the *Theoria motus corporum rigidorum*, and calls it "the third part of my *Mechanica*".
xiii. In the second edition of Poisson's *Traité de Mécanique* [1833: vol. II, Chap. VII].
xiv. The distinction between the "center of gravity" and the "center of inertia", commonly attributed to Euler, already appears in Daniel Bernoulli's *Traité du flux et reflux de la Mer*: "... from which it follows that the point about which these two bodies turn must be placed, so that the centrifugal forces are equal …. .It would be therefore be better to call this point the center of centrifugal forces, or, since in our hypothesis the velocities maintain a constant proportion, center of mass, rather than center of gravity. It is true that these words come down to the same thing when the center of gravity is considered in the usual sense, but what idea do we have of it, when the weight is unequal for different parts of the body? So there is no point that can be called by that name, no matter what definition one gives to it" (*...d'où il suit que le point autour duquel ces deux corps tournent doit être placé, en sorte que les forces centrifuges soient égales ... Il vaudroit donc mieux appeler ce point centre des forces centrifuges, ou bien, puisque les vitesses gardent dans notre hypothese une proportion constante, centre des masses, que centre de gravité. Il est vrai que ces mots reviennent au même, à prendre celui de centre de gravité dans le sens commun: mais quelle idée y peut-on attacher, lorsque la pésanteur est inégale, dans les différentes parties du corps? Il n'y a aucun point alors, qu'on puisse nommer tel, quelque définition qu'on donne à ce mot*) [1741: chap. III, § III]. See also d'Alembert's *Traité de Dynamique* [first edition 1743: § 51, second edition, 1758: § 62].
xv. While the term "tensor" may appear anachronistic in a discussion of eighteenth-century mechanics, there is no doubt that, late in life, Euler clearly understood the meaning of a linear relation between the components of two vectors.
xvi. The analytic expressions of the corresponding equations for statics (the vector sums of the forces and of their moments must be zero at equilibrium) appear, perhaps for the first time, in d'Alembert's *Recherches sur la précession des équinoxes* [1749: Chap. 2]. As usual with d'Alembert, these equations are given in a very complicated form, almost completely opaque for modern scientists.
xvii. For a historical examination of the different points of view of Huygens, Newton and Leibniz on the nature of space see Reichenbach [1924] and Jammer [1954: chap. 4].
xviii. See the studies on the history of astronomy in St. Petersburg during the eighteenth century by Dr. Nina I. Nevskaya [1983a; 1983b]. I am indebted to Prof. G. Mikhailov (Moscow) for this reference.

Johann Bernoulli's Work on the Theory of Gravitation and the Weight of the Atmosphere (pp. 105-121)

i. The far from trivial mistake detected by Bernoulli was contained in *Principia*, Book II, Proposition X, and concerned the motion of projectile in a resisting medium. Johann Bernoulli published his criticism in his "Lettre touchant la maniere de trouver les forces centrales dans des milieux resistans, en raisons composées de leurs densités & des puissances quelconques des vitesses du mobile" [1714a]. The whole story is told in a masterly fashion by D. Whiteside in his comments to *The Mathematical Papers of Isaac Newton* [Newton 1967-80, **8**: 48-58, 312-419]. The mathematical details are examined by Lagrange in his *Theorie des fonctions analytiques* [1797: part III, chap. IV].
ii. Since the prize had not been awarded in 1732 for the same question, in 1734 the amount of the prize was doubled; see [Johann Bernoulli 1752: §§ I and LXVIII].
iii. In modern terms, the invariable plane is simply a plane orthogonal to the total moment of momentum vector. If the system is isolated (like, for example, the Solar system), the orientation of this plane is constant. The invariable plane was discovered by Laplace in 1798, and was used by every writer on theoretical astronomy during the nineteenth century. Historically, its importance lies in the fact that it showed the vectorial character of moment of momentum; see [Caparrini 2001: § 5].

The Kepler Problem from Newton to Johann Bernoulli (pp. 123-134)

i. A detailed history of how Kepler's laws influenced Newton's theory of gravitation is found in Wilson [1970; 1974].

ii. Opinions differ. The debate is mainly centered on the nature and meaning of "proofs" in the early days of mathematical physics. For a review of the question see [Wilson 1994] and [Weinstock 2000]. The different treatments of the same problem by some of the most important mathematicians of the period shed light on the passage from a semi-*geometrical* to a *purely analytical* conception of mechanics.

iii. This theorem is completely equivalent to the conservation of energy for central forces.

iv. The components of the three-dimensional angular velocity vector appears in Euler's paper "Du mouvement de rotation des corps solides autour d'un axe variable" [1765b: § 28]. However, the proof of the vector addition of angular velocities and the clear statement of their geometrical representation by means of segments belong to the Italian mathematician Paolo Frisi [1759: prop. X]. Interestingly, the proof was found independently at the same time by a second Italian mathematician, Tommaso Perelli, professor at the University of Pisa. For more details, see [Caparrini 2002: §§ 2-3].

v. To understand this formula, expressed in Bernoulli's notation (*Opera* **1**, p. 475), the reader should know that here x is the radius vector, z the polar angle, and that a, b, c are constants. Bernoulli's proof is remarkably short, just a few lines.

vi. This formula is given here for the special case of a point of unit mass.

vii. Mention must be made here that of the fact that Bernoulli's paper contains an important result: the explicit formula of the energy integral for a point moving under the action of a central force (*Opera* **1**, p. 474), neatly derived from Newton's theorem XIII.

viii. The expression $\int\phi(r)dr$, where $\phi(r)$ is the central force acting on a mass point, is our potential energy.

ix. Varignon gave his solution of the inverse problem in [1712]. The papers by Varignon on the motion of a point subject to a central force were discussed by M. Blay [1992: 216-21].

x. Another immediate fruit is the equations of motion of a rigid body, obtained in the second part of the paper.

xi. See also "Johann Bernoulli's Work on the Theory of Gravitation and the Weight of the Atmosphere" in this volume.

xii. It is possible to trace back the origin of this point of view on the foundations of mechanics at least to Descartes's *Principia philosophiae* [1644], where dynamics is based on the ideas of size, figure and motion. "Impenetrability" figures prominently among the fundamental properties of matter in d'Alembert's *Traité de Dynamique* [1743: 1]. D'Alembert sought to eliminate the concept of force from mechanics, and impenetrability seemed a more intuitive concept for dealing with motion and impact. A characterization of a "body" not very different from those of Descartes, Bernoulli, d'Alembert and Euler appears, as late as 1833, right at the beginning of the second edition of a very influential textbook, Poisson's *Traité de Mécanique*: "*Matter* is all that can affect our senses in any way whatsoever. *Bodies* are portions of matters that are limited in all ways, and as a consequence have a determined *shape* and *volume*. The *mass* of a body is the quantity of matter that composes it" (*La* matière *est tout ce qui peut affecter nos sens d'une manière quelconque. Le* corps *sont des portions de matière limitées en tous sens, et qui ont, par conséquent, une* forme *et un* volume *déterminés. On appelle* masse *d'un corps, la quantité de matière dont il est composé*) [1833 : 1].

xiii. See, for example, Euler's *Reflexions sur l'espace et le tems* [1750b]. In essence, here Euler notes that inertial motion defines Newton's absolute space and time. This view is tenable even by modern standards, which take into account general relativity. Indeed, according to Einstein, "[In classical mechanics] space must be introduced as the independent cause of the inertial behavior of bodies if one wishes to give the classical principle of inertia (and therewith the classical law of motion) an exact meaning. To have realized this fully and clearly is in my opinion one of Newton's greatest achievements" [Einstein 1953].

Euler's remark that the law of inertia defines a time-scale [1750b: § 20] appears already in d'Alembert's *Traité de Dynamique* [1743: § 11]. See also Neumann's *Ueber die Principien der Galilei-Newton'schen Theorie* [1870] and the *Treatise on Natural Philosophy* by Lord Kelvin and Tait [1890: vol. 1, § 247].

xiv. In the *Anleitung zur Naturlehre* [1862]. See "Euler's *Anleitung zur Naturlehre"* in this volume.

xv. Euler's letter to Johann Bernoulli of 18 [29] October 1740 [Euler 1742], acknowledging receipt of the *Hydraulica* and praising its content, is printed at the very beginning of the *Hydraulica,* showing how proud the old teacher was of his former pupil's judgment. The letter begins: "Previously indeed I highly praised Your Theory of flowing water because of the true and genuine Method which You alone, Most Excellent Sir, first revealed for fully treating Problems of this type. But now, after I had examined another portion of Your Studies, I was thoroughly astounded by the very fluent application of your principles to the solution of the most intricate Problems, because of which most useful and most profound finding Your very distinguished Name will forever be revered among future generations" (*Jam ante quidem, maximi feci Theoriam Tuam aquarum fluentium, propter veram & genuinam Methodum, quam Tu, Vir Excellentissime, primus atque solus aperuisti ad hujus generis Problemata solide pertractanda. Nunc vero, perlecta altera Tuarum Meditationum parte, penitus obstupui faecundissima principiorum Tuorum applicatione ad perplexissima Problemata resolvenda, quo utilissimo pariter ac profundissimo invento Nomen Tuum celeberrimum apud posteros perpetuo erit sacrum*) (trans. T. Carmody and H. Kobus.

xvi. "III. Motive forces are in proportion to the products of masses and accelerative forces. .... IV. The motive force divided by the mass gives the accelerative force, but divided by the latter gives the mass." (III. *Vires motrices, sunt in ratione composita, ex ratione massarum & virium acceleratricium.* ... IV. *Vis motrix divisa per massam, dat vim acceleratricem; per hanc vero divisa, dat massam* [1742: Praefatio, Def. III-IV. English translation by T. Carmody and H. Kobus]).

xvii. Truesdell later amended his statements [1980b].

xviii. These remarks will be published in the commentary to the third volume of the correspondence between Johann I Bernoulli and Pierre Varignon. (I owe this information to Prof. J. Peiffer.)

xix. Voltaire's quatrain: *Son Esprit vit la vérité, / Et son coeur connut la justice, / Il a fait l'honneur de la Suisse / Et celui de l'humanité* ("His spirit saw the truth, / and his heart knew justice, / he does honour to Switzerland / and to humanity"). It is found under Johann Bernoulli's portrait, opposite the title page of the first volume of the *Opera omnia.*

Daniel Bernoulli (pp. 137-154)

i. See [Hankins 1970: 27].

ii. The word *hydraulico-statics* was created by Daniel Bernoulli to denote those problems of *dynamics* of fluids that are solved by means of quasi-static devices; see chap. I, § 15. Thus, for example, to obtain Bernoulli's theorem he imagines a section of the wall containing the fluid suddenly to dissolve.

iii. Apart from questions of priority, there is no doubt that Daniel Bernoulli's relations with his father Johann were not easy. Here is a telling anecdote from Condorcet's eulogy of Daniel: "Johann Bernoulli, his father, regarded these Sciences [mathematics] as the foundation of all others, as a useful instrument in all professions of life; but his manner of teaching had completely repulsed his child. One day, to test his son's strengths, he gave him a small problem, young Daniel carried away it to his room, examined it, solved it, and brought it back happily to his father: he expected praise; you should have solved it right away, was the answer that he received; this answer, the tone, the gesture that accompanied it, appalled the young man, & the memory of this first grief never faded from his memory" (*Jean Bernoulli, son père, regardoit ces Sciences [les mathématiques] comme le fondement de toutes les autres, comme un instrument utile dans toutes les professions de la vie ; mais sa manière de les enseigner eût rebuté tout enfant qui ne seroit pas né pour elles. Un jour, pour essayer les forces de son fils, il lui proposa un petit problème, le jeune Daniel l'emporte dans son cabinet, l'examine, le résout, revient palpitant de joie le rapporter à son père : il s'attendait à des applaudissemens ;* ne devois-tu pas l'avoir resolu sur le champ, *fut la réponse qu'il reçut ; cette reponse, le ton, le geste qui l'accompagnoient, consternèrent le jeune homme, & jamais le souvenir de ce premier chagrin ne s'est effacé de sa mémoire* [Cordorcet 1785 : 82]).

iv. Daniel Bernoulli's own account of this weird experiment appears in his *Oratio physiologica de vita*, written in 1737 but published only in our times [1940/41]; see Daniel Bernoulli's *Werke*, vol. 1, p. 114.

v. Newton's theory of tides in the *Principia* must be traced from Corollaries 19 and 20 to Prop. LXVI in Book I, to Prop. XXIV, XXXVI and XXXVII in Book III.

vi. Maclaurin's theorem says that for a mass of fluid uniformly rotating about an axis under the action of gravity, an oblate spheroid is a possible figure of equilibrium. However, Maclaurin's *De causa physica fluxus et refluxus maris* contains a host of remarkable theorems on the gravitational field of important classes of extended bodies, rigorously demonstrated; they have been studied by Todhunter [1873: vol 1, Chap. IX] and Greenberg [1995: Chap. VIII].

vii. The experiments were described by Socin in [Socin 1760].

viii. Condorcet says this about Daniel's bachelorhood: "He didn't get married. In his youth a very advantageous match had been proposed to him, but the extreme economy of the woman chosen for him made him soon decide to break with her. From that time on he thought no more about marriage except to recall that he had been about to lose in one day all of his liberty & tranquillity for the rest of his life, & to fortify himself against exposure to the same peril" (*Il ne s'est point marié. Dans sa jeunesse on lui proposa un parti très-avantageux, mais l'extrême économie de la femme qu'on lui destinoit, l'eut bientôt décidé à rompre avec elle. Depuis ce temps il n'a plus pensé au marriage que pour se souvenir qu'il avoit été sur le point de perdre en un jour sa liberté & son repos, & pour se fortifier de ne plus s'exposer au même péril* [Condorcet 1785: 102]).

Leonhard Euler: Mathematician, Physicist, Engineer (pp. 157-165)

i. *Ursprünglich eignen Sinn / Lass dir nicht rauben! / Woran die Menge glaubt, / Ist leicht zu glauben* (Johann Wolfgang Goethe, *Zahme Xenien*)

The Principle of Relativity in Euler's Work (pp. 167-178)

i. While Huygens's *Tractatus* referred to here was published only in 1704, there are good reasons to believe that the main part of the work had been completed as early as 1656. Its importance consists in its being a clear and axiomatic formulation of a small but important part of mechanics: the impact of hard bodies. The theory is rigorously founded on the principle of relativity, on some intuitive assumptions of symmetry in impact and on a restricted form of the conservation of momentum.

ii. The vector composition of forces was discovered almost contemporaraneously around 1687 by Newton, Varignon and Lamy. For more details, see [Duhem 1905-06: vol. 2, chap. XVIII, § 6] and [Dugas 1950: Part II, chap. VIII].

The discovery of the vector representation of the moment of a force, of moment of momentum and of angular velocity came only in the second half of the eighteenth century, and was closely related to the work on the dynamics of rigid bodies. The Italian mathematician P. Frisi discovered the vector composition of angular velocities in 1759; he wrote that "In every body two rotations can be composed into one in exactly the same way that two forces, represented by the two sides of a parallelogram, are composed into a third one represented by the diagonal" [1759: Prop. X]. The vectorial composition of moments of forces can be found in a posthumous memoir by Euler, written in 1780. Euler immediately saw the importance of this result: "Therefore the moments about three orthogonal axes can be composed exactly as the simple forces. For if three forces P, Q, R were applied to the point a, acting along the directions af, ag, ah they would form a force equal to $fP+gQ+hR$ acting along the direction az. This marvellous harmony deserves to be considered with the greatest attention, for in general mechanics it can deliver no small development"[1793: § 35]. Prony extended Euler's result to moment of momentum [1800: 110]. The composition of angular velocities also made its appearance in the second edition of Lagrange's treatise of mechanics, now entitled *Mécanique analytique* [1811-15: Vol. 1, Part I, Sect. III,

§ III, Art. 15]. Independently of Euler, Poinsot gave a vectorial theory of the statics of a rigid body [1803]. After about 1815, these results became generally known. For more details, see [Caparrini 2001].

iii. However, since the rigid body considered by Euler *is not* isolated, its center of mass in general *does not* translate uniformly. In Euler's "Découverte" we find the following interesting passage: "Therefore, some motion that has first been impressed on a solid body, & of some forces that are solicited then, to determine its motion at every instant, one will start by considering the body, as if all its mass were united in its center of gravity, & then one will determine by the known principles of Mechanics the motion produced by the soliciting forces; that will be the progressive motion of the body" (*Donc, quelque mouvement qui ait été imprimé d'abord à un corps solide, & de quelques forces qu'il soit sollicité ensuite, pour déterminer son mouvement à chaque instant, on commencera par considérer le corps, comme si toute sa masse étoit réunie dans son centre de gravité, & alors on déterminera par les principes connus de la Mécanique le mouvement produit par les forces sollicitantes ; ce sera le mouvement progressif du corps* [1752b : § VI]) Here Euler obviously refers to the theorem expressed by the formula $\mathbf{F} = m\mathbf{a}_G$, where $\mathbf{F}$ is the resultant of the external forces, m the total mass and $\mathbf{a}_G$ the acceleration of the center of mass. As far as I know, this is the earliest appearance of this result in such generality. According to Truesdell [Truesdell and Toupin 1960: 532], the proof of this formula occurs for the first time in the *Méchanique analitique* [1788: Seconde Partie, Sect. IIIème, § 3].

iv. The most complete and authoritative history of group theory is the book by H. Wussing [1969]. According to Wussing, there are four major sources, four different historical threads, in the development of group theory: Lagrange's theory of equation from 1770, Gauss's *Disquisitiones arithmeticae*, 1801 (composition of quadratic forms, etc.), Klein's Erlangen Program (1874) and the invariant theory of differential equations by Lie, Klein and Poincaré from the 1870s.

v. The recent article by Jeff Suzuki [2007] is a useful résumé of Euler's many contributions to number theory.

vi. However, in the second half of the nineteenth century there was a resurgence of interest in the problem of the "correct" frame of reference in mechanics. These discussions are historically important in the light of the formulation of special relativity by Einstein in 1905. The first to take up the matter was Neumann in 1869, who asserted that for the foundation of mechanics it is necessary to postulate a rigid body "at rest", called the "body Alpha"; in 1886 Lange defined the modern inertial frame of reference by requiring that from such a frame all isolated bodies should appear uniformly translating; Mach [1883] attacked the Newtonian absolute space. For a review of the main theses, see [Jammer 1954: chap. 5]; the classic text by Whittaker discusses the connections with electromagnetism [1951 53], the article by Giorgi [1912] gives a long list of references.

vii. The term *Gruppenpest* was used in the 1920s to refer disparagingly to the abstractedness of the group theoretical point of view on quantum mechanics held by prominent physicists like Wigner and Weyl. On the history of group concept in physics see the article by Bonolis [2004].

Euler, the Principle of Relativity and the Fundamentals of Classical Mechanics (pp. 179-183)

i. Galileo gave a very clear statement of the classical principle of relativity in the *Dialogo sopra i due massimi sistemi del mondo*, [Galileo 1632 in *Opere* VII: 212]. Here he imagined carrying out several experiments (falling stones, projectiles, flying birds and insects, etc.) on a ship moving at constant speed, and argued that it would be impossible to say whether the ship were moving or not from the outcome of these experiments. Indeed, most of the Second Day is spent in discussing various aspects of the principle of relativity. After Galileo, the example of the uniformly translating boat appears in Descartes's *Principia philosophiae* [1644: Part II, §§ 13-15] and in Newton's *Principia mathematica* [1687: Bk. I, Coroll. V to the Third Law]. It was also used as a rhetorical device for introducing the relativity of motion in Huygens's "Tractatus de motu corporum ex percussione" [1704], written around 1656.

ii. It was Mach who severely criticized Newton's absolute space [1883]. His analysis had been partly anticipated by Bishop Berkeley in the first half of the eighteenth century (see [Suchting 1967]). While Mach scored some points (rotations are *kinematically* distinguishable from uniform translations, thanks to the existence of the fixed stars), his general position is untenable: *it is* possible to distinguish rotational from translational motion with respect to an inertial frame of reference, for the former gives rise to centrifugal forces. However, Mach's remarks stimulated the formulation of the so-called "Mach's principle", according to which inertia is determined by the distribution of matter in the universe. (More than likely, this principle would have been rejected by Mach; see [Blackmore 1972].) The status and the exact formulation of Mach's principle is still very much debated. For a general, non-technical discussion see [Sciama 1959].

iii. In his lectures on mechanics [1876], Kirchhoff stated that the second law of motion, today expressed by the vector equation $\mathbf{F} = m\mathbf{a}$, was simply a definition of force. For a description of Kirchhoff's idea, see [Dugas 1950]. The idea that motion is fundamental and force is just a secondary concept had been already put forward by d'Alembert [1743].

iv. Perhaps the first discussion on the experimental or *a priori* character of the second law of motion appears at the beginning of a famous paper by Daniel Bernoulli [1728] in which he tried to demonstrate the parallelogram law for the composition of forces. This paper initiated a fashion in mechanics: it is no exaggeration to say that every single mathematician of some importance in the period 1700-1850 attempted such demonstration . Since many of these "proofs" led to a functional equation, this stream of defective demonstrations gave the impetus to the study of a new branch of pure mathematics; see [Dhombres 1986]. By weakening the basic hypothesis, some of these mathematicians arrived at the composition of forces in non-Euclidean spaces; see [Bonola 1906].

Euler's Work in Physical Optics (pp. 197-209)

i. As is well known, the existence of these "Newtonian black holes" had first been predicted by John Mitchell: "Hence the greatest distance at which it would be possible to distinguish any sensible apparent diameter of a body as dense as the sun cannot well greatly exceed five hundred times ten thousand, that is, five million times the distance of the sun; for if the diameter of such a body was not less than five hundred times that of the sun, its light, as has been shewn above, in art. 16, could never arrive at us" [1784: 50]. Laplace gave the same result in the first edition of his *Exposition*: *Un astre lumineux de même densité que la terre, et dont le diamètre serait deux cents cinquante fois plus grand que celui du soleil, ne laisserait en vertu de son attraction, parvenir aucun de ses rayons jusqu'à nous ; il est donc possible que les plus grands corps lumineux de l'univers, soient par cela même, invisibles* [1796: vol. 2, p. 305]. (Since this paragraph was deleted after the second edition, it does not appear in Laplace's *Oeuvres*.)

ii. The idea first appears in Kepler's appendix to the *Hyperapistes* [1625: *Werke*, vol. 8, p. 419]. It is also discussed in detail in Newton's *Principia* [1687: Bk III, Prop. XLI].

Euler's Writings on Optics, Electricity and Magnetism (pp. 211-223)

i. This summary remained in manuscript form in St Petersburg until it was published in vol. 10, series III of Euler's *Opera omnia* in 2004.

ii. These are letters CXXXVIII-CLIV of E471 [Euler 1772]. The English translation of the letters cited here is by David Brewster (1833). The numbering of the letters is different in this version; we have cited the numbers from the English version for consistency, and indicated the original numbering in the notes.

Pierre Louis Moreau de Maupertuis (pp. 255-268)

i. The *theorems* of central attraction for various powers of the distance r are in Book I of the *Principia*. The third book contains their application to celestial mechanics.

References

ACKERET, JAKOB. 1944. Untersuchung einer nach Euler'schen Vorschlägen (1754) gebauten Wasserturbine. *Schweizerische Bauzeitung* **123**: pp. 9-15.

———. 1957. Vorrede. Pp. VII-LX in *L. Euleri Opera omnia*, ser. II, vol. 15.

AEPINUS, FRANZ ULRICH THEODOR. 1759. *Tentamen theoriae electricitatis et magnetismi. Accedunt dissertationes duae, quarum prior, phaenomenon quoddam electricum, altera, magneticum, explicat.* Petropoli: typis Academiae Scientiarum. English translation, *Essay on the theory of electricity and magnetism*, introduction and notes by R.W. Home, translation by P.J. Connor. Princeton, N.J.: Princeton University Press, 1979.

———. 1761 (1758-59). Descriptio ac explicatio novorum quorundam experimentorum electricorum, *Novi Commentarii academiae Scientiarum Petropolitanae* **7**, pp. 277-302.

AITON, ERIC J. 1989. The Contributions of Isaac Newton, Johann Bernoulli and Jakob Hermann to the Inverse Problem of Central Forces. *Studia Leibnitiana* **17**: pp. 48-58.

ALEMBERT, JEAN LE ROND D'. 2002-. *Oeuvres complètes*. Série I: Mathématiques 1736-56; Série II: Encyclopédie; Série III: Mathématiques 1757-83; Série IV: Littérature, Histoire, Philosophie; Série V: Correspondance. Paris: CNRS Editions.

———. 1749 (1747). Recherches sur la courbe que forme une corde tendue mise en vibration, *Histoire de l'Académie Royale des Sciences et Belles-lettres (Berlin)* **3**, pp. 214-219, 220-249.

———. 1751. Discours préliminaire. Pp. I-XLV, vol. 1, in *Encyclopédie, ou dictionnaire raisonné des sciences, des arts et des métiers, par une Société de Gens de Lettres*, 35 vols., Paris et Neufchastel et Amsterdam, 1751-1780.

———. 1752. *Essai d'une nouvelle théorie de la résistance des fluides*. Paris: David. Facsimile edition, Bruxelles: Culture et civilisation, 1966.

ALGAROTTI, FRANCESCO. 1737. *Il newtonianismo per le dame ovvero Dialoghi sopra la luce e i colori*. In Napoli. French translation, *Le Newtonianisme pour les dames: ou entretiens sur la lumière, sur les couleurs et sur l'attraction*. Traduits de l'Italien par M. Du Perron de Castera. Paris: Montalant, 1738.

ARNDT, HANS WERNER. 1979. J.-H. Lambert et l'esthétique du XVIIIe siècle. Pp. 197-208 in *Colloque international et interdisciplinaire Jean-Henri Lambert: Mulhouse, 26-30 septembre 1977*. Paris: Ophrys.

BALDI, BERNARDINO. 1621. *In mechanica Aristotelis problemata exercitationes: adjecta succincta narratione de autoris vita & scriptis*. Maguntiae: J. Albini.

BEAUJOUAN, GUY. 1957. La science dans l'Occident medieval et chrétien. Pp. 517-582 in vol. 1 of *Histoire générale des sciences*. René Taton, ed. Paris: Presses universitaires de France.

BECCHI, ANTONIO. 2004. *In Q.16.: Leonardo, Galileo e il caso Baldi: Magonza, 26 Marzo 1621*. Venezia: Marsilio.

BELL, JAMES F. 1973. The Experimental Foundations of Solid Mechanics, *Encyclopedia of Physics* **VIa**/1. Berlin, New York: Springer. Reprinted as vol. 1 of *Mechanics of Solids*, edited by C. A. Truesdell, Berlin: Springer, 1984.

———. 1991. The late twentieth-century resolution of a mid-nineteeth-century dilemma generated by the eighteenth-century experiments of Ernst Chladni on the dynamics of rods. *Archive for History of Exact Sciences* **43**: pp. 251-273.

BENVENUTO, EDOARDO and PATRICIA RADELET-DE GRAVE (eds). 1995. *Entre mécanique et architecture/Between mechanics and architecture*. Leipzig: Teubner.

BERNOULLI, DANIEL. 1982–. *Die Werke von Daniel Bernoulli*. Edited by David Speiser. Basel: Birkhäuser.

———. 1724. St. 4. *Exercitationes quaedam mathematicae*. Venetiis: apud Dominicum Lovisam.

———. 1725. St. 8. Discours sur la maniere la plus parfaite de conserver sur Mer l'égalité du mouvement des Clepsidres ou Sabliers, in *Pièces qui ont remporté les prix de l'academie royale des sciences en M.DCC.XXV*, pp. 3-21. Reprinted in *Werke*, vol. 7, pp. 221-239.

———. 1728 (1726). St. 9. Examen principiorum mechanicae, et demonstrationes geometricae de compositione et resolutione virium, *Commentarii academiae Scientiarum Petropolitanae* **1**, pp. 126-142. Reprinted in *Werke*, vol. 3, pp. 119-135. French translation, Examen des principes de la mécanique et démonstrations géométriques de la composition et de la résolution des forces. *Sciences et techniques en perspective* **11** (1986/87): pp. 39-58.

———. 1729a (1727). St. 12. Theoria Nova de Motu Aquarum per Canales quoscunque fluentium, *Commentarii academiae Scientiarum Petropolitanae* **2**, pp. 111-125.

———. 1729b (1727). St. 13. De mutua relatione Centri Virium, Centri Oscillationis et Centri Gravitatis Demonstrationes Geometricae, *Commentarii academiae Scientiarum Petropolitanae* **2**, pp. 208-226. Reprinted in *Werke*, vol. 3, pp. 136-144.

———. 1738 (1732-33). St. 23. Theoremata de Oscillationibus corporum filo flexili connexorum et catenae verticaliter suspensae, *Commentarii academiae Scientiarum Petropolitanae* **6**, pp. 108-122.

———. 1735 (1734). St. 24. Recherches physiques et astronomiques sue le probleme proposé pour la seconde fois par l'Academie Royale des Sciences de Paris : Quelle est la cause physique de l'inclinaison des Plans des Orbites des Planetes par rapport au plan de l'Equateur de la revolution du Soleil autour de son axe ; Et d'où vient que les inclinaisons de ces Orbites sont differentes entre elles, in *Pièces qui ont remporté les prix de l'academie royale des sciences en M.DCC.XXXIV*, pp. 93-122. Reprinted in *Werke*, vol. 3, pp. 303-326.

———. 1740 (1734-34). St. 25. Danielis Bernoulli Demonstrationes Theorematum suorum de Oscillationibus corporum filo flexili connexorum et catenae verticaliter suspensae, *Commentarii academiae Scientiarum Petropolitanae* **7**, pp. 162-173.

———. 1744 (1737). St. 27. De Variatione motuum a Percussione excentrica, *Commentarii academiae Scientiarum Petropolitanae* **9**, pp. 189-206. Reprinted in *Werke*, vol. 3, pp. 145-159.

———. 1737. St. 28. Réflexions sur la meilleure figure à donner aux Ancres, et la meilleur manière de les essayer, in *Pièces qui ont remporté les prix de l'academie royale des sciences en M.DCC.XXXVII*, pp. 49-84.

———. 1747 (1738). St. 29. Commentationes de Immutatione et Extensione Principii Conservationis virium, quae pro Motu Corporum coelestium requiritur, *Commentarii academiae Scientiarum Petropolitanae* **10**, pp. 116-124. Reprinted in *Werke*, vol. 3, pp. 160-169.

———. 1748 (1743). St. 39. Mémoire sur la manière de construire les Boussoles d'inclinaison, Pour faire, avec le plus de précision qu'il est possible, les Observations de l'Inclinaison de l'Aiguille Aimantée, tant sur Mer que sur Terre ; ce qui suppose des Boussoles, qui, étant mises dans un même lieu, donneront sensiblement la même Inclinaison, in *Pièces qui ont remporté les prix de l'academie royale des sciences en M.DCC.XLII*, pp. 3-61. Reprinted in *Werke*, vol. 7, pp. 67-111.

———. 1750 (1745-47). St. 42. Recherches Méchaniques et Astronomiques, sur la Question proposée par l'Académie Royale des Sciences, pour l'année 1745. La meilleure maniere de trouver l'heure en Mer, par observation, soit dans le jour, soit dans les crépusculues, & sur-tout la nuit, quand on ne voit pas l'horisons, in *Pièces qui ont remporté les prix de l'academie royale des sciences en M.DCC.XLII*, pp. 3-78. Reprinted in *Werke*, vol. 7, pp. 241-293.

———. 1750 (1748). St. 43. Remarques sur le principe de la conservation des forces vives pris dans un sens général, *Histoire de l'Académie Royale des Sciences et Belles-lettres (Berlin)* **4**, pp. 356-364. Reprinted in *Werke*, vol. 3, pp. 197-206.

———. 1755a (1753). St. 45. Réflexions et éclaircissemens sur les nouvelles vibrations des cordes exposées dans les mémoires de l'Académie de 1747 & 1748, *Histoire de l'Académie Royale des Sciences et Belles-lettres (Berlin)* **9**, pp. 147-172.

———. 1755b (1753). St. 46. Sur le mêlange de plusieurs especes de vibrations simples isochrones, qui peuvent coexister dans un même système de corps, *Histoire de l'Académie Royale des Sciences et Belles-lettres (Berlin)* **9**, pp. 173-195.

———. 1769 (1754). St. 47. Recherche sur la maniere la plus avantageuse de suppléer à l'action du Vent sur les grands Vaisseaux, soit en y appliquant les Rames, soit en y employant quelqu'autre moyen que ce puisse être, in *Pièces qui ont remporté les prix de l'academie royale des sciences en M.DCC.LIV*, pp. 3-99.

———. 1772 (1757). St. 48. Mémoire sur la maniere de diminuer le Roulis & le Tangage d'un navire, sans qu'il perde sensiblement, par cette diminution, aucune des bonnes qualités que sa construction doit lui donner, in *Pièces qui ont remporté les prix de l'academie royale des sciences en M.DCC.XXXIV*, pp. 3-96.

———. 1738 (1733). *Hydrodinamica, sive de viribus et motibus fluidorum commentarii*. Strasbourg: J. R. Dulsecker. Reprinted in *Hydrodynamik II*, vol. 8 of *Die Werke von Daniel Bernoulli*, edited by Gleb Mikhailov. Basel: Birkhäuser, pp. 93-424.

———. 1740a (1732-33). Theoremata de oscillationibus corporum filo flexili connexorum et catenae verticaliter suspensae, *Commentarii academiae Scientiarum Petropolitanae* **6**, pp. 108-122. Reprinted in *Werke*, vol. 3, pp. 145-159.

———. 1740b (1734-35). Demonstrationes theorematum suorum de oscillationibus corporum filo flexili connexorum et catenae verticaliter suspensae, *Commentarii academiae Scientiarum Petropolitanae* **7**, pp. 162-173. Reprinted in *Werke*, vol. 3, pp. 145-159.

———. 1741 (1740). St. 33. Traité sur le Flux et le Reflux de la Mer, in *Pièces qui ont remporté les prix de l'academie royale des sciences en M.DCC.XL*, pp. 53-191. Reprinted in [NEWTON 1739-42], vol. III, pp. 133-246. Reprinted in *Werke*, vol. 3, pp. 327-438.

———. 1744 (1737). De variatione motuum a percussione excentrica, *Commentarii academiae Scientiarum Petropolitanae* **9**, pp. 189-206. Reprinted in *Werke*, vol. 3, pp. 145-159.

———. 1979. *Correspondance entre Daniel Bernoulli et Jean-Henri Lambert: Annexe aux Actes du colloque J.-H. Lambert*. Edited by P. Radelet-DeGrave and V. Scheuber. Paris: Ophrys.

BERNOULLI, DANIEL and JOHANN II BERNOULLI. 1748 (1746). St. 41. Nouveaux Principes de Méchanique et de Physique, tendans à expliquer la Nature & les Propriétés de l'Aiman, in *Pièces qui ont remporté les prix de l'academie royale des sciences en M.DCC.XLIV-M.DCC.XLVI*, pp. 117-144. Reprinted in *Die Werke von Daniel Bernoulli*, vol. 7, pp. 113-135.

BERNOULLI, JACOB I. 1744a. *Jacobi Bernoulli, Basileensis, Opera*. Edited by Nicholas I Bernoulli with some notes by G. Cramer. 2 vols. Genevæ: Sumptibus Hæedum Cramer & Fratrum Philibert. Facsimile edition, Bruxelles: Culture et Civilisation, 1967.

———. 1993. *Der Briefwechsel von Jacob Bernoulli*. Edited and annotated by André Weil, with essays by von Clifford Truesdell and Fritz Nagel. Basel: Birkhäuser.

———. 1683. Op. CXLVI. *Dissertatio de gravitate aetheris*. Amstelaedami: Wetstenius. Reprinted in *Opera*, vol. 2, pp. 261-364.

———. 1684. Op. IX. Extrait d'une Lettre du Sr. Bernoulli, Ecrite de Bâle à l'Auteur du Journal, sur le démèlé de M. l'Abbé Catelan, avec Mr. Hugens, touchant le Centre d'Oscillation. *Journal des sçavants*, 24 Avril 1684, pp. 142-143. Reprinted in *Opera*, vol. 1, pp. 195-196.

———. 1686. Op. XXIII. Narratio Controversiae inter Dn. Hugenium & Abbatem Catelanum agitatae de Centro Oscillationis, quae loco Animadversasionis esse poterit in Respomsionem Dn. Catelani. *Acta eruditorum*, Julii 1686, pp. 356-360. Reprinted in *Opera*, vol. 1, pp. 277-281.

———. 1691a. Op. XLII. Specimen alterum Calculi differentialis in dimetienda Spirali Logarithmica, Loxodromiis Nautarum, & Areis Triangulorum Sphaericorum: una cum Additamento quodam ad Problema Funicularium, aliisque. *Acta eruditorum*, June1691, pp. 282-290. Reprinted in *Opera*, vol. 1, pp. 442-453.

———. 1691b. Op. XLV. Demonstratio Centri Oscillationis ex natura Vectis, Reperta occasione eorum, quae super hac materia, in *Historia litteraria Rotterodamensi*, recensentur. *Acta eruditorum*, Julii 1691, pp. 317-321. Reprinted in *Opera*, vol. 1, pp. 460-465.

———. 1694. Op. LVIII. Curvatura Laminae Elasticae. Ejus Identitas cum Curvatura

Lintei a pondere inclusi fluidi expansi. Radii Circulorum Osculantium in terminis simplicissimis exhibiti, una cum novis quibusdam Theorematis huc pertinentibus &c. *Acta eruditorum*, Junii 1694, pp. 262-276. Reprinted in *Opera*, vol. 1, pp. 576-600.

———. 1695. Op. LXVI. Explicationes, Annotationes et Additiones ad ea, quae in Actis sup. anni de Curva Elastica, Isochrona Paracentrica & velaria, hinc inde memorata, & partim controversa leguntur ; ubi de Linea mediarum directionum, aliisque novis. *Acta eruditorum*, Decembris 1695, pp. 537-553. Reprinted in *Opera*, vol. 1, pp. 639-663.

———. 1703a. Op. XCVIII. Démonstration générale, Du Centre de Balancement ou d'Oscillation, tirée de la Nature du Levier. Lettre du 13. Mars 1703. *Memoires de l'Académie des Sciences de Paris*, 1703, pp. 78-84. Reprinted in *Opera*, vol. 2, pp. 930-936.

———. 1703b. Op. XCIX. Extrait d'une Lettre, De Mr. Bernoulli, Profess. à Bâle, en date du 11 Sept. 1703, Contenant l'application de sa Règle du Centre de Balancement à toutes sortes de figures. *Memoires de l'Académie des Sciences de Paris*, 1703, pp. 272-283. Reprinted in *Opera*, vol. 2, pp. 937-946.

———. 1704. Op. C. Démonstration du Principe de M. Hugens, touchant le centre de Balancement, & de l'identité de ce centre avec celui de percussion. *Memoires de l'Académie des Sciences de Paris*, 1704, pp. 136-142. Reprinted in *Opera*, vol. 2, pp. 947-953.

———. 1705. Op. CII. Veritable Hypothêse de la Résistance des Solides, avec la Démonstration de la Courbure des corps qui font Ressort. *Memoires de l'Académie des Sciences de Paris*, 1705, pp. 176-186. Reprinted in *Opera*, vol. 2, pp. 976-989.

———. 1713. *Ars Conjectandi. Opus Posthumuum. Accedit Tractatus de Seriebus infinitis, et Epistola Gallicee scripta de ludo Pilae reticularis*. Basileae: Impensis Thurnisiorum, fratrum. Reprinted in *Werke*, vol. 3, pp. 107-286. Facsimile edition, Bruxelles: Culture et Civilisation, 1968. English translation, *The art of conjecturing, together with "Letter to a friend on sets in court tennis"*; translated with an introduction and notes by Edith Dudley Sylla. Baltimore: Johns Hopkins University Press.

———. 1744. Vp. XXVI. Inventio Centri Tensionis. *Opera*, vol. 2, pp. 1105-1108.

BERNOULLI, JACOB II. 1788 (1787). Essai théorique sur les vibrations des plaques élastiques rectangulaires et libres, *Nova Acta academiae Scientiarum Imperialis Petropolitanae* **5**, pp. 197-219.

BERNOULLI, JOHANN I. 1712 (1710). Op. LXXXVI. Extrait de la Réponse de M. Bernoulli à M. Herman, datée de Basle le 7. Octobre 1710. *Memoires de l'Académie Royale des Sciences* (Paris) 1710, pp. 521-533. Reprinted in *Opera*, vol. 1, pp. 470-480.

———. 1714a. Extrait d'une lettre de M. Bernoulli, écrite de Basle le 10. Janvier 1711 touchant la maniere de trouver les forces centrales dans des milieux resistans, en raisons composées de leurs densités & des puissances quelconques des vitesses du mobile, *Memoires de l'Academie royale des Sciences*. Année M.DCCXI., pp. 47-54. Opera, vol. 1, pp. 502-510.

———. 1714b. Op. CXLVI. *Essay d'une nouvelle théorie de la manoeuvre des vaisseaux: avec quelques lettres sur le même sujet*. Basel: chez Jean George König. Reprinted in *Opera*, vol. 2, pp. 261-364.

———. 1729 (1727). Op. CXXXVII. Theoremata selecta pro conservazione virium viva rum demonstranda et experimentis confirmana. Excerpta ex Epistolis datis ad filium

Danielem, *Commentarii academiae Scientiarum Petropolitanae* **2**, pp. 200-207. Reprinted in *Opera*, vol. 3, pp. 124-130.

———. 1752. Op. CXLVI. *Essai d'une nouvelle Physique celeste*. In *Recueil des pièces qui ont remporté les prix de l'Académie Royale des Sciences pour 1734-1737.*. Reprinted in *Opera*, vol. 3, pp. 261-364.

———. 1742a. *Johannis Bernoulli Opera omnia*. Edited by G. Cramer. 4 vols. Lausannæ & Genevæ: Sumptibus Marci-Michaelis Bousquet & Sociorum. Facsimile edition, Hildesheim: Georg Olms, 1968.

———. 1742b. Op. CLXXXVI. Propositiones variae mechanico-dynamicae. In *Opera*, vol. 4, pp. 253-313.

———. 1742c. Op. CLXXXVI. *Hydraulica, nunc primum detecta ac demonstrata directe ex fundamentis pure mechanicis. Anno 1732*. In *Opera*, vol. 4, pp. 387-493. Also in *Commentarii academiae Scientiarum Petropolitanae* **9**, pp. 3-49, 1744 (1737); **10**, pp. 207-260, 1747 (1738).

BERTOLONI MELI, DOMENICO. 2006. *Thinking With Objects: The Transformation of Mechanics in the Seventeenth Century*. Baltimore: Johns Hopkins University Press.

BIARNAIS, MARIE-FRANÇOISE. 1988. Les "Principia Mathematica": défi aux "Principes" cartésiens ou réalité?. *Revue philosophique de Louvain* **86**: pp. 440-466.

BIERMANN, KURT-REINHARD. 1983. Aus der Vorgeschichte der Euler-Ausgabe 1783-1907. Pp. 489-500 in [AA.VV. 1983].

BLACKMORE, JOHN T. 1972. *Ernst Mach: His Work, Life, and Influence*. Berkeley, London: University of California Press.

BLAY, MICHEL. 1992. *La naissance de la mécanique analytique: la science du mouvement au tournant des XVIIe et XVIIIe siècles*. Paris: Presses Universitaires de France.

BOLTZMANN, LUDWIG. 1925a. *Populäre Schriften*. Third edition. Leipzig: Barth.

———. 1925b. Die Grundprinzipien und Grundgleichungen der Mechanik. Pp. 253-307 in [BOLTZMANN 1925a].

BONOLA, ROBERTO. 1906. *La geometria non-euclidea: esposizione storico-critica del suo sviluppo*. Bologna: Zanichelli. English translation, *Non-Euclidean geometry: a critical and historical study of its development*. Authorized English translation, with additional appendices, by H.S. Carslaw. Chicago: Open Court Pub. Co., 1912.

BONOLIS, LUISA. 2004. From the Rise of the Group Concept to the Stormy Onset of Group Theory in the New Quantum Mechanics. A saga of the invariant characterization of physical objects, events and theories. *Rivista del Nuovo Cimento,* **27** (4-5): 1-110.

BORN, MAX. 1909. Die Theorie des starren Elektrons in der Kinematik des Relativitätsprinzips. *Annalen der Physik* **30**: pp. 1-56.

BORN, MAX and EMIL WOLF. 1970. *Principles of Optics: Electromagnetic Theory of Propagation, Interference, and Diffraction of Light*. Fourth edition. Oxford, New York: Pergamon Press.

BOS, HENDRIK JAN MAARTEN. 1972. Huygens, Christiaan. Pp. 597-613 in *Dictionary of scientific biography*, vol. VI, New York : Charles Scribner's Sons.

BOS, HENDRIK JAN MAARTEN and MARTIN J. S. RUDWICK (eds.). 1980. *Studies on Christiaan Huygens: Invited Papers from the Symposium on the Life and Work of Christiaan Huygens*. Lisse: Swets en Zeitlinger.

BOUGUER, PIERRE. 1729. *Essai d'optique, sur la gradation de la lumiere*. A Paris: chez

Claude Jombert. Reprinted, Paris: Gauthier-Villars, 1921.

BRADLEY, JAMES. 1728. A Letter from the Reverend Mr. James Bradley Savilian Professor of Astronomy at Oxford, and F.R.S. to Dr.Edmond Halley Astronom. Reg. &c. Giving an Account of a New Discovered Motion of the Fix'd Stars.A Letter from the Reverend Mr. James Bradley Savilian Professor of Astronomy at Oxford, and F.R.S. to Dr.Edmond Halley Astronom. Reg. &c. Giving an Account of a New Discovered Motion of the Fix'd Stars. *Philosophical Transactions* **35**: pp. 637-661.

BRETSCHER, EGON and JOHN DOUGLAS COCKCROFT. 1955. Enrico Fermi: 1901-1954. *Biographical Memoirs of Fellows of the Royal Society* **1**: pp. 68-78.

BREWSTER, DAVID. 1831. *The Life of Sir Isaac Newton*. London: John Murray.

———. 1855. *Memoirs of the Life, Writings, and Discoveries of Sir Isaac Newton*. 2 vols. Edinburgh: Thomas Constable and Co.

BUCHWALD, JED Z. and BERNARD I. COHEN (eds.). 2001. *Isaac Newton's Natural Philosophy*. Dibner Institute studies in the history of science and technology. Cambridge, Mass.: MIT Press.

BURKHARDT, HEINRICH. 1901-08. Entwicklungen nach oscillirenden Functionen und Integration der Differentialgleichungen der mathematischen Physik. *Jahresbericht der deutschen Mathematiker-Vereinigung* **10**.

CALLAHAN, JAMES J. 1976. The curvature of space in a finite universe. *Scientific American* **235**: pp. 90-100.

CAPARRINI, SANDRO. 1999. On the History of the Principle of Moment of Momentum. *Sciences et Techniques en Perspective* (2)**3**: pp. 47-56.

———. 2002. The Discovery of the Vector Representation of Moments and Angular Velocity. *Archive for History of Exact Sciences* **56**: pp. 151-181.

———. 2007. An Unpublished Letter by Lagrange Concerning the Turin Academy of Science. *Atti della Accademia delle Scienze di Torino* **141**, Classe di Scienze Fisiche, Matematiche e Naturali: pp. 45-52.

CARDANO, GIROLAMO. 1545. *Hieronymi Cardani, Præstantissimi Mathematici, Philosphi, Ac Medici, Artis Magnæ, Sive De Regvlis Algebraicis, Lib. unus : Qui & totius operis de Arithmetica, quod Opvs Perfectvm inscripsit, est in ordine Decimus*. Norimbergæ: per Ioh. Petreium excusum. English translation, *The Great Art: Or, the Rules of Algebra*, translated and edited by T. Richard Witmer, Cambridge, Mass.: M.I.T. Press, 1968.

CASEY, JAMES. 1992. The Principle of Rigidification. *Archive for History of Exact Sciences* **43**: pp. 329-383.

CASSIRER, ERNST. 1906-07. *Das Erkenntnisproblem in der Philosophie und Wissenschaft der neueren Zeit*. Berlin: B. Cassirer.

———. 1911. *Zur Einstein'schen Relativitätstheorie: Erkanntnistheoretische Betrachtungenund*. Berlin: B. Cassirer. English translation, *Substance and function, and Einstein's Theory of Relativity*, authorized translation by W.C. Swabey and M.C. Swabey, Chicago, London: The Open court Publishing Company, 1923.

CAUCHY, AUGUSTIN-LOUIS. 1882-1974. *Oeuvres completes d'Augustin Cauchy*. Publié sous la direction scientifique de l'Academie des sciences et sous les auspices de M. le Ministre de l'instruction publique. 27 vols. Paris: Gauthier-Villars.

———. 1823. Recherches sur l'équilibre et le mouvement intérieur des corps solides ou fluides, élastiques ou non élastiques, *Bulletin de la Société philomathique* **14**, pp. 9-13. Reprinted in *Oeuvres*, ser. II, vol. 2, pp. 300-304.

———. 1827. De la pression ou tension dans un corps solide, *Exercices de mathématiques* **2**, pp. 42-59. Reprinted in *Oeuvres*, ser. II, vol. 7, pp. 60-81.

CHALLIS, CHRISTOPHER EDGAR (ed.). 1992. *A New History of the Royal Mint*. Cambridge: Cambridge University Press.

CHLADNI, ERNST FLORENS FRIEDRICH. 1787. *Entdeckungen über die Theorie des Klanges*. Leipzig: Weidmann. Facsimile edition, Leipzig: Zentralantiquariat d. Dt. Demokrat. Republik, 1980.

———. 1802. *Die Akustik*. Leipzig: Breitkopf & Härtel 1802. Second edition, 1830. Facsimile edition, Hildesheim: Georg Olms, 2004.

CLAIRAUT, ALEXIS CLAUDE. 1749 (1745). Du système du Monde dans les principes de la gravitation universelle. *Memoires de l'Académie des Sciences de Paris*, 1745, pp. 329-364.

COHEN, I. BERNARD. 1969. Hypotheses in Newton's philosophy. Pp. 304-326, vol. 5 in [COHEN and WARTOFSKY 1969].

———. 1971. *Introduction to Newton's "Principia"*. Cambridge: University Press.

———. 1972. Galileo and Newton. Pp. 377-400 in [MACCAGNI 1972].

———. 1974. History and the philosopher of science. Pp. 308-349 in [SUPPE 1974].

———. 1979. Galileo and Newton. Pp. 377-400 in *Colloque international et interdisciplinaire Jean-Henri Lambert: Mulhouse, 26-30 septembre 1977*. Paris: Ophrys.

COHEN, I. BERNARD AND RENÉ TATON (eds.). 1964. *L'aventure des sciences. Mélanges Alexandre Koyré: publiés à l'occasion de son soixante-dixième anniversaire*. 2 vols. Paris: Hermann.

COHEN, ROBERT S. and MARX W. WARTOFSKY (Eds.). 1969. *Proceedings of the Boston Colloquium for the Philosophy of Science (1966/1968)*. 2 vols. Boston Studies in the Philosophy of Science **4-5**. Dordrecht: D. Reidel Publishing Co.

CONDORCET, MARIE JEAN ANTOINE NICOLAS DE CARITAT. 1785. Éloge de M. Bernoulli. *Histoire de l'Academie royale des sciences, avec les mémoires de mathématique et de physique pour la même annee,* année 1782, Histoire, pp. 82-107.

COSTABEL, PIERRE. 1989. Courbure et dynamique. Jean I Bernoulli correcteur de Huygens et de Newton. *Studia Leibnitiana* **l7**: pp. 12-24.

COULOMB, CHARLES AUGUSTIN DE. 1776. Du Essai sur une application des règles de maximis et minimis à quelque problèmes de statique, relatifs à l'archtecture. *Mémoires de mathématique et de physique presentés à l'Académie R. des Sciences par divers savans et lûs dans ses assemblées* **7**, Anné 1773: pp. 343-382

CRAIG, JOHN HERBERT MCCUTCHEON, SIR. 1946. *Newton at the Mint*. Cambridge: University Press.

———. 1963. Isaac Newton and the Counterfeiters. *Notes and Records of the Royal Society of London* **18**: pp. 136-145.

DEBRU, CLAUDE. 1979. Nature et mathématisation des grandeurs intensives. Pp. 187-196 in *Colloque international et interdisciplinaire Jean-Henri Lambert: Mulhouse, 26-30 septembre 1977*. Paris: Ophrys.

DESCARTES, RENE. 1897-1913. *Oeuvres de Descartes*. Publiées par Charles Adam & Paul

Tannery. 12 vols. Paris: Léopold Cerf. Reprint, Paris: Vrin, 1974-1986.

———. 1644. *Renati Des-Cartes Principia philosophiae*. Amsterdam: Apud Ludovicum Elzevirium David. Reprinted in *Oeuvres*, vol. 8, part I.

———. 1664. *Le Monde de M.r Descartes, ou Le Traitté de la Lumiere, & des autres principaux objets des Sens. Avec un Discours du Mouvement Local, & un autre des Fiévres, composez selon les principes du même Auteur*. Paris: Jacques Le Gras, 1664. Reprinted in *Oeuvres*, vol. 11, pp. 1-214.

DHOMBRES, JEAN. 1986. Quelques aspects de l'histoire des équations fonctionnelles liés á l'évolution du concept de fonction. *Archive for History of Exact Sciences* **36**: pp. 91-181.

DICKE, ROBERT HENRY et. al. 1976. New test of the equivalence principle from lunar laser ranging. *Physical Review Letters* **36**: pp. 551-554.

DIEUDONNE , JEAN (ed.). 1978. *Abregé d'histoire des mathématiques: 1700-1900*. 2 vols. Paris: Hermann.

DRAKE, STILLMAN. 1978. *Galileo at Work: His Scientific Biography*. London: Collins Clear-Type Press.

DUGAS, RENE. 1950. *Histoire de la mécanique*. Neuchâtel: Editions du Griffon. English translation, *A History of Mechanics*. Translated by J. R. Maddox. London; Routledge & Kegan Paul, 1957.

———. 1954. *La mécanique au XVIIe siècle: des antécédents scolastiques à la pensée classique*. Paris: Dunod ; Neuchâtel: Editions du Griffon.

ELSKENS, YVES and DAVID SPEISER. 1982. Classical Mechanics of non spherical bodies. *Journal of Mathematical Physics* **123**: pp. 539-546.

EINSTEIN, ALBERT. 1987–. *The Collected Papers of Albert Einstein*. Princeton, N.J.: Princeton University Press.

———. 1915. Erklärung der Perihelbewegung der Merkur aus der algemeinen. Relativitätstheorie. *Preussische Akademie der Wissenschaften* **47**: pp. 831-839. Reprinted in *Collected Papers*, vol. 6, pp. 233-243.

———. 1954. Foreword. Pp. XI-XV in [JAMMER 1954].

———. 1995. Letter to Conrad Habicht of 24 December 1907. *Collected Papers*, vol. 5, p. 82.

EINSTEIN, ALBERT and MICHELE BESSO. 1995 (1913). Manuscript on the Motion of the Perihelion of Mercury. *The Collected Papers of Albert Einstein*, vol. 4, pp. 360-473.

ENESTRÖM, GUSTAF. 1903-05. Der Briefwechsel zwischen Leonhard Euler und Johann I Bernoulli. *Bibliotheca Mathematica* (3) **4**: pp. 344-388 (1903); (3)**5**: pp. 248-291(1904); (3)**6**: pp. 16-87 (1905).

———. 1907-07. Der Briefwechsel zwischen Leonhard Euler und Daniel Bernoulli. *Bibliotheca Mathematica* (3)**7**: pp. 126-156.

———. 1910-13. Verzeichnis der Schriften Leonhard Eulers. *Jahresbericht der deutschen Mathematiker-Vereinigung*, Ergänzungsband 4, 1. Lieferung 1910, 1. Lieferung 1913.

ERLICHSON, HERMAN. 1996. Christiaan Huygens'discovery of the center of oscillation formula. *American Journal of Physics* **64**: pp. 571-574.

EULER, LEONHARD. 1911–. *Leonhardi Euleri Opera omnia*. Series prima: *Opera mathematica*. 29 vols. (Leipzig: Teubner); Series secunda: *Opera mechanica et astronomica*. 31 vols. (Zürich: Orell-Füssli); Series tertia: *Opera physica et miscellanea*. 12 vols. (Basel: Birkhäuser); Series quarta: *Commercium epistolicum et manuscripta*. 4 vols. (in progress), (Basel: Birkhäuser).

———. 1727. E 2. *Dissertatio physica de sono quam annuente numine divino jussu magnifici et sapientissimi philosophorum ordinis pro vacante professione physica ad d. 18. febr. a. MDCCXXVII. in auditorio juridico hora 9. publico eruditorum examini subjicit Leonhardus Eulerus A. L. M. respondente adolescente Ernesto Ludovico Burcardo, phil. cand. Basileae*, Basileae: Thurnisii. Reprinted in *Opera*, ser. III, vol. 3, pp. 183-196.

———. 1728. E 4. Meditationes super problemate nautico, quod illustrissima regia Parisiensis Academia scientarum proposuit, in *Pièce qui a remporté les prix de l'academie royale des sciences en M.DCC.XXVII*, pp. 1-48. Reprinted in *Opera*, ser. II, vol. 20, pp. 1-35.

———. 1736. E 15-16. *Mechanica sive motus scientia analytice exposita*, 2 vols. St. Petersburg: Ex typographia academiae scientiarum. Reprinted in *Opera*, ser. II, vols. 1-2.

———. 1738-40. E17. *Einleitung zur Rechen-Kunst zum Gebrauch des Gymnasii bey der Kayserlichen Academie der Wissenschafften in St. Petersburg*, 2 vols. St. Petersburg: Gedruckt in der Academischen Buchdruckerey scientiarum. Reprinted in *Opera*, ser. III, vol. 2, pp. 1-304.

———. 1739a. E 33. *Tentamen novae theoriae musicae ex certissimis harmoniae principiis dilucide expositae*, St. Petersburg: Ex typographia academiae scientiarum. Reprinted in *Opera*, ser. III, vol. 1, pp. 197-427.

———. 1739b. E 57. Inquisito physica in causam fluxus ac refluxus maris, in *Pièces qui ont remporté les prix de l'academie royale des sciences en M.DCC.XL*, pp. 235-350. Reprinted in [NEWTON 1739-42], vol. III, pp. 283-374. Reprinted in *Opera*, ser. II, vol. 31, pp. 19-124.

———. 1742 (1740). E 64. Letter to Johann I Bernoulli of 18 [29] October, 1740. Partially printed in [Johann Bernoulli 1742]: vol. IV, p. 389. Reprinted in *L. Euleri Opera omnia*, ser. II, vol. 12, p. 1. Reprinted in its entirety in *Opera*, ser. IV, vol. 2, pp. 1 386-389.

———. 1744a. E 65. *Methodus inveniendi lineas curvas maximi minimive proprietate gaudentes sive solutio problematis isoperimetrici latissimo sensu accepti*. Lausannae & Genevae: Bousquet. Reprinted in *Opera*, ser. I, vol. 24.

———. 1744b. E 66. *Theoria motuum planetarum et cometarum: continens methodum facilem ex aliquot observationibus orbitas cum planetarum tum cometarum determinandi*. Berolini: sumtibus Ambrosii Hauds. Reprinted in *Opera*, ser. I, vol. 28, pp. 105–251.

———. 1744c (1736). E 69. De communicatione motus in collisione corporum sese non directe percutientium, *Commentarii academiae Scientiarum Petropolitanae* **9**, pp. 50-76. Reprinted in *Opera*, ser. II, vol. 8, pp. 7-26.

———. 1745 E 77. *Neue Grundsätze der Artillerie enthaltend die Bestimmung der Gewalt des Pulvers nebst einer Untersuchung über der Unterscheid des Wiederstands der Luft in schnellen und langsamen Bewegungen aus dem Englischen des Hrn. Benjamin Robins übersetzt und mit den nöthigen Erläuterungen und vielen Anmerkungen versehen von Leonhard Euler Königlichem Professor in Berlin*. Berlin: bey A. Haude. Reprinted in *Opera*, ser. II, vol. 14, pp. 1-409.

———. 1746a. E 80. *L. Euleri opuscula varii argumenti*, vol. 1. Berolinii: sumtibus Ambr. Haude & Jo. Carol. Speneri.

———. 1746b (1744). E 88. Nova theoria lucis et colorum, in *Opuscula varii argumenti*, vol. 1, pp. 169-244. Reprinted in *Opera*, ser. III, vol. 5, pp. 1-45.

———. 1746c. E103. Recherches physiques sur la cause de la queüe des cometes, de la lumiere boreale, et de la lumiere zodiacale. Originally published in *Mémoires de l'académie des sciences de Berlin*, 1746, pp. 117-140 ; Reprinted in Opera, ser. II, 31, pp. 221-238

———. 1748a. E 101-2. *Introductio in analysin infinitorum*. 2 vols. Lausanne: apud Marcum-Michaelem Bousquet & Socios. Second edition, Lugduni: Apud Bernuset, Delamolliere, 1797. Facsimile edition: Bruxelles: Culture et Civilisation, 1967. Reprinted in *Opera*, ser. I, vols. 8-9. French translation, *Introduction à l'analyse infinitésimale ; traduite du latin en français, avec des notes et des éclaircissements, par J. B. Labey*. Paris: chez Barrois, 1796-97. German translation: *Einleitung in die Analysis des Unendlichen. Aus dem Lateinischen ubersetzt und mit Anmerkungen und Zusatzen begleitet von Johann Andreas Christian Michelsen*. Berlin: Bei G. Reimer, 1835-36. English translation: *Introduction to Analysis of the Infinite; translated by John D. Blanton*. New York, London: Springer-Verlag, 1988-90.

———. 1748b. E 104. Mémoire sur l'effet de la propagation successive de la lumiere dans l'apparition tant des planetes que des cometes, *Histoire de l'Académie Royale des Sciences et Belles-lettres (Berlin)* **2**, pp. 141-181. Reprinted in *Opera*, ser. III, vol. 5, pp. 46-77.

———. 1748c. E 109. Dissertatio de magnete, in *Pièces qui ont remporté les prix de l'academie royale des sciences en M.DCC.XLVIII*, pp. 63-97.

———. 1749a. E 110-111. *Scientia navalis, seu tractatus de construendis ac dirigendis navibus*. Petropoli: typis Academiae Scientiarum. Reprinted in *Opera*, ser. I, vol. 18.

———. 1749b (1747). E 112. Recherches sur le mouvement des corps célestes en général, *Histoire de l'Académie Royale des Sciences et Belles-lettres (Berlin)* **3**, pp. 93-143. Reprinted in *Opera*, ser. II, vol. 25, pp. 1-44.

———. 1749c (1748). E 119. De vibratione chordarum exercitatio, *Nova acta eruditorum*, pp. 512-527. Reprinted in *Opera*, ser. II, vol. 10, pp. 50-62.

———. 1750a (1739). E 127. Explicatio phaenomenorum quae a motu lucis successivo oriuntur, *Commentarii academiae Scientiarum Petropolitanae* **11**, pp. 150-193. Reprinted in *Opera*, ser. III, vol. 5, pp. 78-112.

———. 1750b (1748). E 149. Reflexions sur l'espace et le tems, *Histoire de l'Académie Royale des Sciences et Belles-lettres (Berlin)* **4**, pp. 324-333. Reprinted in *Opera*, ser. III, vol. 2, pp. 376-383.

———. 1751a. E 156. L. *Euleri Opusculorum tomus III. Continens novam theoriam magnetis ab illustr. academia regia scient: Parisina praemio condecoratam a. 1744. Una cum nonnullis aliis dissertationibus analytico-mechanicis*, Berolinii: sumtibus Ambr. Haude & Jo. Carol. Speneri.

———. 1751b. E 109. De motu corporum flexibilium, in *Opuscula varii argumenti*, vol. 3, pp. 88-165. Reprinted in *Opera*, ser. II, vol. 10, pp. 177-232.

———. 1752a (1750). E 176. Exposé concernant l'examen de la lettre de M. de Leibnitz, alléguée par M. le Professeur Koenig, dans le mois de mars 1751 des Actes de Leipzig, à l'occasion du principe de la moindre action, *Histoire de l'Académie Royale des Sciences et Belles-lettres (Berlin)* **6**, pp. 52-62. Reprinted in *Opera*, ser. II, vol. 5, pp. 64-73.

———. 1752b (1750). E 177. Découverte d'un nouveau principe de Mecanique, *Histoire de l'Académie Royale des Sciences et Belles-lettres (Berlin)* **6**, pp. 185-217. Reprinted in *Opera*, ser. II, vol. 5, pp. 81-108.

———. 1752c (1750). E 178. Réflections sur les divers degrés de lumière du soleil et des autres corps célestes, *Histoire de l'Académie Royale des Sciences et Belles-lettres (Berlin)* **6**, pp. 280-310.

———. 1752d (1750). E 181. Recherches sur l'origine des forces, *Histoire de l'Académie Royale des Sciences et Belles-lettres (Berlin)* **6**, pp. 419-447. Reprinted in *Opera*, ser. II, vol. 5, pp. 109-131.

———. 1753. E 187. *Theoria motus lunae exhibens omnes ejus inaequalitates. In additamento hoc idem argumentum aliter tractatur simulque ostenditur quemadmodum motus lunae cum omnibus inaequalitatibus innumeris aliis modis repraesentari atque ad calculum revocari possit*, Petropoli: impensis Academiae Imperialis Scientiarum. Reprinted in *Opera*, ser. II, vol. 23, pp. 64-336.

———. 1755a. E 212. *Institutiones calculi differentialis, cum eius usu in analysi finitorum ac doctrina serierum.* Saint Petersburg: Impensis Academiae Imperialis Scientiarum Petropolitanae. Reprinted in *Opera*, ser. I, vol. 10. Partial English translation: *Foundations of Differential Calculus.* Translated by John D. Blanton. New York: Springer, 2000.

———. 1755b (1754). E 213. Remarques sur les memoires precedens de M. Bernoulli, *Histoire de l'Académie Royale des Sciences et Belles-lettres (Berlin)* **9**, pp. 196-222. Reprinted in *Opera*, ser. II, vol. 10, pp. 233-254.

———. 1755c (1754). E 216. Examen d'une controverse sur la loi de réfraction des rayons de différentes couleurs par rapport à la diversité des milieux transparens par lesquels ils sont transmis, *Histoire de l'Académie Royale des Sciences et Belles-lettres (Berlin)* **9**, pp. 294-309. Reprinted in *Opera*, ser. III, vol. 5, pp. 172-184.

———. 1757a (1753). E 225. Principes généraux de l'état d'équilibre des fluides, *Histoire de l'Académie Royale des Sciences et Belles-lettres (Berlin)* **11**, pp. 217-273. Reprinted in *Opera*, ser. II, vol. 12, pp. 2-53.

———. 1757b (1755). E 226. Principes généraux du mouvement des fluides, *Histoire de l'Académie Royale des Sciences et Belles-lettres (Berlin)* **11**, pp. 274-315. Reprinted in *Opera*, ser. II, vol. 12, pp. 54-91.

———. 1757c (1755). E 227. Continuation des recherches sur la theorie du mouvement des fluides, *Histoire de l'Académie Royale des Sciences et Belles-lettres (Berlin)* **11**, pp. 316-361. Reprinted in *Opera*, ser. II, vol. 12, pp. 92-132.

———. 1758 (1757). E 234. Expériences pour déterminer la réfraction de toutes sortes de liqueurs transparentes, *Histoire de l'Académie Royale des Sciences et Belles-lettres (Berlin)* **12**, pp. 235-266. Reprinted in *Opera*, ser. III, vol. 5, pp. 239 – 264.

———. 1761 (1752). E 258. Principia motus fluidorum, *Novi Commentarii academiae Scientiarum Petropolitanae* **6**, pp. 271-311. Reprinted in *Opera*, ser. II, vol. 12, pp.

133-168.

———. 1762 (1760). E 268. Lettre de M. Euler à M. de la Grange. Recherches sur la propagation des ébranlemens dans une milieu élastique, *Mélanges de philosophie et de mathématique de la Société Royale de Turin pour 1761-62*, pp. 1-10. Partially reprinted in p. 561, vol. 1, in [EULER 1862a]. Reprinted in *Oeuvres de Lagrange* **14**, pp. 178-188. Reprinted in *Opera*, ser. II, vol. 10, pp. 255-263. Partially reprinted in *Opera*, ser. IV, vol. 5.

———. 1765a. E 289. *Theoria motus corporum solidorum seu rigidorum ex primis nostrae cognitionis principiis stabilita et ad omnes motus, qui in hujusmodi corpora cadere possunt, accommodata*. Rostochii et Gryphiswaldiae: litteris et impensis A.F. Röse. Second edition: 1790. Reprinted in *Opera*, ser. II, vols. 3-4.

———. 1765b (1758). E 291. Recherches sur la connoissance mécanique des corps, *Histoire de l'Académie Royale des Sciences et Belles-lettres (Berlin)* **14**, pp. 131-153. Reprinted in *Opera*, ser. II, vol. 8, pp. 178-199.

———. 1765c (1758). E 291. Recherches sur la connoisance mechanique des corps *Histoire de l'Académie Royale des Sciences et Belles-lettres (Berlin)* **14**, pp. 131-153. Reprinted in *Opera*, ser. II, vol. 8, pp. 178-199.

———. 1765d (1758). E 292. Du mouvement de rotation des corps solides autour d'un axe variable, *Histoire de l'Académie Royale des Sciences et Belles-lettres (Berlin)* **14**, pp. 154-193. Reprinted in *Opera*, ser. II, vol. 8, pp. 200-235.

———. 1766a (1764). E 301. De motu corporis ad duo centra virium fixa attracti, *Novi Commentarii academiae Scientiarum Petropolitanae* **10**, pp. 207-242. Reprinted in *Opera*, ser. II, vol. 6, pp. 209-246.

———. 1766b (1762). E 302. De motu vibratorio tympanorum, *Novi Commentarii academiae Scientiarum Petropolitanae* **10**, pp. 243-260. Reprinted in *Opera*, ser. II, vol. 10, pp. 344-358.

———. 1766a (1759). E 305. De la propagation du son, *Histoire de l'Académie Royale des Sciences et Belles-lettres (Berlin)* **15**, pp. 185-209. Reprinted in *Opera*, ser. III, vol. 1, pp. 428 451.

———. 1766b (1759). E 306. Supplément aux recherches sur la propagation du son, *Histoire de l'Académie Royale des Sciences et Belles-lettres (Berlin)* **15**, pp. 210-240. Reprinted in *Opera*, ser. III, vol. 1, pp. 452-483.

———. 1766c (1759). E 307. Continuation des recherches sur la propagation du son, *Histoire de l'Académie Royale des Sciences et Belles-lettres (Berlin)* **15**, pp. 241-264. Reprinted in *Opera*, ser. III, vol. 1, pp. 484-507.

———. 1767a (1765). E 328. De motu corporis ad duo centra virium fixa attracti, *Novi Commentarii academiae Scientiarum Petropolitanae* **11**, pp. 152-184. Reprinted in *Opera*, ser. II, vol. 6, pp. 247-263.

———. 1767b (1751). E 336. Du mouvement d'un corps solide quelconque lorsqu'il tourne autour d'un axe mobile, *Histoire de l'Académie Royale des Sciences et Belles-lettres (Berlin)***16**, pp. 176-227. Reprinted in *Opera*, ser. II, vol. 8, pp. 313-356.

———. 1767c (1760). E 337. Probleme. Un corps étant attiré en raison reciproque quarrée des distances vers deux points fixes donnés, trouver les cas ou la courbe décrite par ce corps sera algébrique, *Histoire de l'Académie Royale des Sciences et Belles-lettres (Berlin)* **16**, pp. 228-249. Reprinted in *Opera*, ser. II, vol. 6, pp. 274-293.

———. 1768-70. E 342-366-385. *Institutionum calculi integralis*, 3 vols., Petropoli: impensis Academiae Imperialis Scientiarum. Reprinted in *Opera*, ser. I, vols. 11-13.

———. 1768-72. E 343. *Lettres à une princesse d'Allemagne sur divers sujets de physique & de philosophie*. 3 vols. Saint Petersbourg: De l'imprimerie de l'Academie Imperiale des Sciences.

———. 1769. E 367. *Dioptricae. Pars prima, continens librum primum. De explicatione principiorum, ex quibus constructio tam telescopiorum quam microscopiorum est petenda*, Petropoli: impensis Academiae Imperialis Scientiarum. Reprinted in *Opera*, ser. III, vol. 3.

———. 1770a. E 387-88. *Vollständige Anleitung zur Algebra*. Petersburg. Reprinted in *Opera*, ser. III, vol. 1.

———. 1770b (1766). E 396. Sectio secunda de principiis motus fluidorum, *Novi Commentarii academiae Scientiarum Petropolitanae* **14**, pp. 270-386. Reprinted in *Opera*, ser. II, vol. 13, pp. 73-153.

———. 1770c. E 386. *Dioptricae. Pars secunda, continens librum secundum, De constructione telescopiorum dioptricorum cum appendice De constructione telescopiorum catoptrico-dioptricorum*, Petropoli: impensis Academiae Imperialis Scientiarum. Reprinted in *Opera*, ser. III, vol. 4.

———. 1771a. E 386. *Dioptricae. Pars tertia, De constructione microscopiorum tam simplicium, quam compositorum*, Petropoli: impensis Academiae Imperialis Scientiarum. Reprinted in *Opera*, ser. III, vol. 4.

———. 1771b (1766). E 409. Sectio tertia de motu fluidorum, *Novi Commentarii academiae Scientiarum Petropolitanae* **15**, pp. 219-360. Reprinted in *Opera*, ser. II, vol. 13, pp. 154-261.

———. 1771c (1770). E 410. Genuina principia doctrinae de statu aequilibrii et motu corporum tam perfecte flexibilium quam elasticorum, *Novi Commentarii academiae Scientiarum Petropolitanae* **15**, pp. 381-413. Reprinted in *Opera*, ser. II, vol. 11, pp. 37-61.

———. 1772 (1766). E 424. Sectio quarta de motu aeris in tubis, *Novi Commentarii academiae Scientiarum Petropolitanae* **16**, pp. 281-425. Reprinted in *Opera*, ser. II, vol. 13, pp. 262-369.

———. 1773. E 426. *Théorie complette de la construction et de la manoeuvre des vaisseaux: mise a la portée de ceux qui s'appliquent a la navigation.* Petersbourg: Académie Imperiale des Sciences. English translation, *A compleat theory of the construction and properties of vessels, with practical conclusions for the management of ships, made easy to navigators*. Translated by Henry Watson. London: printed for P. Elmsley, 1776. Reprinted in *Opera*, ser. II, vol. 21, pp. 80-222.

———. 1776a (1775). E 479. Nova methodus motum corporum rigidorum determinandi, *Novi Commentarii academiae Scientiarum Petropolitanae* **20**, pp. 208-238. Reprinted in *Opera*, ser. II, vol. 9, pp. 99-125.

———. 1776b (1775). E 481. De gemina methodo tam aequilibrium quam motum corporum flexibilium determinandi et utriusque egregio consensu, *Novi Commentarii academiae Scientiarum Petropolitanae* **20**, pp. 286-303. Reprinted in *Opera*, ser. II, vol. 11, pp. 180-193.

———. 1777 (1776). E 487. Réfléxions de Mr. L. Euler. Sur quelques nouvelles expériences optiques, communiquées à l'Académie des Sciences, par Mr. Wilson, *Acta academiae Scientiarum Imperialis Petropolitanae* **1**, pp. 71-77. Reprinted in *Opera*, ser. III, vol. 5, pp. 351-355.

———. 1787. E 579. Calculs sur les ballons aérostatiques, faits par le feu M. Léonard Euler, tels qu'on les a trouvés sur son ardoise, après sa mort arrivée le 7 septembre 1783, *Memoires de l'academie royale des sciences pour 1784*, pp. 264-268. Reprinted in *Opera*, ser. II, vol. 16, pp. 165-169.

———. 1793 (1780). E 658. De momentis virium respectu axis cuiuscunque inveniendis; ubi plura insignia symptomata circa binas rectas, non in eodem plano sitas, explicantur, *Nova Acta Academiae Scientarum Imperialis Petropolitanae* **7**, pp. 191-204. Reprinted in *Opera*, ser. II, vol. 9, pp. 387-398.

———. 1849a. *Leonhardi Euleri commentationes arithmeticae collectae*. 2 vols. Petropoli: Typis ac impensis Academiae Imperialis ScientiarumVoss.

———. 1849b. E 792. Tractatus de numerorum doctrina capita sedecim, quae supersunt. Pp. 503-575, vol. 2, in *Opera postuma mathematica et physica anno MDCCCXLIV detecta*. 2 vols. Petropoli: Eggers; Leipzig: Voss. Facsimile edition, Nendeln: Kraus Reprint, 1969. Reprinted in *Opera*, ser. I, vol. 4, pp. 182-283.

———. 1862a. E 802. *Opera postuma mathematica et physica anno MDCCCXLIV detecta*. 2 vols. Petropoli: Eggers; Leipzig: Voss. Facsimile edition, Nendeln: Kraus Reprint, 1969.

———. 1862b. E 822. Letter from Euler to Lagrange of 24 June 1760. Pp. 561-563, vol. 1, in [EULER 1862a]. Reprinted in *Oeuvres de Lagrange* **14**, pp. 193-198. Reprinted in *Opera*, ser. IV, vol. 5, pp. 441-444.

———. 1862c. E 842. Anleitung zur Natur-Lehre, worin die Gründe zu Erklärung aller in der Natur sic ereignenden Begebenheiten und Veränderungen festgesetzet werden. Pp. 449-560, vol. 2, in [EULER 1862a]. Reprinted in *Opera*, ser. III, vol. 1, pp. 16-180.

———. 1892a. Letter to Lagrange of 23 October 1759. *Oeuvres de Lagrange*, vol. 14, p. 164-170. Reprinted in *Opera*, ser. IV, vol. 5, pp. 423-456.

———. 1892b. Letter to Lagrange of 9 November 1762. *Oeuvres de Lagrange*, vol. 14, p. 200-204. Reprinted in *Opera*, ser. IV, vol. 5, pp. 448-450.

———. 2004. Descriptio ac explicatio novorum quorundam experimentorum electricorum. Auctore F. U. T. Aepino, *Opera*, ser. III, vol. 10, pp. CXXI-CXXIV.

EULER, JOHANN ALBRECHT. 1769 (1762). A 28. Sur la réfraction des fluides, *Histoire de l'Académie Royale des Sciences et Belles-lettres (Berlin)* **18**, pp. 279-342. Reprinted in *L. Euleri Opera omnia*, ser. III, vol. 5, pp. 300-350.

FARADAY, MICHAEL. 1852. On the Physical Character of the Lines of Magnetic Force. *London, Edinburgh and Dublin Philosophical Magazine and Journal of Science* **3**: pp. 401-428. Reprinted in [Faraday 1855b], pp. 407-443.

———. 1855a. On Some Points of Magnetic Philosophy. *London, Edinburgh and Dublin Philosophical Magazine and Journal of Science* **9**: pp. 81-113. Reprinted in [Faraday 1855b], pp. 528-565.

———. 1855b. *Experimental Researches in Electricity. Vol. III*. London: Richard Taylor and William Francis. Facsimile reprint, New York: Dover Publications, 1965.

———. 1932-36. *Faraday's Diary: Being the Various Philosophical Notes of Experimental Investigation by Michael Faraday, During the Years 1820-1862 and*

Bequeathed by Him to the Royal Institution of Great Britain; Now, by Order of the Managers, Printed and Published for the First Time, Under the Editorial Supervision of Thomas Martin; With a Foreword by Sir William H. Bragg. 8 vols. London: G. Bell and Sons.

FELLMANN, EMIL A. 1971. Die Marginalnoten von Leibniz in Newtons *Principia mathematica. Verhandlungen der Schweizerischen Naturforschenden Gesellschaft. Wissenschaftlicher Teil*, 151. Jahresversammlung, Freiburg, 1971: pp. 169-173.

———. 1975-76. Newtons *Principia. Jahresbericht der Deutschen Mathematiker-Vereinigung* **77**: pp. 107-137.

———, ED. 1983. *Leonhard Euler, 1707-1783: Beiträge zu Leben und Werk.* Basel: Birkhäuser.

FEYNMAN, RICHARD. 1965. *The Character of Physical Law*. Cambridge, Mass.: M.I.T. Press.

FIERZ, MARKUS. 1954. Über den Ursprung und die Bedeutung der Lehre Isaac Newtons vom absoluten Raum. *Gesnerus* **12**: pp. 61-120.

FLECKENSTEIN, JOACHIM OTTO. 1949. Johann und Jacob Bernoulli. Supplement to *Elemente der Mathematik*, Beiheft 7 (November 1949). Basel: Birkhäuser.

———. 1969. Editor's Introduction to "Fragen der heliozentrischen Bahnbestimmung". In *Die Werke von Jakob Bernoulli*, vol. 1, pp. 118-124.

FORMEY, JOHANN HEINRICH SAMUEL. 1778. Éloge de Lambert. *Nouveaux Mémoires de l'Académie Royale des Sciences et Belles-Lettres (Berlin)* **9** : 72-90.

FRANKLIN, ALLAN. 1976. Principle of Inertia in the Middle Ages. *American Journal of Physics* **44**: pp. 529-545.

FRANKLIN, BENJAMIN. 1751. *Experiments and Observations on Electricity. Made at Philadelphia in America, by Mr. Benjamin Franklin, and Communicated in Several Letters to Mr. P. Collinson, of London, F. R. S*. London: printed and sold by E. Cave, at St. John's Gate.

FRASER, CRAIG G. 1997. *Calculus and Analytical Mechanics in the Age of Enlightenment.* Variorum collected studies series. Aldershot, (Great Britain); Brookfield, VT: Variorum.

———. 1983. L. Lagrange's early contributions to the principles and methods of mechanics. *Archive for History of Exact Sciences* **28**: pp. 197-241.

———. 1985. D'Alembert's Principle: The Original Formulation and Application in Jean d'Alembert's *Traité de Dynamique. Centaurus* **34**: pp. 31-61, 145-159.

FUETER, RUDOLF. 1941. Vorwort des Herausgebers. Pp. VIII-XLIV in *L. Euleri Opera omnia*, ser. I, vol. 4.

FRISI, PAOLO. 1759. Contient : Problematum præcessionis æquinoctiorum, nutationis terrestris axis, aliarumque vicissitudinum diurni motus geometrica solutio; de atmosphæra cælestium corporum dissertatio physico-mathematica; de natura, & motu ætheris. Pp. 1-81 in *Paulli Frisii Mediolanensis Dissertationum varium*, vol. I. Lucae: Apud V. Junctinium.

FUSS, PAUL-HEINRICH VON. 1843. Correspondance mathématique et physique de quelques célèbres géomètres du XVIIIème siècle: précédée d'une notice sur les travaux de Léonard Euler, tant imprimés qu'inédits. 2 vols. Petersbourg: Imprimerie de l'Académie impériale des sciences.

GALILEO GALILEI. 1890-1909. *Le Opere di Galileo*. 20 vols. Florence: G. Barbèra Editore.

———. 1610. *Sidereus, nuncius: magna, longeqve admirabilia spectacula pandens, suspiciendaque proponens vnicuique praesertim vero philosophis, atque astronomis quae a Galileo Galileo ... perspicilli nuper a sereperti beneficio sunt obseruata in lvnae facie, fixis innvmeris, lacteo circvlo stellis nebvlosis, apprime vero in qvatvor planetis circa Iovis stellam disparibus interuallis, atq; periodis, celeritate mirabili circumuolutis; quos, nemini in hanc vsq; diem cognitos, nouissime auctor deprehendit primus, atque Medicea sidera nvncvpandos decrevit.* Venetiis: Apud Thomam Baglionum. Reprinted in *Opere*, vol. III/1, pp. 53-96. English translation by Albert Van Helden, Oakland, Ca.: Octavo, 2000.

———. 1899. Letter to Lodovico Cigoli of 26 June 1612. *Opere*, vol. XI, pp. 340-343. English translation in [Panofsky 1954], p. 34-37.

———. 1632. *Dialogo di Galileo Galilei Linceo matematico sopraordinario dello studio di Pisa: E filosofo, e matematico primario del serenissimo gr. duca di Toscana. Dove ne i congressi di quattro giornate si discorre sopra i due massimi sistemi del mondo tolemaico, e copernicano; proponendo indeterminatamente le ragioni filosofiche, e naturali tanto per l'una, quanto per l'altra parte.* Fiorenza: Per Gio: Batista Landini. Facsimile edition, Florence: L.S. Olschki, 1999. Reprinted in *Opere*, vol. VII, pp. 21-520. English translation, *Dialogue Concerning the Two Chief World Systems, Ptolemaic & Copernican*, translated by Stillman Drake, Berkeley: University of California Press, 1953.

———. 1638. *Discorsi e dimostrazioni matematiche intorno à due nuove scienze, attenenti alla mecanica & i movimenti locali. Con una appendice del centro di gravità d'alcuni solidi.* Leyden: Appresso gli Elsevirii. Facsimile edition, Bruxelles: Culture et Civilization, 1966. Reprinted in *Opere*, vol. VIII, pp. 39-448. English translation, *Dialogues Concerning Two New Sciences*, translated by Henry Crew and Alfonso de Salvio, New York: Macmillan, 1914.

GALLUZZI, PAOLO. 1979. *Momento: studi galileiani.* Roma: Edizioni dell'ateneo & Bizzarri.

GIORGI, GIOVANNI. 1912. Il problema del moto assoluto nelle leggi fondamentali della dinamica. *Rendiconti del Circolo Matematico di Palermo* **34**: pp. 301-332.

GRANT, ROBERT. 1852. *History of physical astronomy: from the earliest ages to the middle of the nineteenth century, comprehending a detailed account of the theory of gravitation by Newton, and its development by his successors ; with an exposition of the progress of research of all the other subjects of celestial physics.* London: H.G. Bohn. Reprint, New York, London: Johnson Reprint, 1966.

GRANT, EDWARD. 1964. Motion in the Void and the Principle of Inertia in the Middle Ages. *Isis* **55**: pp. 265-292.

GRAY, GEORGE JOHN. 1888. *Bibliography of the works to Sir Isaac Newton: together with a list of books illustrating his life and works.* Cambridge: Macmillan & Bowes. Second edition, revised and enlarged, Cambridge: Bowes and Bowes, 1907. Reprint, London: Dawsons, 1966. Facsimile reprint of the second edition, Connecticut: Martino Fine Books, 2002.

GREENBERG, JOHN LEONARD. 1995. *The Problem of the Earth's Shape from Newton to Clairaut: The Rise of Mathematical Science in Eighteenth-Century Paris and the Fall of "Normal" Science*. Cambridge; New York: Cambridge University Press.

GUICCIARDINI, NICCOLO. 1999. *Reading the Principia: the Debate on Newton's Mathematical Methods for Natural Philosophy from 1687 to 1736*. Cambridge: Cambridge University Press.

HALL, ALFRED RUPERT. 1992. *Isaac Newton, Adventurer in Thought*. Oxford, UK; Cambridge, Mass., USA: Blackwell.

HALLEY, EDMUND. 1683. A Theory of the Variation of the Magnetical Compass by the Most Ingenious Mr. Ed. Halley. *Philosophical Transactions of the Royal Society* **XIII**, 208-221.

HANKINS, THOMAS L. 1970. *Jean d'Alembert: Science and the Enlightenment*. Oxford: Clarendon Press. Second edition, New York: Gordon and Breach, 1990.

HARNACK, ADOLF VON. 1900. *Geschichte der Königlich Preußischen Akademie der Wissenschaften zu Berlin*. 4 vols. Berlin: Reimer. Facsimile edition, Hildesheim: Georg Olms, 1970.

HELMHOLTZ, HERMANN LUDWIG FERDINAND VON. 1847. *Über die Erhaltung der Kraft : eine physikalische Abhandlung, vorgetragen in der Sitzung der physikalischen Gesellschaft zu Berlin am 23sten Juli 1847 von H. Helmholtz*. Berlin: G. Reimer.

HERIVEL, JOHN. 1965. *The Background to Newton's Principia: A Study of Newton's Dynamical Researches in the Years 1664-84*. Oxford: Clarendon Press.

———. 1988. L'influence de Descartes sur Newton en dynamique. *Revue philosophique de Louvain* **86**: pp. 467-484.

HERMANN, JACOB. 1712. Extrait d'une lettre de M. Hermann a M. Bernoulli datée de Padoüe le 12 Juillet 1710. *Memoires de l'Academie Royale des Sciences*, 1710. Reprinted in *Johannis Bernoulli Opera omnia*, pp. 85.

———. 1716. *Phoronomia, sive de viribus et motibus corporum solidorum et fluidorum libri duo*. Amsterdam: apud R. & G. Wetstenios.

HERTZ, HEINRICH. 1894. *Die Prinzipien der Mechanik: in neuem Zusammenhange dargestellt*. In *Gesammelte Werke*, vol. 3, Leipzig: Barth. English translation, *The Principles of Mechanics, Presented in a New Form*. Macmillan & Co.: London, 1899.

HOUZEL, CHRISTIAN. 1978. Fonctions elliptiques et intégrales abéliennes. Pp. 1-113, vol. 2, in [DIEUDONNE 1978].

HUYGENS, CHRISTIAAN. 1988-1950. *Les Œuvres complètes de Christiaan Huygens*. 22 vols. Dutch Society of Science. The Hague: Nijhoff.

———. 1669. Regles du movement dans la rencontre des corps. *Journal des Sçavans*, 18 Mars 1669, pp 19-24. Reprinted in *Mémoires de l'Académie royale des sciences depuis 1666 jusqu'à 1699* (1729-1734), pp. 341-343. Reprinted in *Oeuvres*, vol. 16, pp. 179-181. Latin translation, A Summary Account of the Laws of Motion, Communicated by Mr. Christian Hugens [*sic*] in a Letter to the R. Society, and Since Printed in French in the Iournal Des Scavans of March 18, 1669. *Philosophical Transactions* **4** (1669), pp. 925-928.

———. 1673. *Horologium oscillatorium, sive de motu pendulorum ad horologia aptato demonstrationes geometricae*. Paris: Muguet. Reprinted in *Oeuvres*, vol. 18, pp. 69-368. English translation, *Christiaan Huygens' The pendulum clock or Geometrical Demonstrations concerning the motion of pendula as applied to clocks*. Translated with notes by Richard J. Blackwell, introduction by H. J. M. Bos, Ames: Iowa State Univ. Press, 1986.

———. 1690a. *Traite de la Lumiere: Où sont expliquées Les causes de ce qui luy arrive Dans la Reflexion, & dans la Refraction ; Et particulierement Dans l'etrange Refraction du Cristal D'Islande ; Avec un Discours de la Cause de la Pesanteur*. Leyden: Chez Pierre Vander Aa, Marchand Libraire. Reprinted in *Oeuvres*, vol. 19, pp. 451-537. Facsimile edition, Bruxelles: Culture et civilisation, 1967. English translation, *Treatise on Light: In which are Explained the Causes of that which Occurs in Reflexion, & in Refraction and Particularly in the Strange Refraction of Iceland Crystal*. Rendered into English by Silvanus P. Thompson, London: Macmillan and Co.

———. 1690b. Remarques de M. Huygens sur la Lettre précédente a sur le récit de M. Bernoulli dont on y fait mention.

———. 1698. *Cosmotheoros, sive De Terris Coelestibus earumque ornatu conjecturae*. The Hague: Apud Adrianum Metjens. Reprinted in *Oeuvres*, vol. 21, pp. 653-821. English translation, *The celestial worlds discover'd, or, Conjectures concerning the inhabitants, plants and productions of the worlds in the planets*. London: Printed for Timothy Childe, 1698.

———. 1704. Tractatus de motu corporum ex percussione. In *Opuscula postuma*, pp. 369-398. Reprinted in *Oeuvres*, vol. 16, pp. 29-91. English translation, Christiaan Huygens' The Motion of Colliding Bodies. Translated by Richard J. Blackwell, *Isis* **68**: pp. 574-597.

———. 1716. *Phoronomia, sive de viribus et motibus corporum solidorum et fluidorum libri duo*. Amsterdam: apud R. & G. Wetstenios.

IGNOTUS, P. ET AL. 1961. *The Logic of 'Personal Knowledge': essays presented to Michael Polanyi on his seventieth birthday*, 11th March 1961. London: Routledge & Paul.

JACOBI, CARL GUSTAV JACOB. 1881–1891. *Gesammelte Werke*. 8 vols. Berlin: Reimer

———. 1866. *Vorlesungen über Dynamik: nebst fünf hinterlassenen Abhandlungen*. Berlin: Reimer. Reprinted in *Werke*, supplementary volume.

JAKI, STANLEY L. 1979. The Cosmological Letters of Lambert and his Cosmology. Pp. 291-300 in *Colloque international et interdisciplinaire Jean-Henri Lambert: Mulhouse, 26-30 septembre 1977*. Paris: Ophrys.

JAMMER, MAX. 1954. *Concepts of Space: The History of Theories of Space in Physics*. Cambridge, Mass.: Harvard University Press.

KANT, IMMANUEL. 1755. *Allgemeine Naturgeschichte und Theorie des Himmels oder Versuch von der Verfassung und dem mechanischen Ursprunge des ganzen Weltgebäudes: nach Newtonschen Grundsätzen abgehandelt*. Königsberg; Leipzig: Petersen .

———. 1790. *Critik der Urtheilskraft*. Berlin; Libau: Lagarde und Friederich.

KEPLER, JOHANNES. 1938-1988. *Gesammelte Werke*. 21 vols. Im Auftrag der Deutschen Forschungsgemeinschaft und der Bayerischen Akademie der Wissenschaften begründet von Walther von Dyck und Max Caspar; fortgesetzt von Franz Hammer; herausgegeben von der Kepler-Kommission. Munich: Beck.

———. 1625. *Tychonis Brahei Dani Hyperaspistes, adversvs Scipionis Cleramontii Cæsennatis Itali, doctoris & equitis Anti-Tychone, quo libro doctrina præstantissima de parallaxibus, deque nouorum siderum in sublimi æthere discursionibus, repetitur, confirmatur, illustratur ; cum indice rerum memorabilium*. Frankfurt am Main: Apud

Godefridum Tampachium Tambach, Gottfried. Reprinted in *Werke*, vol. 8, pp. 263-437.

———. 1627. *Tabulæ Rudolphinæ, quibus astronomicæ scientiæ, temporum longinquitate collapsæ restauratio continetur illustratur....* Ulm: Typis J. Saurii. Reprinted in *Werke*, vol. 10.

KEYNES, JOHN MAYNARD. 1946. Newton the Man. Pp. 27-34 in [Royal Society 1946].

KIRCHHOFF, GUSTAV ROBERT. 1876. *Vorlesungen über mathematische Physik. Bd. 1: Mechanik.* Leipzig: Teubner.

KLEIN, FELIX. 1926-27. *Vorlesungen über die Entwicklung der Mathematik im 19. Jahrhundert.* 2 vols. Berlin: Springer-Verlag.

KLEVER, WIM (ed.). 1997. *Die Schwere der Luft in der Diskussion des 17. Jahrhunderts.* Wiesbaden: Harrassowitz.

KOESTLER, ARTHUR. 1959. *The Sleepwalkers: A History of Man's Changing Vision of the Universe.* London: Hutchinson; New York: Macmillan.

———. 1964. *The Act of Creation.* New York: Macmillan.

KÖNIG, SAMUEL. 1751. De universali principio aequilibrii & motus in vi viva reperto, deque nexu inter vim vivam & actionem, utriusque minimo, dissertatio. *Nova Acta eruditorum*, 1751, pp. 125-135, 162-176. Reprinted in *L. Euleri Opera omnia*, ser. II, vol. 5, pp. 303-324.

———. 1752. *Appel au public du Jugement de l'Académie royale de Berlin, sur un fragment de lettre de Mr. de Leibnitz cité par Mr. Koenig.* A Leide: de l'impr. d'Elie Luzac fils.

KOYRÉ, ALEXANDRE. 1939. *Etudes Galiléennes.* 3 vols. Paris: Hermann. Second edition, 1966. English translation: *Galileo Studies*. Translated from the French by John Mepham. Atlantic Highlands: Humanities Press; Hassocks: Harvester Press, 1978.

———. 1965. *Newtonian Studies.* London: Chapman & Hall.

———. 1957. *From the Closed World to the Infinite Universe.* Hideyo Noguchi Lectures, vol. 7. Baltimore: Johns Hopkins Press)].

KRONIG, RALPH DE LAER and ISIDOR ISAAC RABI. 1927. The Symmetrical Top in Undulatory Mechanics. *Physical Review* **29**: pp. 262-269.

LAGRANGE, JOSEPH-LOUIS. 1867-92. *Oeuvres de Lagrange.* 14 vols. Publiées par les soins de m. J.-A. Serret (and G. Darboux) sous les auspices de Son Excellence le ministre de l'instruction publique. Paris: Gauthier-Villars. Facsimile edition of vols. 3 and 8-14, Paris, Gabay, 2007.

———. 1759. Recherches sur la nature, et la propagation du son. *Miscellanea Taurinensia*, p. 1-112. Reprinted in *Oeuvres*, vol. 1, pp. 39-148.

———. 1762. Application de la méthode précédente à la solution de différens Problèmes de Dynamique. *Miscellanea Taurinensia. Mèlanges de philosophie et de mathématiques de la Societé royale de Turin pour les années 1760-1761*, p. 11-172. Reprinted in *Oeuvres*, vol. 1, pp. 151-316.

———. 1780 (1778). Sur le probleme de la détermination des orbites des Cometes d'après trois observations. Premier Mémoire. *Nouveaux Mémoires de l'Académie Royale des Sciences et Belles-Lettres (Berlin)* **9**, pp. 111-123. Reprinted in *Oeuvres*, vol. 4, pp. 439-451.

———. 1782 (1780). Théorie de la libration de la Lune & des autres Phénomenes qui dépendent de la figure non sphérique de cette Planete, *Nouveaux Mémoires de l'Académie des Sciences et Belles-Lettres (Berlin), année 1780* , pp. 203-309. Reprinted in *Oeuvres*, vol. 5, pp. 5-122.

———. 1788. *Méchanique analitique.* Facsimile edition, Paris: Gabay, 1989. Paris: chez la veuve Desaint. Second edition, much augmented, *Mécanique analytique.* 2 vols. Paris: 1811-15.

———. 1797. *Theorie des fonctions analytiques, contenant les principes du calcul differentiel, degages de toute consideration d'infiniment petits ou d'evanouissans, de limites ou de fluxions, et reduits a l'analyse algebrique des quantites finies.* Leipzig: Brockhaus. Second edition revised and augmented, Paris: de l'imprimerie de la Republique. Third edition, Paris: Bachelier, 1847. Reprinted in *Oeuvres*, vol. 9.

LAMBERT, JOHANN HEINRICH. 1758. *Les proprietés remarquables de la route de la lumière, par les airs et en général par plusieurs milieux refringens spheriques et concentriques avec la solution des problèmes, qui y ont du rapport, comme sont les refractions astronomiques et terrestres, et ce qui en depend.* A la Haye: chez H. Scheurleer, F.Z. & compagnie. Second edition, La Haye: chez Nicolaas van Daalen, 1750. Facsimile edition, text révised by Patricia Radelet-De Grave; with a preface by David Speiser, Paris: A. Brieux, 1977. German translation, *Merkwürdigste Eigenschaften der Bahn des Lichts durch die Luft und überhaupt durch verschiedene sphärische und concentrische Mittel: nebst der Auflösung verschiedener Aufgaben, welche sich darauf beziehen, als die astronomische und Erdstrahlenbrechung und was davon abhängt.* Berlin: Bey Haude und Spener, 1772. Reprint of the German edition, 1773.

———. 1760. *Photometria sive de mensura et gradibus luminis, colorum et umbrae.* Augustae Vindelicorum: sumptibus Viduae Eberhardi Klett. Partial German translation, *Lambert's Photometrie: Photometria sive de mensura et gradibus luminis, colorum et umbrae (1760).* Translation and commentary by E. Anding. 3 vols.Ostwald's Klassiker der exakten Wissenschaften, vols. 31-33. Leipzig: W. Engelmann, 1892. English translation, *Photometry, or, On the measure and gradations of light, colors, and shade.* With introductory monograph and notes by David L. DiLaura. New York: Illuminating Engineering Society of North America, 2001.

———. 1946-48. *Johannis Henrici Lamberti Opera mathematica: Commentationes arithmeticae, algebraicae et analyticae.* 2 vols. Edited by Andreas Speiser. Turici: Füssli.

LAPLACE, PIERRE SIMON, MARQUIS DE. 1878-1912. *Oeuvres complètes de Laplace.* 14 vols. Publiées sous les auspices de l'Académie des sciences, par MM. les secrétaires perpétuels. Paris: Gauthier-Villars. Facsimile edition of vols. 1 to VII, Paris, Gabay, 1995-2007.

———. 1779. Suite des Recherches sur plusieurs points du système du monde. *Histoire de l'Academie royale des sciences, avec les mémoires de mathématique et de physique pour la même annee,* année 1776, Mémoires, pp. 177-267. Reprinted in *Oeuvres*, vol. 9, pp. 281-310.

———. 1785. Théorie des attractions des sphéroïdes et de la figure des planètes. *Histoire de l'Academie royale des sciences, avec les mémoires de mathématique et de physique pour la même annee,* année 1782, Mémoires, pp. 113-197. Reprinted in *Oeuvres*, vol.

10, pp 339-419.

———. 1789. Mémoire sur la théorie de l'anneau de Saturne. *Histoire de l'Academie royale des sciences, avec les mémoires de mathématique et de physique pour la même annee,* année 1787, Mémoires, pp. 249-267. Reprinted in *Oeuvres*, vol. 11, pp. 273-292.

———. 1796. *Exposition du système du monde.* 2 vols. Paris: de l'imprimerie du Cercle-Social. Second edition, revised and augmented, Paris: Crapelet, 1799. Third edition, Paris: Courcier, 1808. Fourth edition, Paris: Courcier, 1813. Fifth edition, Paris: Bacheler, 1824. Reprinted in *Oeuvres*, vol. 6.

LEIBNIZ, GOTTFRIED WILHELM, FREIHERR VON. 1850-63. *Leibnizens mathematische Schriften.* 7 vols. Edited by C.I. Gerhardt. Halle: Schmidt. Facsimile edition, Bruxelles: Olms, 1971.

———. 1684a. Demonstrationes novae de resistentia solidorum. *Acta eruditorum*, Julii 1684, pp. 319-325. Reprinted in *Mathematische Schriften*, vol. 6, pp. 106-112.

———. 1684b. Nova Methodus pro Maximis et Minimis, itemque tangentibus, quae nec fractas, nec irrationales quantitates moratur, & singulare pro illis calculi genuus. *Acta eruditorum*, Octobris 1684, pp. 319-325. Reprinted in *Leibnizens mathematische Schriften*, vol. 5, pp. 220-226.

———. 1855. Letter to Jacob I Bernoulli of 24 September [4 October] 1690. Pp. 13-20, vol. 2 of *Leibnizens mathematische Schriften*. Reprinted in *Der Briefwechsel von Jacob Bernoulli*, pp. 59-67.

L'HOSPITAL, GUILLAUME FRANCOIS ANTOINE DE. 1696. *Analyse des infiniment petits, pour l'intelligence des lignes courbes*. Paris: de l'Imprimerie Royale. Facsimile edition, Paris : ACL, 1988.

LOVE, AUGUSTUS EDWARD HOUGH. 1892-93. *A Treatise on the Mathematical Theory of Elasticity*. Cambridge: University Press.

LÜTZEN, JESPER. 1983. Euler's Vision of a General Partial Differential Calculus for a Generalized Kind of Function. *Mathematics Magazine* **56**: pp. 299-306.

———. 1995. Interactions between mechanics and differential geometry in the 19th century. *Archive for History of Exact Sciences* **49**: pp. 1-72.

MACCAGNI, CARLO (Ed.). 1972. *Saggi su Galileo Galilei*. Firenze: G. Barbèra.

MACH, ERNST. 1883. *Die Mechanik in ihrer Entwickelung, historisch-kritisch dargestellt*. Leipzig: Brockhaus. English translation, *The science of mechanics: a critical and historical exposition of its principles*. Translated from the second German edition by Thomas J. McCormack. Chicago: Open Court Pub. Co., 1893.

MACLAURIN, COLIN. 1741 (1740). De causa physica fluxus et refluxus maris, in *Pièces qui ont remporté les prix de l'academie royale des sciences en M.DCC.XL*, pp. 195-234. Reprinted in [NEWTON 1739-42], vol. III, pp. 247-282.

———. 1748. *An Account of Sir Isaac Newton's Philosophical Discoveries, in four Books*. London: printed for the author's children: and sold by A. Millar, and J. Nourse; G. Hamilton and J. Balfour, and A. Kincaid at Edinburgh; J. Barry at Glasgow, and J. Smith at Dublin. Second edition, London: Millar, 1750. Third edition, London: printed for J. Nourse, W. Strahan, J. and F. Rivington, W. Johnston, D. Wilson …etc., 1775. French translation, *Exposition des découvertes philosophiques de M. le Chevalier Newton*. Ouvrage traduit de l'Anglois par La Virotte, Paris: Durand, 1749. Latin translation, *Expositio philosophiae Newtonianae*. Ex ed. Paris. anni 1749 in Latinum conversa a Georgio Falck, Vindobonae: Trattner, 1761. Facsimile edition, New York: Johnson, 1968. Facsimile edition, Hildesheim: Olms, 1971. Facsimile edition, Bristol: Thoemmes Continuum, 2004.

MALABRANCHE, NICOLAS. 1958-1978. *Œvrres complètes*. 20 vols. André Robinet, ed. Paris : Vrin

MALTESE, GIULIO. 1992. *La storia di F=ma: la seconda legge del moto nel 18. secolo*. Florence: Olschki.

———. 2001. *Da F = ma alle leggi cardinali del moto: lo sviluppo della tradizione newtoniana nella meccanica del '700*. Milano: Hoepli.

MANUEL, FRANK EDWARD. 1963. *Isaac Newton, Historian*. Cambridge, Mass.: Belknap Press of Harvard University Press.

———. 1968. *A Portrait of Isaac Newton*. Cambridge, Mass.: Belknap Press of Harvard University Press. Second edition, New York: Da Capo Press, 1990.

MARCOLONGO, ROBERTO. 1918. *Il problema dei tre corpi da Newton (1686) ai nostri giorni*. Milano: Hoepli.

MÀRGANI, MARGHERITA. 1921-22. Sull'autenticità di una lettera attribuita a G. Galilei. *Atti della Reale Accademia delle Scienze di Torino* **57**: pp. 556-568.

MARIOTTE, EDME. 1673. *Traitté de la percussion ou chocq des corps, dans lequel les principales règles du mouvement, contraires à celles que Mr. Descartes et quelques autres modernes ont voulu établir, sont démonstrées par leurs véritables causes*. Paris: Michallet. Second edition, 1676.

———. 1686. *Traite du mouvement des eaux et des autres corps fluides*. Paris: Michallet. Second edition, 1676. Second edition, 1700. Third edition, 1718. English translation, *The motion of water and other fluids : being a treatise of hydrostaticks, written originally in French by the late Monsieur Marriotte ; divided into five parts and translated into English; together with a little treatise of the same author, giving practical rules for fountains or jets d'eau by J.T. Desaguliers ; by whom are added several annotations for explaining the doubtful places*, Erschienen: London: J. Senex and W. Taylor, 1718.

MARSHAK, ROBERT EUGENE, FAYYAZUDDIN RIAZUDDIN and CIARAN P. RYAN, (Eds.). 1969. *Theory of Weak Interactions in Particle Physics*. New York: Wiley-Interscience.

MASCHERONI, LORENZO. 1785. *Nuove ricerche sull'equilibrio delle volte*. Bergamo: per Francesco Locatelli. Second edition, Milano: Silvestri, 1829.

MATHÚNA, DIARMUID Ó. 1995. Jacob II Bernoulli and the problem of the vibrating plate. Pp. 165-177 in [BENVENUTO and RADELET-DE GRAVE (eds.) 1995].

MAUPERTUIS, PIERRE-LOUIS-MOREAU DE. 1738. *La figure de la terre, déterminée par les observations de messieurs de Maupertuis, Clairaut, Camus, le Monnier, de l'Académie Royale des Sciences, & de M. l'abbé Outhier, correspondant de la même académie, accompagnés de M. Celsius, professeur d'astronomie à Upsal, faites par ordre du roy au cercle polaire*. Paris: de l'Imprimerie Royale. English translation, *The Figure of the Earth: Determined from Observations Made by Order of the French King, at the Polar Circle*, London: Printed for T. Cox, C. Davis, J. and P. Knapton, and A. Millam, 1738.

———. 1742 (1740). Loi du repos des corps, *Mémoires de l'Académie Royale des Sciences (Paris)* **47**, pp. 170-176. Reprinted in *L. Euleri Opera omnia*, ser. II, vol. 5, pp. 268-273.

———. 1748a (1744). Accord de différentes loix de la Nature, qui avoient jusqu'ici paru incompatibles, *Mémoires de l'Académie Royale des Sciences (Paris)* **56**, pp. 417-426. Reprinted in *L. Euleri Opera omnia*, ser. II, vol. 5, pp. 274-281.

———. 1748b (1746). Les lox du mouvement et du repos deduites d'un principe metaphysique, *Histoire de l'Académie Royale des Sciences et Belles-lettres (Berlin)* **2**, pp. 267-294. Reprinted in *L. Euleri Opera omnia*, ser. II, vol. 5, pp. 282-302.

———. 1752. *Les Oeuvres De Mr. De Maupertuis*. A Dresde: Ge. Conr. Walther.

MAXWELL, JAMES CLERK. 1873. *A Treatise on Electricity and Magnetism*. 2 vols. Oxford: Clarendon Press.

MICHELL, JOHN. 1784. On the Means of Discovering the Distance, Magnitude, &c. of the Fixed Stars, in Consequence of the Diminution of the Velocity of Their Light, in Case Such a Diminution Should be Found to Take Place in any of Them, and Such Other Data Should be Procured from Observations, as Would be Farther Necessary for That Purpose. By the Rev. John Michell, B. D. F. R. S. In a Letter to Henry Cavendish, Esq. F. R. S. and A. S.. *Philosophical Transactions of the Royal Society of London* **74**: pp. 35-57.

MIKHAILOV, GLEB. 1983. Leonhard Euler und die Entwicklung der theoretischen Hydraulik im zweiten Viertel des 18 Jahrunderts. Pp. 229-241 in *Colloque international et interdisciplinaire Jean-Henri Lambert: Mulhouse, 26-30 septembre 1977*. Paris: Ophrys.

MOODY, ERNEST A. 1951. Laws of Motion in Medieval Physics. *The Scientific Monthly* **72**: pp. 18-23.

MORE, LOUIS TRENCHARD. 1934. *Isaac Newton: a Biography, 1642-1727*. New York; London: Scribner.

MUSSCHENBROEK, PETRUS VAN. 1729. Introductio ad cohaerentiam corporum firmorum. Pp. 421-472 in *Petri Van Musschenbroek A.L.M. Med. & Phil. D. Phil. & Mathes. Profess. in Acad. Ultraj. Physicae Experimentales, Et Geometricae, De Magnete, Tuborum Capillarium Vitreorumque Speculorum Attractione, Magnitudine Terrae, Cohaerentia Corporum Firmorum Dissertationes: Ut Et Ephemerides Meteorologicae Ultrajectinae*. Lugduni Batavorum: Apud Samuelem Luchtmans.

NAUENBERG, MICHAEL. 2001. Newton's perturbation methods for the three-body problem and their application to lunar motion. Pp. 189-224 in [BUCHWALD and COHEN 2001].

NAVIER, CLAUDE LOUIS MARIE HENRI. 1864. *Résumé des leçons données à l'Ecole des Ponts et Chaussées sur l'application de la mécanique à l'établissement des constructions et des machines*. 2 vols. Third edition. Paris: Dunod.

NEUMANN, CARL. 1870. *Ueber die Principien der Galilei-Newton'schen Theorie: Akademische Antrittsvorlesung gehalten in der Aula der Universität Leipzig am 3 November 1869*. Leipzig: Teubner.

NEVSKAYA, NINA I. 1983a. *Peterburgskaya astronomičeskaya škola XVIII veka* (The Petersburg Astronomical School of the 18th Century). Leningrad: Nauka.

———. 1983b. Euler als Astronom. Pp. 363-371 in[AA.VV. 1983].In [AA.VV. 1983]: 363-371.

NEWTON, ISAAC. 1967-80. *The Mathematical Papers of Isaac Newton*. 8 vols. Edited by D. T. Whiteside, with the assistance in publicaton of M. A. Hoskin and A. Prag.

Cambridge: Cambridge University Press.
———. 1958. *Isaac Newton's Papers & Letters on Natural Philosophy and Related Documents*. Edited, with a general introduction by J. Bernard Cohen assisted by Robert E. Schofield. With explanatory prefaces by Marie Boas, Charles Coulston Gillispie, Thomas S. Kuhn. Cambridge, Mass.: Harvard University Press. Second edition, 1978.
———. 1959-77. *The Correspondence of Isaac Newton.* 7 vols. Edited by H. W. Turnbull (vols. 1-3), J.F. Scott (vol. 4), A. Rupert Hall and Laura Tilling (vols. 5-7). Cambridge: Cambridge University Press.
———. 1686. Letter to Halley of 20 June 1686. In *Correspondence*, vol. 3, pp. 435-441.
———. 1687. *Philosophiae naturalis principia mathematica.* London: Royal Society. Second edition, Cambridge: Printed by Cornelius Crownfield, 1713. Third edition, London: apud Guil. & Joh. Innys..., 1726.
———. 1701. Scala graduum caloris: Calorum descriptiones & signa. *Philosophical Transactions* **22**: pp. 824-829. Reprinted in [Newton 1958], pp. 259-264.
———. 1704. *Opticks: or a treatise of the reflexions, refractions, inflexions and colours of light. Also two treatises of the species and magnitude of curvilinear figures.* London: Sam. Smith and Benj. Facsimile edition, Bruxelles: Culture et Civilisation, 1966.
———. 1707. *Arithmetica universalis: sive de compositione et resolutione arithmetica liber. Cui accessit Halleiana æquationum radices arithmetice inveniendi methodus.* Cambridge: typis academicis. London: impensis Benj. Tooke. English translation: *Universal arithmetick: or, a treatise of arithmetical composition and resolution. To which is added, Dr. Halley's method of finding the roots of æquations arithmetically.* Translated from the Latin by the late Mr. Raphson, and revised and corrected by Mr. Cunn. London: printed for J. Senex; W. Taylor, T. Warner, and J. Osborn, 1720. Critical edition, based on the original manuscript: *Newton's Lectures on Algebra During 1673-1683*, in *Mathematical Papers*, vol. V, pp. 54-517.
———. 1728. *The Chronology of Antient Kingdoms Amended: To Which is Prefix'd, a Short Chronicle from the First Memory of Things in Europe, to the Conquest of Persia by Alexander the Great. With Three Plates of the Temple of Solomon.* Dublin: printed by S. Powell, for George Risk, George Ewing, and William Smith.
———. 1733. *Observations Upon the Prophecies of Daniel, and the Apocalypse of St. John: In two parts.* London: printed by J. Darby and T. Browne.
———. 1739-1742. *Philosophiae naturalis principia mathematica. Auctore Isaaco Newtono ... Perpetuis commentariis illustrata, communi studio PP. Thomæ Le Seur & Francisci Jacquier.* 3 vols. Geneva: typis Barrillot & Filii. Reprint, Coloniæ Allobrogum: sumptibus C. & A. Philibert, 1760. Reprint, Glasgow: ex prelo academico, typis A. et J.M. Duncan, 1822. Reprint, Glasgow: G. Brookman, 1833.
———. 1962. *Unpublished Scientific Papers of Isaac Newton: A Selection from the Portsmouth Collection in the University Library, Cambridge.* Chosen, edited, and translated by A. Rupert Hall and Marie Boas Hall. Cambridge: Cambridge University Press.
———. 1972. *Isaac Newton's Philosophiae naturalis principia mathematica. The third edition, 1726, with variant readings.* Assembled and edited by Alexandre Koyré and I. Bernard Cohen with the assistance of Anne Whitman. 2 vols. Cambridge: Cambridge University Press.

———. 1999. *The Principia: mathematical principles of natural philosophy*. A new translation by I. Bernard Cohen and Anne Whitman, assisted by Julia Budenz, preceded by a guide to Newton's *Principia* by I. Bernard Cohen. Berkeley; London: University of California Press.

NORDTVEDT JR., KENNETH. 1968a. Equivalence Principle for Massive Bodies. I. Phenomenology. *Physical Review* **169**: pp. 1014-1016.

———. 1968b. Testing Relativity with Laser Ranging to the Moon. *Physical Review* **170**: pp. 1186-1187.

———. 1996. Post-Newtonian Gravitational Effects in Lunar Laser Ranging. *Physical Review D* 7: pp. 2347-2356.

OESCHGER, JOHANNES (ed.). 1960. *Briefe von und nach Basel aus fünf Jahrhunderten*. Basel: J.R. Geigy.

OLDENBURG, HENRY. 1667. To Observe the Declinations and Variations of the Compass or Needle from the Meridian Exactly, in as Many Places as They Can, and in the Same Places, Every Several Voyages, *Philosophical Transactions of the Royal Society* **II**: pp. 435-438.

PANOFSKY, ERWIN. 1954. *Galileo as a Critic of the Arts*. The Hague, M. Nijhoff.

———. 1956. Galileo as a Critic of the Arts: Aesthetic Attitude and Scientific Thought. *Isis* **47**: pp. 3-15.

PARENT, ANTOINE. 1713. De la veritable méchanique des resistances relatives des solides, et réfléxions sur le sistême de M. Bernoulli de Bâle. Pp. 187-201, vol. 3, in *Essais et recherches de mathematique et de physique. Nouvelle edition augmentée d'un troisiéme volume, & d'un tiers au moins en chacun des deux premiers*. 3 vols. A Paris: chez Jean de Nully.

PAULI, WOLFGANG. 1921. Relativitättheorie. In *Encyklopädie der mathematischen Wissenschaften : mit Einschluss ihrer Anwendungen*. Vol. 5, *Physik*, Part 2. Edited by Arnold Sommerfeld, Leipzig: B.G. Teubner, 194-1922, pp. 539-775. [Issued November 15, 1921]. English translation, *Theory of Relativity*, with supplementary notes by the author, translated by G. Field. London: Pergamon, 1958.

PEMBERTON, HENRY. 1728. *A View of Sir Isaac Newton's Philosophy*. London: Printed by S. Palmer.

PFEIFFER, HENRI. 1979. Les études perspectives et chromatologiques de Lambert. Pp. 307-312 in *Colloque international et interdisciplinaire Jean-Henri Lambert: Mulhouse, 26-30 septembre 1977*. Paris: Ophrys.

PINGRÉ, ALEXANDRE GUY. 1783-84. *Cométographie; ou, Traité historique et théorique des comètes*. 2 vols. Paris: Imprimerie Royale.

PLANCK, MAX. 1928. *Einführung in die theoretische Physik. Bd. 1: Einführung in die allgemeine Mechanik*. Fourth edition. Leipzig : S. Hirzel.

POINCARE, HENRI. 1911. *Leçons sur les hypothèses cosmogoniques*. Paris: Hermann. Second edition, 1913.

POINSOT, LOUIS. 1803. *Élémens de Statique*. A Paris: Chez Calixte-Volland.

POISSON, SIMEON-DENIS. 1809. Mémoire sur la variation des constants arbitraries dans les questions de Mécanique. *Journal de l'Ecole Polytechnique*, cahier XV, vol. 8: pp. 266-344.

———. 1811. *Traité de Mécanique*. 2 vols. Paris: Courcier. Second edition, Paris: Bachelier, 1833. Third edition, Bruxelles: Société Belge de Librairie, 1838. English translation, *A treatise of mechanics, by S. D. Poisson. Translated from the French, and elucidated with explanatory notes by the Rev. Henry H. Harte*. London: Longman and Co., 1842.

———. 1809. Mémoire sur la variation des constants arbitraries dans les questions de Mécanique. *Mathematics Magazine* **56**: pp. 299-306.

PRINGSHEIM, ERNST. 1910. *Vorlesungen über die Physik der Sonne*. Leipzig und Berlin: B. G. Teubner.

PRONY, GASPARD-CLAIR-FRANÇOIS-MARIE RICHE DE. 1800. Mécanique philosophique, ou Analyse raisonnée des diverses parties de la science de l'équilibre et du mouvement. *Journal de l'Ecole Polytechnique*, cahiers VII-VIII, vol. 3.

RAIMI, RALPH A. 1976. The First Digit Problem. *American Mathematical Monthly* **83**: pp. 521-538.

RADELET-DE GRAVE, PATRICIA and DAVID SPEISER. 1975. Le "De Magnete" de Pierre de Maricourt. traduction et commentaires. *Revue d'Histoire des Sciences* **28**: pp. 193-233.

———. 1987a. Introduction to Daniel Bernoulli's memoirs on general mechanics, in *Die Werke von Daniel Bernoulli*, vol. 3, pp. 1-118.

———. 1987b. Introduction to Daniel Bernoulli's memoirs on the mechanics of the planetary system and of the tides, in *Die Werke von Daniel Bernoulli*, vol. 3, pp. 241-302.

RADELET-DE GRAVE, PATRICIA. 1982. *Les lignes magnétiques du XIIIème siècle au milieu du XVIIIème siècle*. Cahiers d'histoire et de philosophie des sciences, Nouvelle série, no. 1. Paris: Centre de Documentation Sciences Humaines.

———. 1994. Commentaire sur le magnetisme, in *Die Werke von Daniel Bernoulli*, vol. 7, pp. 1-64.

RADICATI DI BROZOLO, LUIGI. A. (ed.). 1999. *"Principio di Secol Nuovo": Saggi su Galileo*. Pisa: Cassa di Risparmio di Pisa.

RAVETZ, JEROME RAYMOND BERNARD. 1961. Vibrating string and arbitrary functions. Pp. 71-88 in [AA. VV 1961]

REICHENBACH, HANS. 1924. Die Bewegungslehre bei Newton, Leibniz und Huyghens. *Kantstudien* **29**: pp. 416-438. English translation, The theory of motion according to Newton, Leibniz and Huyghens, in *Modern Philosophy of Science: Selected Essays by Hans Reichenbach*, translated and edited by Maria Reichenbach, London, Routledge & Kegan Paul, 1959, pp. 46-66.

REIFF, RICHARD AUGUST and ARNOLD SOMMERFELD. 1902. Standpunkt der Fernwirkung. Die Elementargesetze. Pp. 4-22 in vol. V, Part 2 of *Encyklopädie der Mathematischen Wissenschaften*. Leipzig: B. G. Teubner.

RIEMANN, GEORG FRIEDRICH BERNHARD. 1892. *Gesammelte mathematische Werke und wissenschaftlicher Nachlass*. Edited by Heinrich Weber, with assistance of Richard Dedekind. Second edition. Leipzig: Teubner. Reprint, New York: Dover Publications, 1953. Facsimile edition, Berlin ; London: Springer, 1990.

———. 1868. Ueber die Darstellbarkeit einer Function durch eine trigonometriche Rehie, *Abhandlungen der K. Gesellschaft der Wissenschaften in Göttingen* **13**, pp. 87-132. Reprinted in *Werke*, pp. 227-271.

ROSENFELD, LEON. 1964-65. Newton and the law of gravitation. *Archive for History of Exact Sciences* **2**: pp. 365-386.

———. 1969. Newton's views on aether and gravitation. *Archive for History of Exact Sciences* **6**: pp. 29-37.

(ROYAL SOCIETY). 1946. *Newton Tercentenary Celebrations 15-19 July 1946*. Cambridge: Published for the Royal Society at the University Press.

RUFFNER, JAMES A. 1964. Reinterpretation of the genesis of Newton's "Law of Cooling". *Archive for History of Exact Sciences* **2**: pp. 138-152.

SAINT-VENANT, ADHÉMAR JEAN CLAUDE BARRÉ DE. 1864. Historique abrégé des recherches sur la résistance et sur l'élasticité des corps solides. Pp. xc-cccxj, vol. 1 in [NAVIER 1864].

SALVINI, SALVINO. 1717. *Fasti consolari dell' Accademia fiorentina*. Firenze: Nella Stamperia di S.A.R., Per Gio: Gaetano Tartini, e Santi Franchi.

SCHRAMM, MATTHIAS. 1985. *Natur ohne Sinn? Das Ende des teleologischen Weltbildes*. Graz: Styria.

SCIAMA, DENNIS W. 1959. *The Unity of the Universe*. London: Faber and Faber.

SEGNER, JOHANN ANDREAS VON. 1758. *De adfrictu solidorum in motu constitutorum*. Halle: Litteris Joannis Friderici Grunerti.

SETTLE, THOMAS B. 1961. An Experiment in the History of Science. *Science* **133**: pp. 19-23.

SHAPIRO, IRWIN I., CHARLES C. COUNSELMAN, III and ROBERT W. KING. 1976. Verification of the principle of equivalence for massive bodies. *Physical Review Letters* **36**: pp. 555-558.

SHEA, WILLIAM R. (ed.). 1988a. *Revolutions in Science: Their Meaning and Relevance*. Papers presented at the Seventh Conference of the International Union for the History and Philosophy of Science, April 18-22, 1988, held at the University of Coimbra, Portugal. New Canton, MA: Science History Publications.

———. 1988b. The Unfinished Revolution: Johann Bernoulli (1667-1748) and the Debate Between Cartesians and the Newtonians. Pp. 70-92 in [Shea 1988a].

SMITH, ROBERT. 1738. *A Compleat System of Opticks in Four Books viz. a Popular, a Mathematical, a Mechanical, and a Philosophical Treatise. To Which are Added Remarks upon the Whole*. 2 vols. Cambridge: Printed for the author. German translation, *Vollständiger Lehrbegriff der Optik nach Herrn Robert Smiths Englischen mit Aenderungen und Zusäszen*. Translation by Abraham Gotthelf Kästner. Altenburg: Richter 1755. French translation, *Traité d'optique par M. Smith, professeur d'astronomie et de philosophie expérimentale à Cambridje, traduit de l'anglais et considérablement augmenté*. Translated by Nicolas-Claude Duval-Le-Roy. Brest: chez Romain Malassis, 1767.

SNOBELEN, STEPHEN D. 2001. "God of Gods, and Lord of Lords": The Theology of Isaac Newton's General Scholium to the Principia. *Osiris* (2)**16**: pp. 169-208.

SOCIN, ABEL. 1670. Tentamina electrica in diversis morborum generibus, quibus accedunt levis electrometri Bernoulliani adumbratio, et quorundam experimentorum instituendorum ratio, *Acta Helvetica. Physico-Mathematico-Anatomico-Botanico-medica* **4**, pp. 214-230.

SOMMERFELD, ARNOLD. 1947. *Vorlesungen über theoretische Physik: Bd. 1, Mechanik.* Third edition. Leipzig: Akad.-Verl. Geest & Portig. English translation, *Lectures on Theoretical Physics: Vol. !, Mechanics.* New York: Academic Press, 1952.

SPEISER, ANDREAS. 1932. *Die mathematische Denkweise.* Zürich, Leipzig, Stuttgart: Rascher.

———. 1939. *Die Basler Mathematiker.* Basel: Helbing & Lichtenhahn.

SPEISER, DAVID. 1962. Editor's Introduction to the *Commentationes Opticae*, in *L Euleri Opera omnia*, ser. III, vol. 5, pp. VII-LVI.

———. 1994. The Symmetries of the Battistero and the Torre Pendente in Pisa. *Annali della Scuola Normale Superiore, Classe di Lettere*, Serie III, vol. XXIV: pp. 511-564.

SPIESS, OTTO. 1936. *Basel anno 1760: nach den Tagebüchern der ungarischen Grafen Joseph und Samuel Teleki.* Basel: Birkhäuser.

———. 1955. Einleitung aus der Briefwechsel mit dem Marquis de l'Hôpital, in *Die Briefwechsel von Johann Bernoulli*, vol. 1, pp. 97-157.

STECK, MAX. 1943. *Bibliographia Lambertiana: Ein Führer durch das gedruckte und ungedruckte Schrifttum von Johann Heinrich Lambert, 1728-1777.* Berlin: Lüttke. Reprint, Hildesheim: Gerstenberg, 1970.

STRAUB, HANS. 1970. Bernoulli, Daniel. *Dictionary of Scientific Biography* **2:** 36-46.. New York: Charles Scribner's Sons.

SUCHTING, W. A. 1967. Berkeley's Criticism of Newton on Space and Motion. *Isis* **58**: pp. 186-197.

SUPPE, FREDERICK (Ed.). 1974. *The Structure of Scientific Theories.* Urbana: University of Illinois Press.

SUZUKI, JEFF. Euler and Number Theory: A Study in Mathematical Invention. Pp. 363-383 in *Leonhard Euler: Life, Work and Legacy.* Robert E. Bradley and C. Edward Sandifer, eds. Amsterdam: Elsevier.

SYNGE, JOHN LIGHTON. 1956. *Relativity: The Special Theory.* Amsterdam: North Holland.

SZABÓ, ISTVÁN. 1979. *Geschichte der mechanischen Prinzipien.* Second edition. Basel: Birkhäuser.

TATON, RENE A. 1976. Inventaire chronologique de l'oeuvre d'Alexis-Claude Clairaut (1713-1765). *Revue d'histoire des sciences et de leurs applications* **29**: pp. 97-122.

———, (ed.). 1957-69. *Histoire générale des sciences.* 2 vols. Paris: Presses universitaires de France.

———, (ed.). 1982. *Huygens et la France : table ronde du Centre national de la recherche scientifique, Paris, 27-29 mars 1979.* Paris: J. Vrin.

———, (ed.). 1978. *Roemer et la vitesse de la lumière.* Paris: Vrin.

TAYLOR, BROOK. 1714 (1713). De motu nervi tensi. *Philosophical Transactions* **28**, No. 337: pp. 26-32.

THIÉBAULT, DIEUDONNÉ. 1813. *Mes souvenirs de 20 ans de séjour a Berlin.* Paris.

THOMSON, WILLIAM (LORD KELVIN) and PETER GUTHRIE TAIT. 1890. *Elements of Natural Philosophy.* 2 vols. Second edition. Cambridge: Cambridge University Press.

TIMOSHENKO, STEPHEN PROKOFIEVITCH. 1953. *History of Strength of Materials. With a Brief Account of the History of Theory of Elasticity and Theory of Structures.* New York: McGraw-Hill.

TRIP, SCATO. 1705. *Disputatio Philosophica de Vacuo ...* . Groningen: Ex Officina Cornelii Barlinck-hof.

TODHUNTER, ISAAC. 1865. *A History of the Mathematical Theory of Probability from the Time of Pascal to that of Laplace.* Cambridge and London: Macmillan and Co. Reprint, New York: G.E. Stechert & Co., 1931. Facsimile edition, New York: Chelsea Pub. Co, 1949. Facsimile edition, Bristol: Thoemmes Press, 1993. Facsimile edition, Honolulu: University Press of the Pacific, 2004.

———. 1873. *A History of the Mathematical Theories of Attraction and Figure of the Earth, from the Time of Newton to that of Laplace.* 2 vols. London: Macmillan and Co. Reprint, New York: Dover Publications, 1962.

TODHUNTER, ISAAC and KARL, PEARSON. 1886-93. *A History of the Theory of Elasticity and of the Strength of Materials, from Galilei to the Present Times.* Edited and completed by Karl Pearson. 2 vols. in 3 pts. Cambridge: Cambridge University Press. Reprint, New York: Dover Publications, 1960.

TRUESDELL, CLIFFORD AMBROSE. 1954. Rational fluid mechanics (1687-1765); in *L. Euleri opera omnia*, ser. II, vol.12, pp. VII, CXXV.

———. 1956. Editor's Introduction. I. The first three sections of Euler's treatise on fluid mechanics (1766). II. The theory of aerial sound (1687-1788); III. Rational fluid mechanics (1765-1788). In *L. Euleri Opera omnia*, ser. II, vol.13, pp. VII-CXVII.

———. 1960a. *The Rational Mechanics of Flexible or Elastic Bodies (1638 – 1788).* In *L. Euleri Opera omnia*, ser. II, vol. 11.

———. 1960b. A Program Toward Rediscovering the Rational Mechanics of the Age of Reason. *Archive for History of Exact Sciences* **1**: pp. 1-36. Reprinted in [Truesdell 1968a]: 85-137.

———. 1964. Whence the Law of Moment of Momentum?. Pp. 588-612 in [AA.VV. 1964]. Reprinted in [Truesdell 1968a]: 239-271.

———. 1967. Reactions of Late Baroque Mechanics to Success, Conjecture, Error, and Failure in Newton's *Principia. The Texas Quarterly*, Autumn, 238-258. Reprinted in [Truesdell 1968a], pp. 138-183.

———. 1968a. *Essays in the history of mechanics*. Berlin; New York: Springer-Verlag.

———. 1968b. The Creation and Unfolding of the Concept of Stress. Pp. 184-238 in [Truesdell 1968a].

———. 1976. History of Classical Mechanics. *Die Naturwissenschaften* 63, 2 (February 1976): part I, to 1800, pp. 53-62; part II, the 19th and 20th centuries, pp. 119-130.

———. 1980a. *The Tragicomic History of Thermodynamics, 1822-1854.* New York, Heidelberg, Berlin: Springer.

———. 1980b. Review of István Szabó's *Geschichte der mechanischen Prinzipien. Centaurus* **23**: pp. 163-175. Reprinted in [Truesdell 1984]: 254-265

———. 1984. *An Idiot's Fugitive Essays on Science: Methods, Criticism, Training, Circumstances.* New York, Berlin, Heidelberg, Tokyo: Springer.

TRUESDELL, CLIFFORD AMBROSE and RICHARD A. TOUPIN. 1960. The Classical Field Theories . Pp. 226-793, vol. 3, part. 1, in *Handbuch der Physik*, edited by S. Flügge, Berlin, New York: Springer.

VARIGNON, PIERRE. 1703a. Du mouvement en général par toutes sortes de courbes ; & des forces centrales, tant centrifuges que centripetes, necessaires aux corps qui les décrivent. *Histoire de l'Académie royale des sciences avec les mémoires de mathématiques et de physique tirés des registres de cette Académie*, Anné 1700: pp. 83-101.

———. 1703b. Des forces centrales, ou des pesanteurs necessaires aux planetes pour leur faire décrire les orbes qu'on leur a supposés jusqu'icy. *Histoire de l'Académie royale des sciences avec les mémoires de mathématiques et de physique tirés des registres de cette Académie*, Anné 1700: pp. 218-237.

———. 1712. Des forces centrales inverses. *Histoire de l'Académie royale des sciences avec les mémoires de mathématiques et de physique tirés des registres de cette Académie*, Anné 1710: pp. 533-544.

VILLAGGIO, PIERO. 1999. Galileo fondatore della teoria meccanica della forma. Pp. 229-249 in [RADICATI DI BROZOLO 1999].

VIVIANI, VINCENZO. 1717. Racconto istorico della vita di Galileo Galilei. Pp. 397-431 in [SALVINI 1717]. Reprinted in *Opere di Galileo*, vol. XIX, pp. 597-632.

VOLLGRAFF, J. A. 1950. Biographie de Christiaan Huygens. In *Œuvres de Huygens*, vol. XXII, pp. 383-778.

VOLTAIRE. 1753. *Diatribe Du Docteur Akakia, Médecin du Pape. Decret De L'Inquisition; Et Rapport Des Professeurs De Rome, Au sujet d'un Prétendu Président.* Rome [i.e. Leyden]: [Luzac]. Édition critique, avec une introduction et un commentaire par Jacques Tuffet, Paris: Nizet, 1967.

———. 1965. Letter to Johann Bernoulli, 30 January 1740, in Voltaire, *Correspondence*, ed. Theodore Bestermann, 1965, vol. II, p. 255.

WAERDEN, BARTEL L. VAN DER. 1983. Eulers Herleitung des Drehimpulssatzes. Pp. 271-281 in [FELLMANN (ED.) 1983].

WAGNER, FRITZ. 1976. *Isaac Newton im Zwielicht zwischen Mythos und Forschung: Studien zur Epoche der Aufklärung*. Freiburg: Alber.

WATSON, GEORGE NEVILLE. 1922. *A Treatise on the Theory of Bessel Functions*. Cambridge: Cambridge University Press.

WEIL, ANDRÉ. 1983. L'œuvre arithmétique d'Euler. Pp. 111-132 in [AA.VV. 1983].

———. 1984. *Number Theory: An Approach Through History from Hammurapi to Legendre*. Boston: Birkhäuser.

WEINSTOCK, ROBERT. 2000. Inverse-Square Orbits in Newton's *Principia* and Twentieth-Century Commentary Thereon. *Archive for History of Exact Sciences* **55**: pp. 137-162.

WESTFALL, RICHARD S. 1980. *Never at Rest: A Biography of Isaac Newton*. Cambridge: Cambridge University Press.

WEYL, HERMANN. 1923. *Mathematische Analyse des Raumproblems: Vorlesungen gehalten in Barcelona und Madrid*. Berlin: Springer.

———. 1949. *Philosophy of Mathematics and Natural Science*. Revised and augmented English edition based on a translation by Olas Helmer. Princeton: Princeton University Press.

WHITE, LYNN TOWNSEND. 1969. *The Expansion of Technology, 500-1500*. London: Collins Clear-Type Press.

WHITTAKER, EDMUND TAYLOR. 1951-53. *A History of the Theories of Aether and Electricity*. 2 vols. London: Thomas Nelson. Reprint, New York: Humanities Press, 1973. Reprint, New York: Dover Publications, 1989.

WILSON, CURTIS A. 1970. From Kepler's Laws, So-called, to Universal Gravitation: Empirical Factors. *Archive for History of Exact Sciences* **6**: pp. 89-170.

———. 1974. Newton and Some Philosophers on Kepler's "Laws". *Journal of the History of Ideas* **33**: pp. 231-258.

———. 1985. The great inequality of Jupiter and Saturn: from Kepler to Laplace. *Archive for History of Exact Sciences* **33**: pp. 15-290.

———. 1992. Euler on action-at-a-distance and Fundamental Equations in Continuum Mechanics. Pp- 399-420 in *The Investigation of Difficult Things: Essays on Newton and the History of the Exact Sciences in Honour of D. T. Whiteside*, Peter Harmon and Alan Shapiro, eds. Cambridge: Cambridge University Press.

———. 1994. The Newton's Orbit Problem: A Historian's Response. *The College Mathematics Journal* **25**: pp. 193-200.

———. 2001. Newton on the Moon's variation and apsidal motion: the need for a newer "new analysis". Pp. 139-188 in [BUCHWALD and COHEN 2001].

WOLF, R. 1859. *Biographien zur Kulturgeschichte der Schweiz*. Zurich.

WUSSING, HANS. 1969. *Die Genesis des abstrakten Gruppenbegriffes: ein Beitrag zur Entstehungsgeschichte der abstrakten Gruppentheorie.* Berlin: Deutscher Verlag der Wissenschaften. English translation, *The Genesis of the Abstract Group Concept: A Contribution to the History of the Origin of Abstract Group Theory*, translation by Abe Shenitzer. Erschienen: Cambridge, Mass.: MIT Press, 1984.

YOUSCHKEVITCH, ADOL'F PAVLOVICH. 1976-77. The Concept of function up to the Middle of the 19th Century. *Archive for History of Exact Sciences* **16**: pp. 35-85.

———. 1979. Lambert et Léonard Euler. Pp. 211-223 *Colloque international et interdisciplinaire Jean-Henri Lambert: Mulhouse, 26-30 septembre 1977.* Paris: Ophrys.

ZEHE, HORST. 1997. "Un Thermometre qui marquera toujours la Veritable agitations des parties de l'Air". Pp. 171-186 in [KLEVER 1997].

ZINSSER JUDITH P. 2001. Translating Newton's 'Principia': The Marquise du Châtelet's Revisions and Additions for a French Audience. *Notes and Records of the Royal Society of London* **55**: pp. 227-245.

Illustration Credits

Portraits

Galileo Galilei (p. 16). Portrait by Justus Sustermans (1597-1681). Courtesy of the National Maritime Museum, Royal Observatory Greenwich.

Jacob Bernoulli (p. 70). Portrait by Nicolaus Bernoulli (1662-1716). Courtesy of the Naturhistorisches Museum, Basel.

Johann I Bernoulli (p. 104). Courtesy of the Naturhistorisches Museum, Basel.

Daniel Bernoulli (p. 136). Courtesy of the Naturhistorisches Museum, Basel.

Leonhard Euler (p. 156). Courtesy of the Naturhistorisches Museum, Basel.

Christiaan Huygens (p. 40). Courtesy of the Collection Huygensmuseum Hofwijck, Voorburg, The Netherlands.

Johannes Kelpler (p. 122). From *Opera Omnia*, vol. 1, Ch. Frisch, ed. Frankfurt: Heyder & Zimmer, 1858. Public domain.

Johann Heinrich Lambert (p. 234). Drawing by Pierre Roch Vigneron. Public domain.

Pierre Louis Moreau de Maupertuis (p. 254). Engraving by Konrad Westermayr (1765-1834) Public Domain.

Isaac Newton (p. 50). From *Optice: sive de reflexionibus, refractionibus, inflexionibus et coloribus lucis Libri Tres*, Lausannae & Genevae: Sumpt. Marci-Michaelis Bousquet & Sociorum, 1740. Courtesy of the Accademia delle Scienze, Turin. Public domain.

Frontispieces

Daniel Bernoulli, *Hydrodinamica, sive de viribus et motibus fluidorum commentarii*, 1738 (p. 141). Courtesy of the Accademia delle Scienze, Turin.

Jacob Bernoulli, Curvatura Laminae Elasticae, 1694 (p. xii). Courtesy of the Accademia delle Scienze, Turin.

Johann I Bernoulli, *Hydraulica, nunc primum detecta ac demonstrata directe ex fundamentis pure mechanicis. Anno 1732* (p. 143).Courtesy of the Accademia delle Scienze, Turin.

Leonhard Euler, *Mechanica sive motus scientia analytice exposita*, vol. I, 1736 (p.166). Courtesy of the Accademia delle Scienze, Turin.

Galileo Galilei, *Discorsi e dimostrazioni matematiche intorno à due nuove scienze, attenenti alla mecanica & i movimenti locali*, 1638 (p.18). Courtesy of the Accademia delle Scienze, Turin.

Christiaan Huygens, *Horologium oscillatorium, sive de motu pendulorum ad horologia aptato demonstrationes geometricae*, 1673 (p.42). Courtesy of the Accademia delle Scienze, Turin.

Johann Heinrich Lambert. *Les proprietés remarquables de la route de la lumière, par les airs...*, 1758 (p. 241). Courtesy of Alain Brieux.

Index of Names

Index of Subjects

SPRINGER NATURE

GPSR Compliance

The European Union's (EU) General Product Safety Regulation (GPSR) is a set of rules that requires consumer products to be safe and our obligations to ensure this.

If you have any concerns about our products, you can contact us on ProductSafety@springernature.com

In case Publisher is established outside the EU, the EU authorized representative is:

Springer Nature Customer Service Center GmbH
Europaplatz 3
69115 Heidelberg, Germany

The manufacturer's authorised representative in the EU is Springer Nature Customer Service Centre GmbH, Europaplatz 3, 69115 Heidelberg, Germany. If you have any concerns regarding our products, please contact ProductSafety@springernature.com

Printed and bound by CPI Group (UK) Ltd, Croydon, CR0 4YY
28/07/2026
02179207-0009